Editors

Wolfgang Hillebrandt
Rudolf Kuhfuß
Ewald Müller
MPI für Astrophysik
Karl-Schwarzschild-Str. 1, D-8046 Garching, FRG

James W. Truran
Astronomy Department, University of Illinois
Urbana-Champaign, West Springfield, Urbana, IL 61801, USA

ISBN 3-540-18279-9 Springer-Verlag Berlin Heidelberg New York
ISBN 0-387-18279-9 Springer-Verlag New York Berlin Heidelberg

Printed in Germany

Printing: Druckhaus Beltz, Hemsbach/Bergstr.;
Bookbinding: J. Schäffer GmbH & Co. KG., Grünstadt
2153/3140-543210

Preface

One evening, when I was contemplating as usual the celestial vault, whose aspect was so familiar to me, I saw, with inexpressible astonishment, near the zenith, in Cassiopeia, a radiant star of extraordinary magnitude. Struck with surprise, I could hardly believe my eyes ...

Thus spoke Tycho Brahe of his discovery of a supernova in Cassiopeia in 1572. Thirty-two years later, Kepler and also Galileo observed a supernova in Ophiuchus in 1604. There has followed a period of almost 400 years during which no supernova event has been seen to occur in our Milky Way Galaxy.

It is thus clear why this is such an extraordinarily exciting time for researchers in nuclear astrophysics. The occurrence of Supernova 1987A in the Large Magellanic Cloud, a mere 180,000 light years away, has provided active researchers with an opportunity they had only dared to dream might occur in their lifetimes. We can now test our elaborate theoretical models of the evolution of massive stars, of the gravitational collapse of stellar cores, of the characteristics of the emerging neutrino spectrum, of the nature of the remnant neutron star or black hole, of explosive nucleosynthesis, of the supernova explosion mechanism itself, and of the evolution of the supernova light curve and spectrum. With the detection of neutrinos from Supernova 1987A, we have already witnessed the birth of extra-solar-system neutrino astronomy, and exciting opportunities for gamma-ray astronomy may lie ahead. Discussions of many of these issues are to be found in these proceedings.

We thank the authors for their great efforts to assist us in providing proceedings of this exciting conference on a rapid time scale.

We wish to express our sincere thanks to the Max-Planck-Society for financial support and for the hospitality of the Ringberg Castle, Tegernsee, where this workshop was held. The proceedings also benefited from a recent attempt to coordinate the activities of various institutes in Belgium, France, and Germany in the field of nuclear astrophysics (Programme International de Cooperation Scientifique, PICS). We acknowledge strong support from agencies of the three countries.

Garching, July 1987 W. Hillebrandt, R. Kuhfuß, E. Müller, J.W. Truran

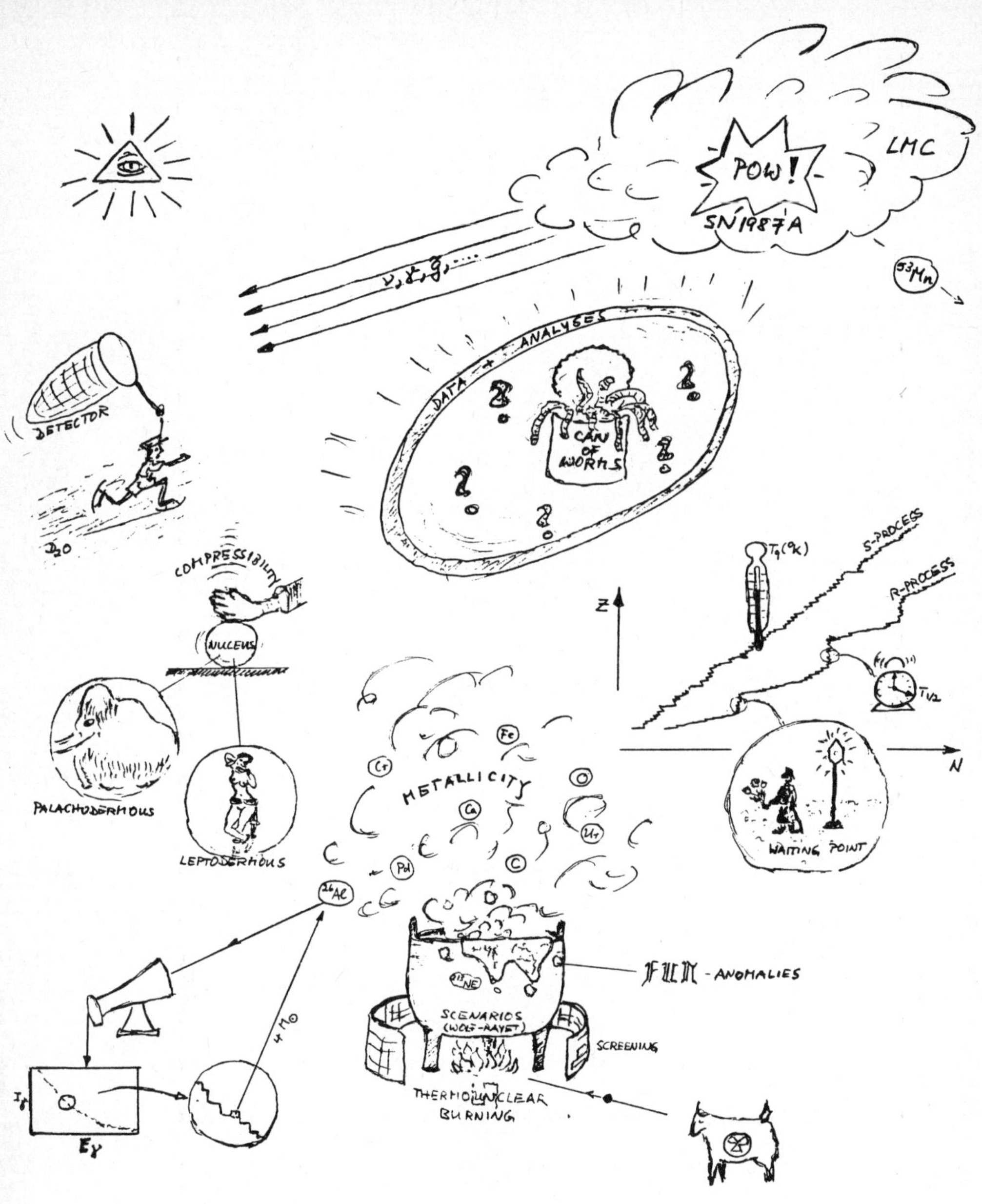

POW!
LMC
SN1987A
^{53}Mn
DATA + ANALYSES
CAN OF WORMS
DETECTOR
D_2O
COMPRESSIBILITY
NUCEUS
PALACHODERMOUS
LEPTODERMOUS
METALLICITY
Fe
Cr
Ca
Ur
Pd
C
^{26}Al
$T_9(^9K)$
S-PROCESS
R-PROCESS
Z
N
$T_{1/2}$
WAITING POINT
FUN - ANOMALIES
^{22}NE
SCENARIOS (WOLF-RAYET)
SCREENING
THERMONUCLEAR BURNING
$4\,M_\odot$
I_γ
E_γ

Table of Contents

I. Thermonuclear Reaction Rates in Stars: Experimental and Theoretical Approaches

New Experimental Approaches in Nuclear Astrophysics

C. Rolfs 2

Direct Cross Section Measurements Towards Thermal Energies

H.W. Becker 18

A New γ–Ray Detector for Studies of Capture Reactions Involving Radioactive Nuclei

F.B. Waanders 29

Coulomb Dissociation as a Source of Information on Radiative Capture Processes of Astrophysical Interest

H. Rebel 38

New Experimental Results for Nuclear Reactions in Explosive Hydrogen Burning

M. Wiescher, J. Görres, L.O. Lamm, C.P. Browne, B.W. Filippone and B. Vogelaar 54

Beta-Decay Half-Lives of Very Neutron-Rich Nuclei and Their Consequences for the Astrophysical r-Process

K.-L. Kratz 68

Experimental Studies of Thermal Effects During s-Process Nucleosynthesis

F. Käppeler 79

Thermonuclear Reactions at High Temperatures and Densities

J.W. Truran, F.-K. Thielemann and M. Arnould 91

Thermonuclear Functions

H.J. Haubold, A.M. Mathai and W.J. Anderson 102

A Microscopic Approach to Reactions of Astrophysical Interest

P. Descouvemont and D. Baye 111

The ETFSI Approach to the Nuclear Mass Formula

J.M. Pearson, F. Tondeur and A.K. Dutta 126

Nuclear Matter Compressibility from Low Energy Nuclear Physics

M.M. Sharma 135

II. Stellar Evolution, Nucleosynthesis and Isotopic Anomalies in Meteorites

Early Nucleosynthesis, Chemical Evolution of Galaxies and Particle Physics
J. Audouze 148

Chemodynamical Models of Galactic Evolution
A. Burkert and G. Hensler 159

Abundance Patterns in Some Old Stars
B. Baschek 174

Evolution of Wolf-Rayet Stars
N. Langer 180

Advanced Phases of Nucleosynthesis in Very Massive Stars
M.F. El Eid, N. Prantzos and N. Langer 187

Overshooting and Electron-Positron Pair Instability
G. Meynet and A. Maeder 195

s-Process Production in the Central Helium Burning of Large Masses ($M \geq 15M_{\odot}$)
D. Bencivenni, V. Castellani and A. Chieffi 204

On the Synthesis of the Proton-Rich Nuclei
M. Rayet 210

Studies of Non-Local and Time-Dependent Convection
R. Kuhfuß 222

Nucleosynthesis in Explosions of High Metallicity Supermassive Objects
W. Hillebrandt, F.-K. Thielemann and N. Langer 233

Isotopic Anomalies and Wolf-Rayet Stars
J.B. Blake and D.S.P. Dearborn 243

The ^{26}Al γ–Ray Line: A Status Report
N. Prantzos 250

A Possible Relationship Between Extinct ^{26}Al and ^{53}Mn in Meteorites and Early Solar Activity
G.J. Wasserburg and M. Arnould 262

The Contamination of Cometary Globules by the Ejecta of Nearby Massive Stars
J.-P. Arcoragi 277

III. Supernovae and SN 1987A

Binary Systems as Supernova Progenitors (Some Frequency Estimates)

A. Tornambè, F. Matteucci, I. Iben Jr. and K. Nomoto 284

On Stellar Models for the Progenitor of Supernova 1987A

J.W. Truran and A. Weiss 293

A Few Comments on the Evolutionary History of SN 1987A Before Explosion

A. Renzini 305

Model Calculations for Scattering Dominated Atmospheres and the Use of Supernovae as Distance Indicators

P. Höflich 307

Synthetic Spectra for Supernovae II

W. Spies, P. Hauschildt, R. Wehrse, B. Baschek and G. Shaviv 316

Monte Carlo Methods for Neutrino Transport in Type-II Supernovae

H.-T. Janka 319

Neutrinos from SN 1987A: Remarks on Possible Interpretations

W. Hillebrandt 335

List of Participants 347

I. Thermonuclear Reaction Rates in Stars: Experimental and Theoretical Approaches

NEW EXPERIMENTAL APPROACHES IN NUCLEAR ASTROPHYSICS*

C.Rolfs

Institut für Kernphysik, Universität Münster, Münster, W.Germany

1. INTRODUCTION

Charged-particle-induced nuclear reactions play a crucial role in the understanding of primordial and stellar nucleosynthesis as well as of the evolution of various astrophysical scenarios [1,2]. The experimental investigations of such reactions, in principle to be carried out over a wide range of energies and to as low an energy as is technically feasible (subcoulomb energies), require often the use of novel experimental approaches. This report describes some new techniques used in such investigations. Other new techniques are discussed by H.W.Becker and F.B.Waanders in these proceedings.

2. IMPLANTED TARGETS

2.1. Carbon targets

The $^{12}C(\alpha,\gamma)^{16}O$ reaction is one of the most important processes in nuclear astrophysics [1,2]. Enormous experimental efforts have gone into studies of this reaction [Ref.3 and references therein], where formidable problems are encountered. The problems arise from the combination of a low γ-ray yield, in the nb and pb region, and a high neutron-induced γ-ray background (Fig. 1a) arising mainly from the $^{13}C(\alpha,n)^{16}O$ reaction [$\sigma(\alpha,n)/\sigma(\alpha,\gamma) \simeq 10^7$]. As a consequence, the measurements (e.g., of γ-ray angular distributions) at subcoulomb energies required [3] the use of high α-beam currents (up to 700 μA) and ^{12}C targets depleted in ^{13}C to reduce the neutron-induced background and capable of withstanding these currents. Such targets were produced [4] by the implantation technique (Figs. 1b and 1c).

A large number of factors affect the quality of the implanted targets and limit the use of the implantation technique. Phenomena such as ion collection, saturation, sputtering, diffusion, solubility and blistering depend on both the implanted ion and the backing material and determine the

---

*Supported in part by the Deutsche Forschungsgemeinschaft (Ro429/15-2) and the Friedrich Flick Förderungsstiftung.

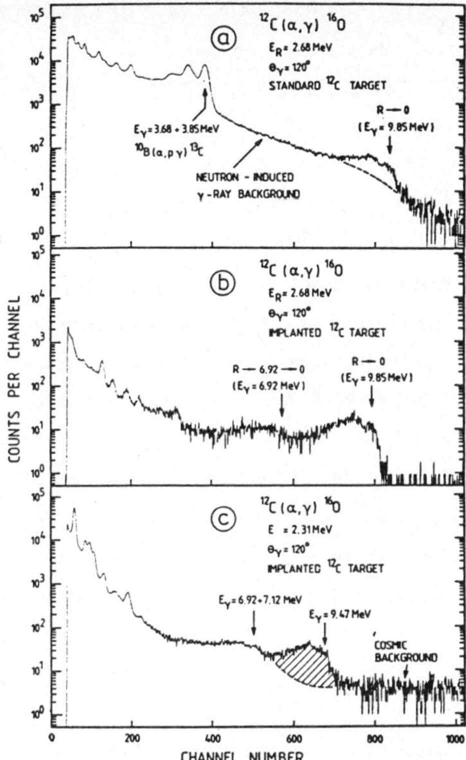

Fig. 1. Gamma-ray spectra obtained [3] with a NaI(Tl) crystal at the strong and narrow resonance at E_R = 2.68 MeV using (*a*) a standard target of normal isotopic composition and (*b*) an implanted ^{12}C target. A significant improvement in the signal-to-noise ratio for the capture transitions is noted. At energies away from this resonance, the capture cross section is much smaller and the analysis of the R → 0 γ-ray intensity (shaded area in (*c*)) depends to some extent on the assumption of background subtraction. Due to the low ^{13}C content in the implanted targets ($^{13}C/^{12}C \approx 10^{-4}$), high-resolution Ge(Li) detectors could be used safely in other experiments [3].

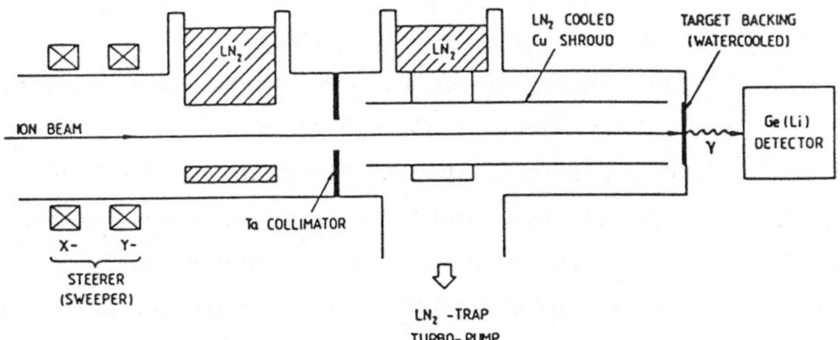

Fig. 2. Schematic diagram of the experimental set-up used in the production and investigation of implanted targets [4]. The liquid-nitrogen (LN_2) trap before the Ta collimator is about 1 m long.

maximum dose to be implanted and the implantation energy. Since no general understanding of these phenomena exists, the production of implanted targets for each element (nuclide) must be treated as a special case.

The 350 kV accelerator at the Institut für Kernphysik in Münster was used as an isotope separator providing various ion beams in the energy range of 30 to 350 keV. The magnetically analyzed ion beam (mass resolution $M/\Delta M \simeq 150$) passed trough a Ta collimator and was focused on the backing to a beam spot of about 1 cm diameter (Fig. 2). The backing was directly watercooled. A LN_2 cooled copper tube extended from the Ta collimator to within 5 mm of the backing. This tube together with the backing formed the Faraday cup for beam integration. From the integrated ion-beam currents, the dose of the ions D_{in} (in atoms/cm^2) incident on the backing was deduced. Magnetic steerers were placed at a distance of about 2 m from the target and scanned the ion beam over the backings. In this way, the incident ions were collected nearly homogeneously over a large area of the backings ($\simeq$ 3 to 4 cm^2).

The distribution of the implanted ions was investigated using the γ-ray yields of (p,γ) reactions induced on the target nuclides. For this investigation a Ge(Li) detector was positioned at 0° close to the target (Fig. 2).

In the case of ^{12}C implanted targets, the content and distribution of the implanted ^{12}C zone was determined by use of the $^{12}C(p,\gamma)^{13}N$ reaction at an incident proton energy of E_o = 330 keV (e.g., Fig. 3). Due to the smooth cross section of this reaction, the intensity and energy distribution of the isotropic γ-transition contains the desired information [4]. As seen in Fig. 3a, the ^{12}C nuclides are nearly homogeneously distributed from the surface of the target to a depth of $\Delta = 94 \pm 4$ keV at half maximum, and the area of the curve in Fig. 3a yielded an implanted dose of $(10 \pm 3) x 10^{18}$ atoms/cm^2. The target depth and the implanted dose were investigated as a function of incident dose. The results for an incident energy of E_C = 70 keV are displayed in Fig. 3b. They show that the incident ^{12}C doses are found nearly 100% in the implanted backings (i.e., no losses) and the thickness of the carbon zone increases nearly linearly with incident dose, i.e., the thickness is not given by the range of the incident ions in the backing material. For implantation in Au at doses $D_{in} \gtrsim 2x10^{18}$ atoms/cm^2, nearly pure ^{12}C targets are obtained, i.e., the carbon ions form their own target layer independent of the Au backing and thus there are no saturation effects (in contrast e.g. to nitrogen-implanted targets, section

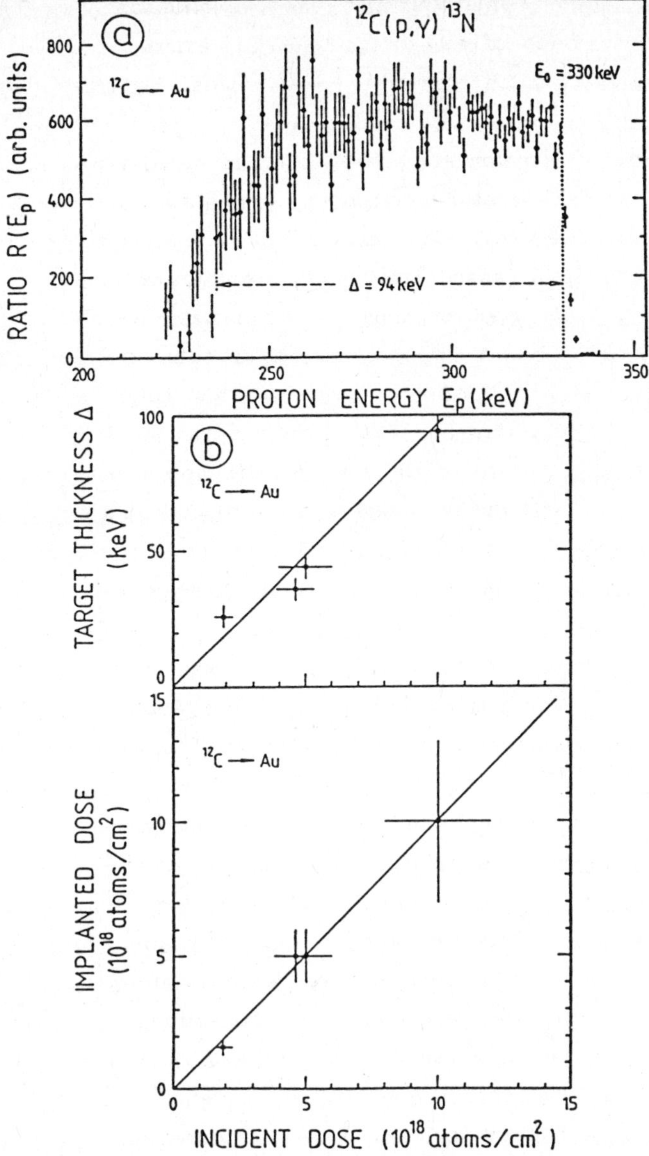

Fig. 3. Shown are results for the ^{12}C implantation into Au at E_C = 70 keV [4]. (*a*) Relevant section of the γ-ray spectrum of the $^{12}C(p,\gamma)^{13}N$ reaction as obtained with the Ge(Li) detector at an incident proton energy of E_o = 330 keV. The γ-ray yields have been corrected for the energy dependence of the capture cross section and the original γ-ray energy scale has been transformed into the equivalent proton energy scale E_p within the implantation zone. (*b*) The target thickness Δ and the implanted dose are displayed as a function of incident dose. The lines through the data points assume a linear relationship.

2.2). Similar conclusions were reached for implantation into Ta backings.

The resulting targets had a ^{13}C depletion [3] of about two orders of magnitude compared to a standard target, a limit set by the mass resolution of the accelerator facility. The depletion was sufficiently low to reduce significantly the neutron-induced background (Figs. 1b and 1c). The targets were found to be very stable against high beam loads [3].

With these implanted targets the $^{12}C(\alpha,\gamma)^{16}O$ reaction has been studied at $E_{c.m.}$ = 0.94 to 2.84 MeV involving NaI(Tl) crystals and, for the first time, germanium detectors [3]. An R-matrix fit to the data of the groundstate-transition led to $S_{E1}(E_o = 0.30\ \text{MeV}) \simeq 0.20$ MeV b for the E1 capture amplitude and $S_{E2}(E_o) \simeq 0.10$ MeV b for the E2 capture amplitude. The inclusion of cascade transitions resulted finally in $S_{tot}(E_o) = 0.32 \pm 0.18$ MeV b. The inclusion of elastic scattering data in the R-matrix fit yielded $S_{tot}(E_o) = 0.31 \pm 0.11$ MeV b [5], a factor of four higher compared to the value recommended in 1975. Other analyses such as the hybrid R-matrix model led to $S_{tot}(E_o) \simeq 0.22$ MeV b [3], a factor of three higher than the recommended value. Clearly, additional work is needed to reduce the present uncertainties of this astrophysically important reaction.

2.2. Nitrogen targets

Hydrogen burning in massive stars of second generation is governed by the CNO cycles [1,2]. The $^{14}N(p,\gamma)^{15}O$ reaction is a member of the chain of reactions involved in the cycles. It is the slowest reaction in the main CN cycle and thus controls the energy generation in the cycles. Several investigators have obtained data for this reaction at various energy regions [Ref.6 and references therein]. However, the results did not provide a coherent picture of the reaction mechanisms involved and thus did not allow for a reliable determination of the reaction rate at stellar energies.

The γ-ray work of Bailey and Hebbard [7] at $E_p \simeq 0.2$ to 1.1 MeV indicated that the predominant contribution to S(0) was the direct capture (DC) process into the 6.18 and 6.79 MeV states with $S_{6.18}(0) \simeq 1.0$ keV b and $S_{6.79}(0) \simeq 1.4$ keV b. Taking into account weaker transitions in the capture process, a total S(0) factor of 2.75 keV b was reported [7]. Based on this and other work Fowler et al. [8] recommended the value of S(0) $\simeq$ 3.32 keV b for stellar model calculations. However, the results of Bailey and Hebbard for the DC→6.18 and DC→6.79 MeV transitions lead to spectroscopic factors of $C^2S \simeq 0.27$ and 0.49 for the 6.18 and 6.79 MeV states, while stripping reactions reveal values of $C^2S \simeq 0.04$ to 0.16 and 0.27 to 0.47, respectively. The γ-ray studies [7] were complicated by the poor energy resolution of NaI(Tl) detectors in combination with high background yields in particular from the reaction $^{15}N(p,\alpha\gamma)^{12}C$. As a consequence, the capture processes into excited states of ^{15}O could only be detected via their high-energy secondary γ-ray transitions. Thus, improved studies of this reaction required ^{15}N-depleted ^{14}N targets. However, ^{14}N gas depleted in ^{15}N from the normal 0.36% abundance is not commercially available.

For this reason, such targets were produced [4] by ^{14}N implantation into Ta and Au backings (Fig. 4). The ^{14}N distribution in the Ta and Au backings was investigated using the thick-target yield curves obtained at the E_R = 278 keV resonance of $^{14}N(p,\gamma)^{15}O$ (Γ = 1.1 keV). A "target profile" obtained after a ^{14}N irradiation dose of $8x10^{18}$ atoms/cm^2 in Ta at E_N = 200 keV is shown in Fig. 5a. As seen in the figure the ^{14}N atoms are nearly homogeneously distributed from the surface of the Ta to a depth of 52 keV at half maximum. This is consistent with the range of ^{14}N ions in Ta. The target thickness Δ and the number of implanted ^{14}N ions have been determined as a function of the amount of incident ^{14}N. The results (Fig. 5b) show that the implantaion reaches a saturation above a concentration of incident ^{14}N ions of about $5x10^{18}$ atoms/cm^2, which corresponds to a target stoichiometry nearly equal to that of the compound Ta_2N_3. All implantations in Ta as well as in Au, carried out at E_N = 50 to 200 keV, reached a saturation in implanted ^{14}N; however, the implanted ^{14}N concentration at saturation in Au was a factor of 4 to 5 less than for implantation into Ta. Thus, Au is much less suitable as an implantation material for nitrogen ions than Ta. Similar to the ^{12}C targets (section 2.2), the ^{15}N in these targets was found to be reduced by about two orders of magnitude from the natural abundance level. This ^{15}N reduction was sufficient for improved studies of the $^{14}N(p,\gamma)^{15}O$ reaction over a wide range of proton energies [6]. The implanted targets were extremely stable against high intensity proton beams.

Analyses of the data obtained for capture into excited states of ^{15}O led to the following results [6]:

(i) The S(0) value, 1.41 keV b, for capture into the 6.79 MeV state is in excellent agreement with the previous result [7].

(ii) The reported transition into the 6.18 MeV state, making apparently another major contribution to the total S(0) value [7], has been found to be of minor importance: $S_{6.18}(0)$ = 0.14 keV b; this finding also reconciles the spectroscopic factors deduced from the DC process and stripping reactions. (iii) The total S(0) value of 1.65 keV b for all transitions into

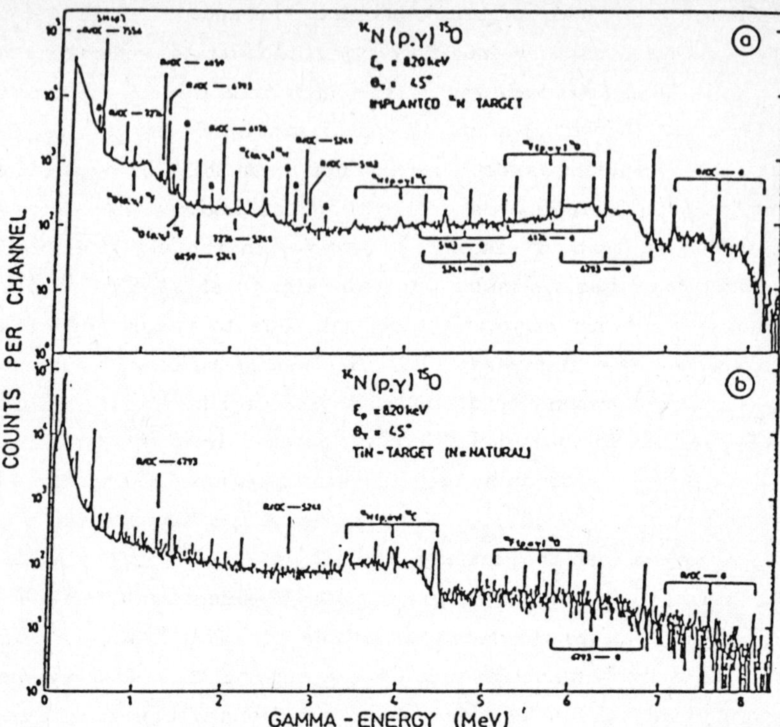

Fig. 4. Gamma-ray spectra obtained [6] with a Ge(Li) detector using (*a*) an implanted ^{14}N target and (*b*) a TiN target of normal nitrogen isotopic composition. A significant improvement in the signal-to-noise ratio for the capture transitions below E_γ = 4.4 MeV is noted. The identification of the observed γ-ray peaks is also given (B = background lines).

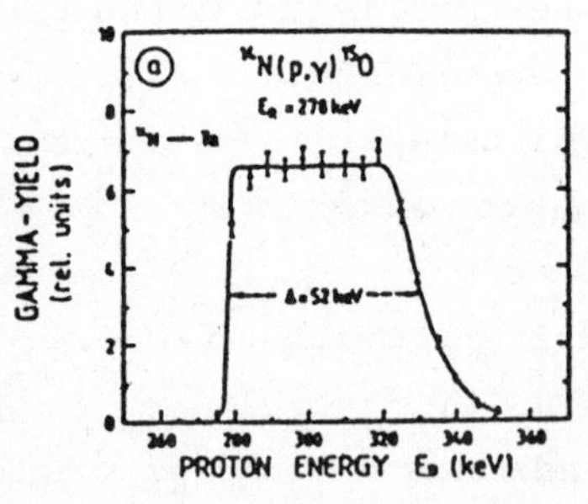

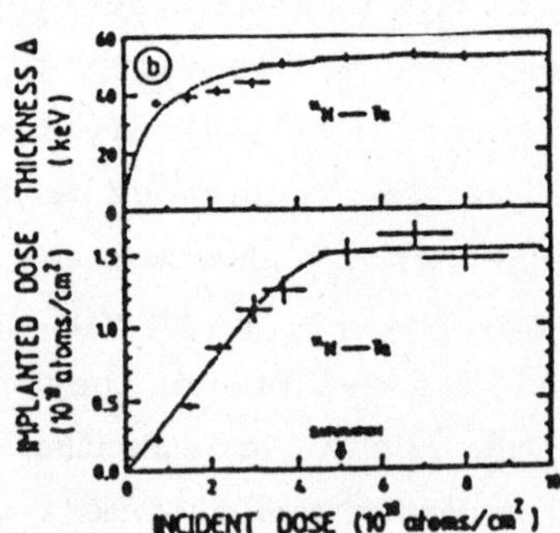

Fig. 5. (*a*) Thick-target γ-ray yield curve of the 278 keV resonance of $^{14}N(p,\gamma)^{15}O$ obtained with a ^{14}N-implanted Ta sheet. (*b*) The target thickness Δ and the number of implanted ions are displayed as a function of incident dose for E_N = 200 keV. The data show a saturation in implantation at about D_{in} = $5x10^{18}$ atoms/cm^2. The curves through the data points are to guide the eye [4].

excited states is a factor of two lower than the recommended value.

The data for capture into the ^{15}O ground state were analysed [6] in terms of the known (broad) resonances including the DC process. The best fit to the data, shown as a dotted curve in Fig. 6, did not reproduce the data and the resulting spectroscopic factor of the ^{15}O ground state was inconsistent with stripping data. Including the high-energy wing of the E_R = -504 keV subthreshold resonance (s-wave) in the analysis resulted in a significantly improved fit (solid curve in Fig. 6) and led to a spectroscopic factor consistent with stripping reactions. Due to the large partial widths in both the particle and γ-ray channels (nearly maximum "pole" strength), this subthreshold resonance makes a major contribution to the S(0) value, 1.55 keV b, although it is relatively far located from the proton threshold. The contribution of this subthreshold resonance has to date been neglected in the analyses.

It is noted that the capture process into the 6.79 MeV state and the capture process into the high-energy wing of this state make the dominant contribution of 93% to the total S(0) value of 3.20±0.54 keV b. This result shows that in light nuclei a single state can be of crucial importance for stellar burning rates. The summed contributions of all transitions to the S(E) factor are displayed as a solid curve in Fig. 7. Also shown are the data from previous work. It is seen that the results of Lamb and Hester and of Pixley are in excellent overlap with the curve and that the apparent discrepancy between the two data sets (assuming an energy-independent non-resonant S(E) factor) is removed. The data of Bailey and Hebbard at $E_p \geq 0.35$ MeV are higher than the curve, probably due to problems discussed above. The energy dependence of the data of Duncan and Perry is in fair agreement with the curve except for the absolute scale.

In summary, although there are two major changes in the previously suggested reaction mechanisms (i.e., a negligible contribution of the DC→6.18 MeV process and a significant contribution of the E_R = - 504 keV subthreshold resonance), they cancel each other predominantly. Thus, the total S(0) value is essentially identical with the recommended value. However, the present work improved the understanding of the reaction mechanisms involved and thus of the extrapolated S(0) value for this astrophysically important reaction.

2.3. Sodium targets

The $^{22}Na(p,\gamma)^{23}Mg$ reaction is of current interest to γ-ray astronomy for investigating hot and explosive astrophysical scenarios [9]. Due to the

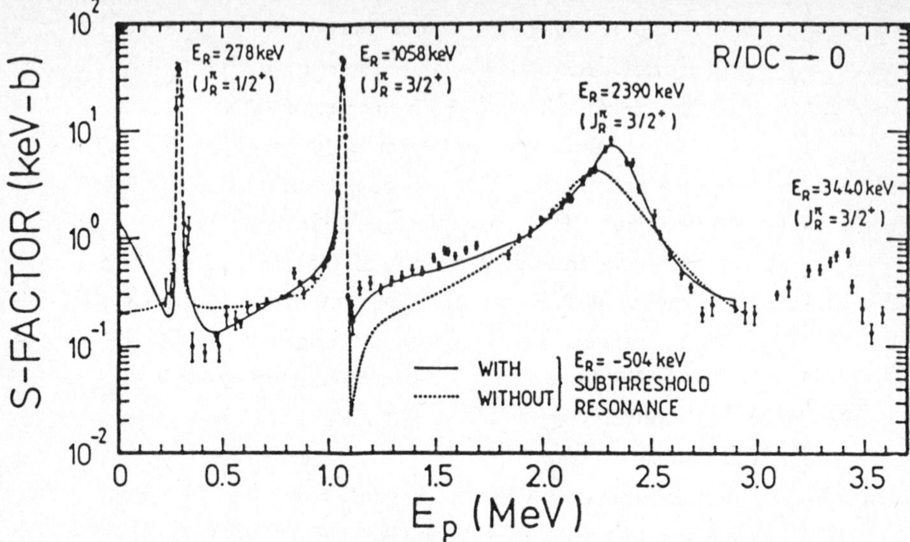

Fig. 6. The $^{14}N(p,\gamma)^{15}O$ data for the capture transition into the ^{15}O ground state are displayed in the form of the astrophysical S(E) factor [6]. The solid (dotted) curve is the result of an R-mátrix fit including (excluding) the E_R = -504 keV subthreshold resonance. The improved fit (solid curve) led to an extrapolated value of S(0) = 1.55 keV b.

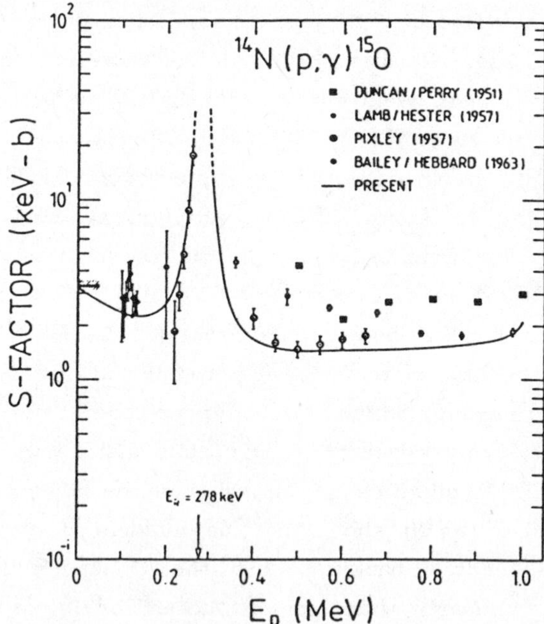

Fig. 7. S(E)-factor data of $^{14}N(p,\gamma)^{15}O$ are compared with previous results [6]. The solid curve with its dotted band represents the present results including the experimental uncertainties. The dashed curve at $E_p \leq 0.2$ MeV represents the extrapolation of the present data to zero energy. Also shown is the previously recommended S(0) value (dashed area).

^{22}Na half-life of 2.60 y, a reasonably thick target (1 to 2 keV thick at E_p = 1 MeV) will have a radioactivity of about 10 mCi (E_γ = 0.51 and 1.27 MeV). To study the capture γ-rays in the environment of such a "hot" radioactive target, a new γ-ray detector is presently being constructed (see contribution by F.B.Waanders), which is based on the γ-ray detection via D(γ,n)H neutron counting [9]. Radioactive ^{22}Na material is commercially available. Ten mCi of ^{22}Na in a target spot of 0.5 cm^2 would yield at E_p = 1 MeV a target thickness of 0.6, 1.5 and 2.4 keV for a pure ^{22}Na target, a ^{22}NaCl and a ^{22}NaTa target, respectively. Of course, the target should be as pure in ^{22}Na as possible and the target thickness should be larger than the beam-energy resolution ξ (here: $\xi \simeq$ 0.5 keV). Since Na metal deposited on a backing will form a compound with the backing, the best choice might be a target material deposited in the oxide or chloride form. However, minimizing background contributions due to the target partner requires a heavier partner, achievable possibly with ^{22}Na implanted in Ta or other high-Z backings. Such implanted targets might also be more stable against high beam loads than evaporated targets.

This possibility for the production of ^{22}Na targets was investigated by testing the implantation of ^{23}Na ions in various heavy backings at E_{Na} = 30 and 50 keV [4]. The ^{23}Na distribution in the embedded targets was studied using the thick-target yield curves at the E_R = 309 keV resonance of ^{23}Na(p,γ)^{24}Mg ($\Gamma \leq$ 20 eV). Sample ^{23}Na distributions in Ni, Zr, Nb, Ta and Au backings are shown in Fig. 8 as obtained for incident doses D_{in} = 8.6, 1.25, 0.56, 1.3 and 1.4 atoms/cm^2, respectively. As seen in the figure the ^{23}Na nuclides are distributed from the surface of the backing to a mean depth of Δ(Zr, Nb) $\simeq$ 11 to 12 keV and Δ(Ta, Au) $\simeq$ 14 to 15 keV, consistent with the range of ^{23}Na ions in the backings. In all these backings, the distributions reveal a more or less pronounced concentration near the surface of the backing, which is not understood. The exception to these features is Ni showing a nearly homogeneous distribution from the surface to a depth, which is significantly larger than the range of ^{23}Na ions in Ni. The data obtained for Ni show that the incident doses are found nearly 100% in the implanted backings and the thickness of the ^{23}Na zone increases nearly linearly with incident dose. For $D_{in} \geq$ 1x10^{18} atoms/cm^2, nearly pure ^{23}Na targets are obtained, i.e., the ^{23}Na ions form their own target layer independent of the Ni backing and thus there are no saturation effects, an analog situation to the ^{12}C targets discussed in section 2.1. For the other backings a saturation concentration is found near D_{in} = 0.5x10^{18} atoms/cm^2 (Fig. 8). It is found that -aside from Ni- the

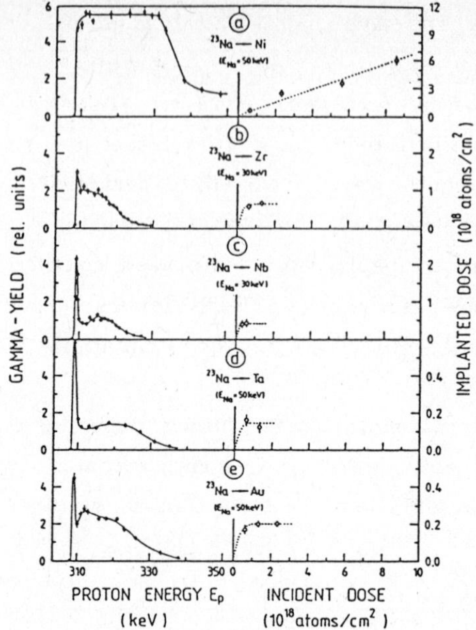

Fig. 8. Thick-target yield curves of the E_R = 309 keV resonance in $^{23}Na(p,\gamma)^{24}Mg$ are displayed on the left side of the figure showing sample ^{23}Na distributions implanted in Ni, Zr, Nb, Ta and Au backings [4]. On the right side the implanted dose is shown as a function of incident dose. With the exception of Ni, all backings exhibit a saturation concentration at about D_{in} = 0.5×10^{18} atoms/cm^2. The solid and dotted curves through the data points are to guide the eye.

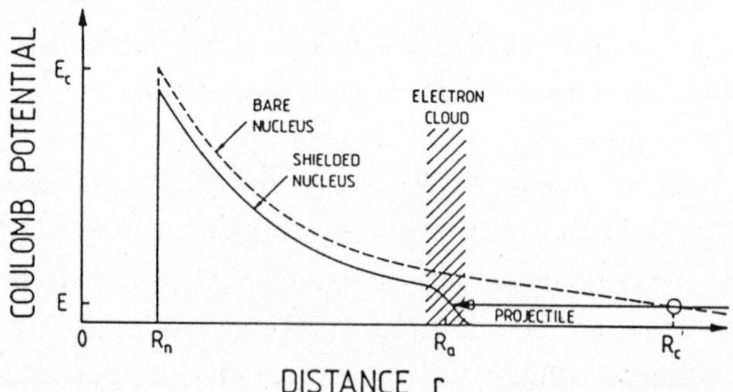

Fig. 9. Shown in an exaggerated and idealized way is the effect of the atomic electron cloud on the Coulomb potential of a bare nucleus [10]. This potential is reduced at all distances and goes essentially to zero beyond the atomic radius R_a. The effect of this electron shielding on an incident projectile is always to increase the penetrability through the barrier and thus also the cross section (or equivalently the S(E) factor).

backings Au and Ta are much less suitable as implantation materials for sodium ions than Zr and Nb. Thus, the requirements of a high ^{23}Na concentration and of a homogeneous ^{23}Na distribution in the backing appear to be best fulfilled by Ni, and to a lesser extend by Zr and Nb. The ^{23}Na-implanted Ni targets were found to deteriorate by about 35% for an accumulated charge below 1 coulomb and to remain stable thereafter, while the other ^{23}Na-implanted backings deteriorated continuously with a rate of about 9% per coulomb [4]. The thick Ni layers (evaporated on Ta) were found to be extremely free of target contaminations (investigated for $Z \leq 14$).

In summary, the implantation technique provides a good means for the production of isotopically enriched targets capable of withstanding high beam loads and thus for their use in reaction studies relevant to nuclear astrophysics.

3. ATOMIC SCREENING

In the usual treatment of charged-particle-induced nuclear reactions at subcoulomb energies it is assumed that the Coulomb potential of the target nucleus as seen by the projectile is that resulting from a bare nucleus of charge number Z_1 and thus would extend to infinity (Fig. 9). However, for nuclear reactions studied in the laboratory the target nuclei are usually in the form of atoms or molecules (i.e., gas or solid targets) and the projectiles are usually in the form of positively charged ions (here assumed to have an effective charge of $Z_{eff} = Z_2$, e.g. $^1H^+$). The atomic (or molecular) electron cloud surrounding the target nucleus acts as a screening potential. As a result, the total potential goes to zero outside the atomic (molecular) radius R_a. An incoming projectile sees no repulsive Coulomb force until it penetrates beyond the atomic (molecular) radius; thus, it effectively sees a reduced Coulomb barrier. In a simple approximation the total Coulomb potential within the atomic radius is $\Phi_{tot} = (Z_1e/r) - (Z_1e/R_a)$. The effective height E_c^* of the Coulomb barrier seen by the incoming projectile (Fig. 9) is then given by the equation $E_c^* = Z_1Z_2e^2(1/R_n - 1/R_a)$, where R_n is the nuclear interaction radius. The equation shows that the effect of the electron shielding on the height of the Coulomb barrier is in the ratio of nuclear to atomic radii, that is $R_n/R_a \simeq 10^{-5}$. In general, this shielding correction is negligible. However, at low projectile energies, when the classical turning point R_c of an incoming projectile for the bare nucleus is near or outside the atomic radius (Fig. 9), the magnitude of the shielding effect becomes significant. Since the classical turning point is related to the projectile energy E by the relation $E = Z_1Z_2e^2/R_c$, the condition $R_c \geq R_a$ leads to energies of $E \leq U_e = Z_1Z_2e^2/R_a$, where the shielding effects cannot be disregarded. Setting the atomic radius equal to the radius of the innermost electrons of the target (or projectile) atom, i.e., $R_a = R_H/Z_i$ with the Bohr radius R_H, the relevant energies U_e from the above equation are listed in Table 1 for representative examples. These energies are all quite low and thus the

Table 1
Atomic screening effects on charged-particle-induced nuclear reactions[a)]

System	U_e (keV)	Enhancement ratio f E/U_e = 1	10	100	1000	Experiment E/U_e
d+d	0.027	1.7×10^{24}	16.5	1.10	1.003	107
$d+{}^3He$	0.11	2.0×10^{26}	20.9	1.11	1.003	64
${}^3He+{}^3He$	0.22	7.4×10^{41}	131	1.18	1.006	73
$p+{}^7Li$	0.24	6.4×10^{22}	14.0	1.09	1.003	125
$p+{}^{11}B$	0.68	1.3×10^{23}	14.4	1.09	1.003	32
$\alpha+{}^{12}C$	2.0	3.2×10^{58}	868	1.25	1.007	450
${}^{12}C+{}^{12}C$	5.9	3.2×10^{144}	1.9×10^7	1.76	1.016	338

a)For details, see [10] and references therein.

shielding effects might appear to be effectively unimportant. However, the penetration through a shielded Coulomb barrier at projectile energy E is equivalent to that of bare nuclei at energy $E_{eff} = E + U_e$ (Fig. 9). The shielding effect reduces the Coulomb barrier and eases the penetration of the Coulomb barrier. Thus, it increases the cross sections. Assuming a constant astrophysical S(E) factor over a relatively small energy interval (e.g., thin target), the enhancement ratio f in cross sections is [10] $f \simeq \exp(-2\pi\eta(E_{eff}))/\exp(-2\pi\eta(E))$. The examples given in Table 1 indicate that for energies $E/U_e \geq 1000$ the shielding effects are negligible and laboratory experiments can be regarded as measuring essentially the cross section for penetration through a Coulomb barrier for the bare nuclei. However, at energies $E/U_e \leq 100$ the shielding effects cannot be disregarded and become important for the understanding of the low-energy data and thus for their theoretical extrapolation to zero energy. The examples (Table 1 and Fig. 10) illustrate also that with improved experimental techniques (see contribution by H.W.Becker) direct yield measurements have been carried out towards, and in some cases even below, the region $E/U_e = 100$.

An improved theoretical approach within the Born-Oppenheimer approximation has been carried out recently [10]. These studies verified

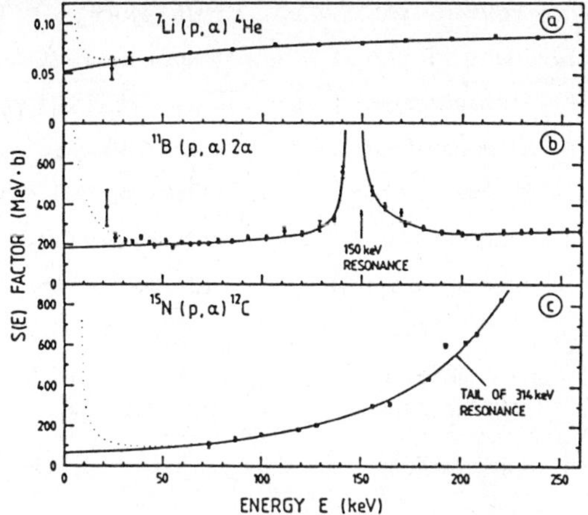

Fig. 10. Shown are the low-energy data of the (p,α) reactions on ^{7}Li, ^{11}B and ^{15}N. The solid curves through the data points in (*a*) and (*b*) are to guide the eye, while the solid curve in (*c*) is the result of a Breit-Wigner fit to the data. The dotted curves represent the predicted enhancements due to atomic effects [10].

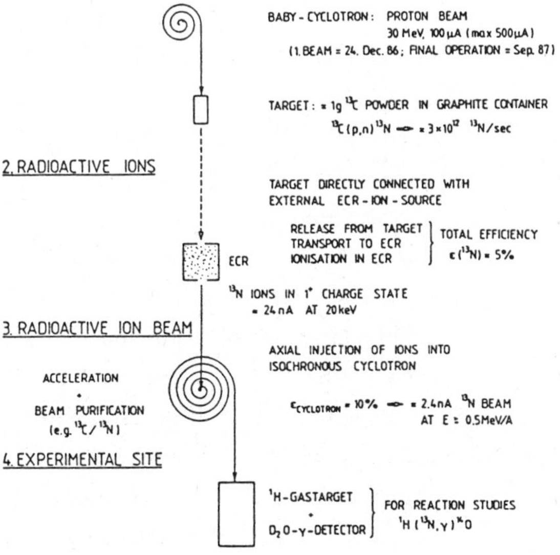

Fig. 11. Schematic diagram of the experimental procedures for the production of radioactive ion beams at Louvain-la-Neuve [12].

essentially the results of the simple theoretical considerations discussed above and indicated that a significant enhancement of the cross sections can occur already at beam energies, which are about a factor 100 higher than the electron binding energies. Thus, the laboratory measurements of charged-particle-induced cross sections must be corrected for the shielding effects of the orbital electrons around the target and projectile nuclei in order to arrive at the cross sections for bare nuclei, which is the needed information for nuclear astrophysics. It should be pointed out that relatively small effects at energies $E/U_e = 100$ can become significantly amplified in the extrapolation to zero energy. Thus, for reliable extrapolations the possible influence of the electron shielding on nuclear cross sections at low energies must be well understood. Experimental work for several nuclear reactions at lower energies than those shown in Fig. 10 is in progress, where the electron screening effects should become clearer visible. A better understanding of these screening effects might eventually also be of importance for the calculation of the screening effects in a stellar plasma [10].

4. HOT/EXPLOSIVE BURNING

In the hot/explosive burning phase of stellar evolution, nuclear burning times can be measured in seconds or less. If the lifetime of radioactive nuclei (or nuclei in isomeric states) are comparable to or longer than the burning times, they will become involved in the nuclear burning processes [1,2]. Laboratory studies of such reactions (predominantly of the type (p,γ), (p,α), and (α,γ); for example $^{13}N(p,\gamma)^{14}O$, $T_{1/2} \simeq 10$ min, hot CNO-cycle) require new experimental techniques [11]. An important question in the experimental approaches is the relative merit of radioactive targets and radioactive ion beams. A decision on the most advantageous way of studying such a reaction is found to be a funçtion of the half-life of the radioactive nuclei [2,11]. If the half-life is shorter than about 1 hour, then the radioactive beam experiment is more advantageous. If it is longer, then a radioactive target experiment is better (e.g., ^{22}Na, section 2.3). It should be noted that this conclusion is independent of the reaction cross section and of the production rate of the radioactive nuclides (e.g., Fig. 11). However, in the radioactive target technique (and to some extend in the radioactive ion beam technique) there is the problem of "hot" targets (and e.g. "hot" beam collimators), which might influence the above conclusion. For γ-ray spectroscopy measurements, the new D_2O-detector might offer here a solution (see contribution by F.B.Waanders). Other new detection techniques in such hot environments are presently developed in several laboratories [11].

Recently, a technique has been proposed [12] for the production of radioactive ion beams, which is novel and unique (compared to other techniques [11]) and shown schematically in Fig. 11. Specific numbers are given in the figure for a ^{13}N radioactive ion beam. Using a windowless hydrogen gas target and the presently constructed D_2O-γ-detector (total efficiency $\simeq 10^{-3}$), the expected ^{13}N beam current will lead near the $E_{c.m.} \simeq$ 0.55 MeV resonance of $^{13}N(p,\gamma)^{14}O$ to a count rate of about 1 event per second.

It is clear that much research and technical development is needed on all aspects of the problem before nuclear reactions involving short-lived radioactive nuclei can routinely be carried out in the laboratory.

REFERENCES

1. W.A.Fowler, Rev.Mod.Phys. 56(1984)149
2. C.Rolfs, H.P.Trautvetter and W.S.Rodney, Rep.Progr.Phys. 50(1987)3
3. A.Redder, H.W.Becker, C.Rolfs, H.P.Trautvetter, T.R.Donughue, T.C.Rinckel, J.W.Hammer and K.Langanke, Nucl.Phys. A462(1987)385
4. S.Seuthe, H.W.Becker, A.Krauss, A.Redder, C.Rolfs, U.Schröder, H.P.Trautvetter, K.Wolke, S.Wüstenbecker, R.W.Kavanagh and F.B.Waanders, Nucl.Instr.Meth. (submitted)
5. R.Plaga, H.W.Becker, A.Redder, C.Rolfs, H.P.Trautvetter and K.Langanke, Nucl.Phys. A465(1987)291
6. U.Schröder, H.W.Becker, G.Bogaert, J.Görres, C.Rolfs, H.P.Trautvetter, R.E.Azuma, C.Campbell, J.D.King and J.Vise, Nucl.Phys. A467(1987)240
7. G.M.Bailey and D.F.Hebbard, Nucl.Phys. 46(1963)529; 49(1963)666
8. W.A.Fowler, G.R.Caughlan and B.A.Zimmerman, Ann.Rev.Astron.Astrophys. 5(1967)525; 13(1975)69
9. C.Rolfs and R.W.Kavanagh, Nucl.Instr.Meth. A247(1986)507
10. H.J.Assenbaum, K.Langanke and C.Rolfs, Zeitsch.Phys. (submitted)
11. L.Buchmann and J.M.d'Auria, *Accelerated Radioactive Beams Workshop*, Parksville, Canada (TRIUMF TRI-85-1)
12. J.Vervier, Universite Catholique de Louvain, Louvain-La-Neuve, private communication (1987)

DIRECT CROSS SECTION MEASUREMENTS TOWARDS THERMAL ENERGIES*

H.W.Becker

Institut für Kernphysik, Universität Münster, Münster, W.Germany

1. INTRODUCTION

It is essential for the understanding of the structure and evolution of the stars in the quiescent burning phases to know the cross sections of the relevant reactions in the thermal energy region. However, for the charged particle induced reactions it is in the most cases impossible to measure this cross sections directly because the influence of the coulombbarrier leads to such small cross sections that with present techniques we are far of beeing able to measure such weak processes. Therefore the experimental investigations have to be extended to energies as low as possible, to provide data for an extrapolation to stellar energies [1].

In principle such measurements are carried out in the well known standard way with an accelerator to provide an ion-beam with the appropriate energy, a target with the nuclei of interest and a detector to measure which and how probable a reaction takes place. This contribution reports about efforts to build and tune each of this components for the special requirements of astrophysical low energy measurements. This requirements can only fullfilled with a dedicated accelerator providing very high beam currents and a target which can stand this beams intensities. In addition a detection technique with high detection efficiency for the reaction products but not for any kind of background had to be developed.

The first part of this contribution discusses the lay out and features of the experimental techniques, the following sections present some examples of reaction cross section measurements carried out with this experimental set up.

2.EXPERIMENTAL EQUIPMENT

2.1 Accelerator

Many low energy accelerators provide an energy of a few hundred keV for the ions, thus in running at a few keV or between ten to fifty keV they

* Supported in part by the Deutsche Forschungsgemeinschaft (Ro429/15-2) and the Friederich Flick Förderungsstiftung.

come at the end of their operating region and the limits for the beam transport system. In this region their beam currents, which are in the optimal cases up to 500 μA are reduced to a few μA.
Therefore an accelerator has been build at the Dynamitron Tandem Laboratory at Bochum for the special purposes of the astrophysical low energy measurements [2]. Fig.1 shows a shematic view of this accelerator.

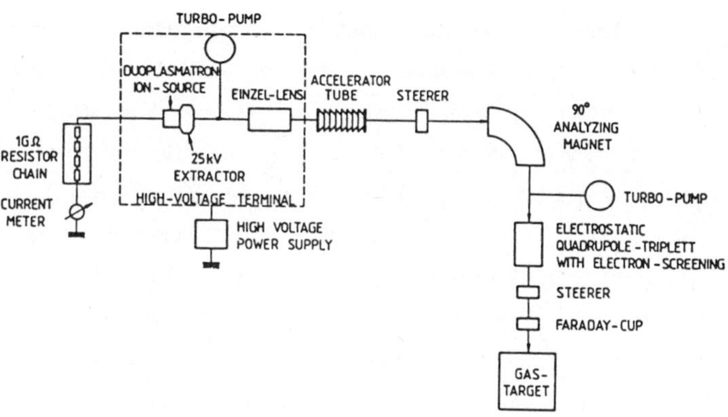

Fig.1 Schematic diagram of the main components of the 100 kV accelerator facility at Bochum.

The ions are produced in a duoplasmatron ion source, which can provide up to 10 mA beam current. After the usual extraction from the source with 25 kV the ions are focused by an einzel lens through the small acceleration tube, acceleratet to ground and deflected by the 90° magnet through slits to provide a sufficient energy determinaton. From here the beam is transported by an electrostatic quadrupol triplet and a pair of magnetic steerers into the experimental set up. The total path length of the ions from the source to the target is only 6 m. In order to keep the ion currents and the beam optics nearly constant over the full energy range of the accelerator, the extraction voltage is kept constant, for ion energies below 25 keV the beam has to be decelerated.
Ion beams with high intensities and this low energies are difficult to focuse because the effects of space charge tend to blow up the beam. Therefore it is necessary to add electrons to the beam at the front and exit of the triplet by installing scimmers, where the beam produces itself secondary electrons.
Several additional improvements [2] finaly led to the following features of

this accelerator. It can provide protons, ^{3}He and ^{4}He with analysed currents of about 1 mA stable over periods of a week. The stability of energy is ± 20eV.

2.1 Energy calibration of the accelerator

At low energies far below the Coulombbarrier the cross section depends exponential of the energy according to the penetrability through the Coulombwall. This leads in addition to the problem of less and less reaction events to the difficulty of determing the effective beam energy very pecisely. This measurements become at low energies as important as the measurement of the reaction events itself. For the example of the D+D reaction an uncertainty of ± 500 V in the absolute beam energy results at 70 keV in an error in the cross section of ± 2%, at an energy of 10 keV the same uncertainty leads to an error of ± 40%.

Therefore different methods for the energy calibration were applied. In general, for this low energies it is possible to measure the accelerating voltages with sufficient accuracy and to decuce from that the beam energy. A first value for the energy can be obtained by adding the readings of the accelerating powersupplies, namely the extraction and the accelerating voltages. In addition a precise and constant resistor chain were build and calibrated together with a digital ampreremeter at the Physikalisch-Technische- Bundesanstalt with an accuracy of 5×10^{-4}. The two methods agreed very well.

Due to the importance of the information of the beam energy a new and independent method of accelerator calibration appeared desirable; the usual way of calibrating via resonances in nuclear reaction is not possible since there are no well known resonances in this energy region. Therefore a time of flight technique has been developed to measure the velocities of the beam ions and deduce from that the energy.

In principle, the beam is passing two detectors (Fig.2), where the secondary electrons produced by the ions in a thin Carbon foil are accelerated by a grid, deflected out of the beam direction by an electrostatic mirror and detected by multichannel plates. To avoid an absolute determination of the distance of the foils and the flight time, the stopdetector is precisely movable in beam direction, thus the change in flight time is measured in dependence of a change in flight distance. This method is very suitable for low energy beams because the velocities of the ions are low enough to be measured with sufficient precision. However, this technique has the disadvantage, that the evaluated beam energy has to be

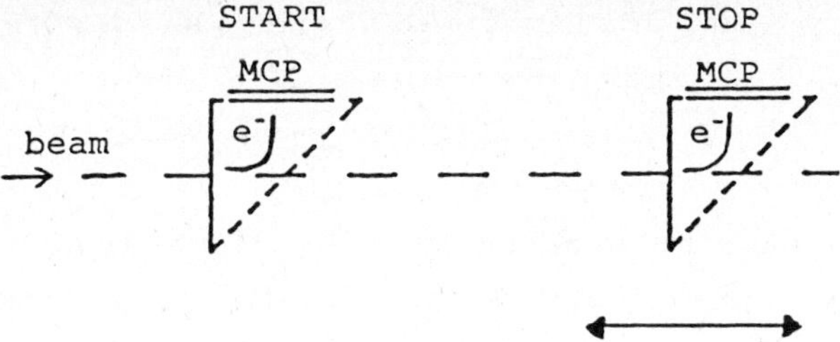

Fig.2 Schematic diagram of the time of flight set-up for the energy determination of the accelerator beam.

corrected for the energy loss in the first foil, which is difficult to determin with the required precision. The results of a first analysis of the data seem to confirm the energy calibration via the resistor chain.

2.3 Target and detection technique

The measurements of the reaction between the very light ions such as hydrogen or helium enforces the use of a gas target, which has to be windowless, an entrance window leads to an energy loss of the beam and a high uncertainity in the effective beam energy in the target. In addition gas targets have several advantages for the low energy measurements, namely the purity and stability against high beam loads. The gastarget

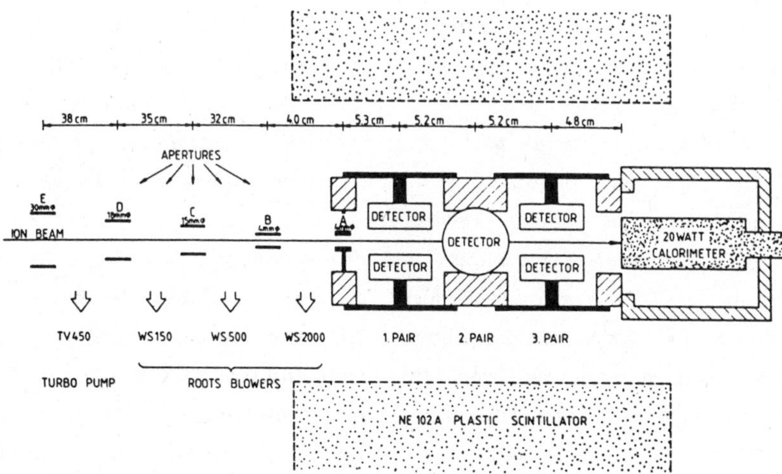

Fig.3 Schematic diagram of the relevant parts of the windowless gas target system an the experimental set-up.

sytem used is described in detail elsewhere[3], Fig.3 shows only the relevant parts. The beam enters the rectangular shaped targetchamber via five entrance collimators, which serve also as reduction stages for the target pressure. In the chamber, which is filled with the target gas, six detectors are mounted very close to the center of the beam to yield solid angles as large as possible. The detectors itself were covered with thin foils to protect them from the intense elastic scattering. Due to the energy loss of the beam in the target the yield in each detector pair corresponds to a slightly different effective beam energy, this effects as well as the effect of the large solid angle can be nicely studied by the simple variation of the pressure in the targetchamber [2].
The beam is finaly stopped in a calorimeter to determin the intensity of the beam by measuring its power, because the normal charge integration is due to charge exchange effects in the target gas not reliable.

3. THE D+D AND THE ^{3}He+D REACTIONS

The deuterium induced reactions on deuterium and ^{3}He are relevant for the primordial nucleoshynthesis during the early stages of the universe, especially in the production of the light nuclei between deuterium and ^{7}Li [4]. Beside this they are also important for the concept and design of future fusion reactors, which are supposed to run at temperatures, corresponding to energies of 1 to 30 keV. For this reason, these reactions have been studied already long time ago, but recent evaluations indicate the possibility of large systematic errors, mainly due the problems of determing the effective beam energy in the foil-contained gastargets, used in that studies.
The D+D reaction leads to a neutron and a ^{3}He or to a proton and a ^{3}T, the ^{3}He+D reaction to a proton and a ^{4}He. Fig 4. shows spectra of the two reactions at the lowest energy, measured in the present investigations [2]. The peaks of the reaction products are clearly visible, in the case of the D+D reaction the ^{3}He particles could not be obseved, since they are stopped in the detector foils for beam energies below 12 keV. The shape of the lines is given by the varying kinematics and solid angles for different reaction locii in front of the detector.
Fig.5 and 6 display the results in terms of the astrophysical S-factor in comparison to previous work and the compilation of Fowler [7]. This graph containes not only measurements at energies below 100 keV at the small maschine at Bochum, but also results at higher energies obtained at the 350 kV machine at Münster. Furthermore the data are also based on studies of

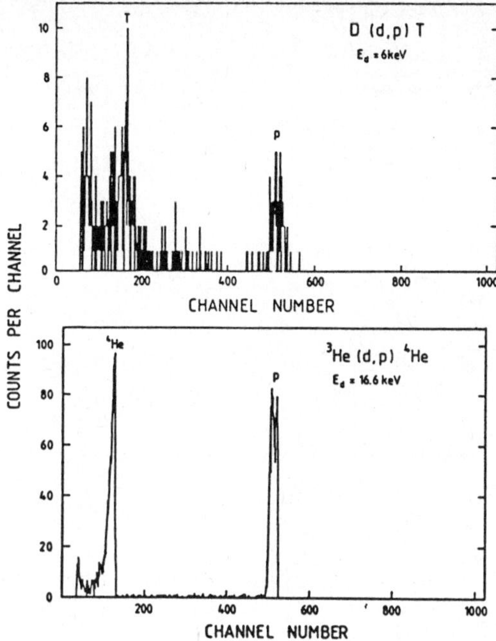

Fig.4 Spectra of the D(d,p)T and the ^{3}He(d,p)^{4}He reactions at the lowest energy measured.

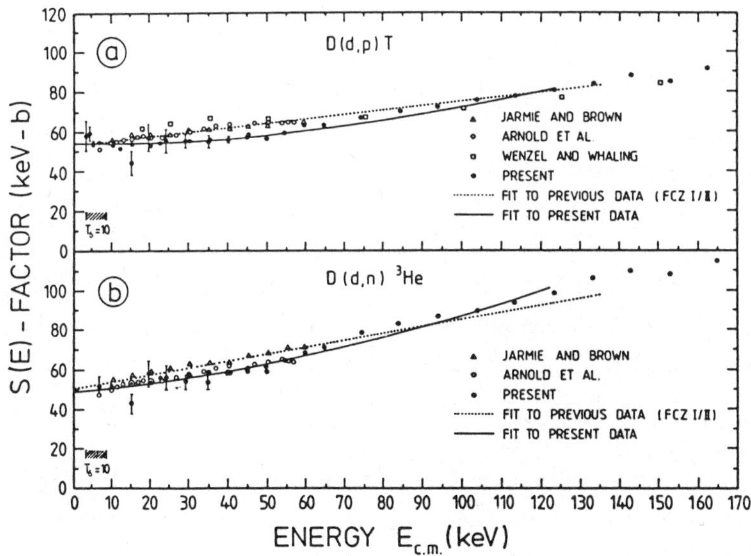

Fig.5 S(E)-factor data for the D+D fusion reactions are compared with previous results and a previous fit ([2] and references therein).

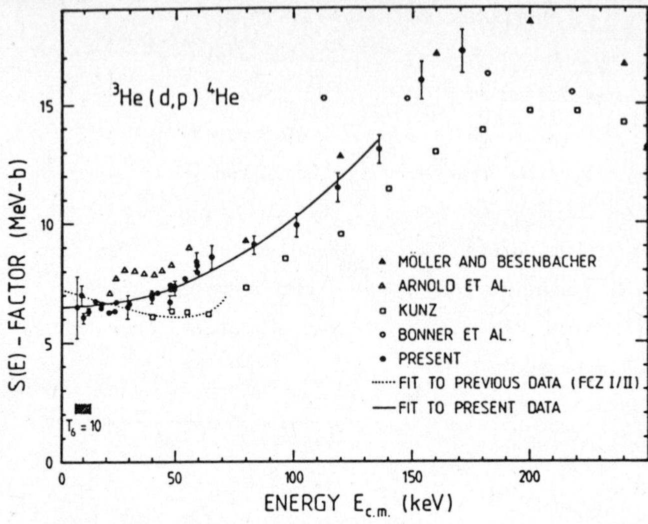

Fig.6 S(E)-factor data for the ^{3}He(d,p)^{4}He fusion reaction are compared with previous results and a previous fit ([2] and references therein)

the angular distributions of the reaction products [2], not discussed here.

The present data are in the energy dependence as well as in absolute cross section in good agreement with previous work. The fit to the S-factor, presented by the solid line, leads to similar values in the thermal energy region as they were compiled by Fowler, Caughlan and Zimmerman [5]. From that picture the cross section of the D+D reactions appear to be well known and established.

The situation is different for the ^{3}He+D reaction (fig.6), where already the previous work gave different results mainly in the absolut cross section but also in the energy dependence of the S-factor. The present data lie at low energies between the values of previous studies, at higher energies additional experimental information appear to be desirable. The discreapancies could be explained in part by differences in the determination of the beam energy.

To demonstrate the capability of the accelerator and the experimental set-up it should be noted, that e.g. the D+D spectrum (fig.4) could be obtained in only one hour running time. It is possible and might be interesting e.g. in future studies of atomic screening effects [6] to extend the data even into the subthermal energy region.

4. THE $^3He(^3He,2p)^4He$ REACTION

The $^3He(^3He,2p)^4He$ reaction is involved in the hydrogen burning p-p-chain, where after the production of ^3He and ^4He this reaction is in competition to the $^3He+^4He$ reaction. Therefore a good understanding of this reaction is extremly important for the solar neutrino problem.

Fig.7 displays the results of previous work, a polynomial fit to the data and a theoretical calculation based essentialy on a reaction mechanism, which assumes the tunneling of one neutron from one nucleus to the other leading to a ^4He and the decay of a diproton. This curve has been normalized in absolute scale to the data.

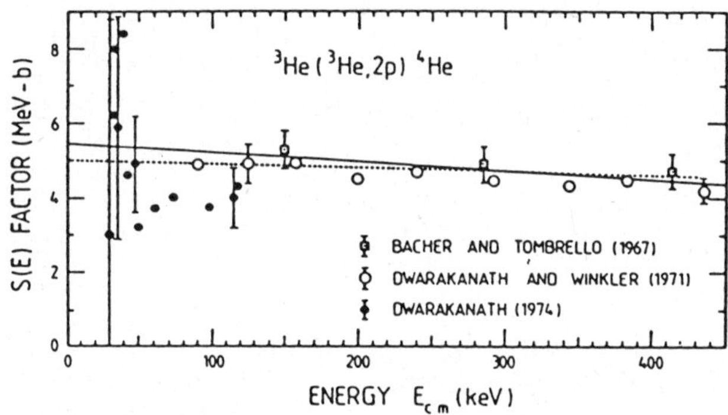

Fig.7 Astrophysical S(E)-factor of the $^3He(^3He,2p)^4He$ reaction as obtained in previous work. The solid curve is a polinomial fit to the data and the dotted curve a theoretical calculation normalized to S(0) = 5.0 Mev·barn ([9] and references therein)

The measurements of this reaction have been extended to energies down to 30 keV, but the lowest data point have a large error up to 200%. This most recent studies have been stimulated by the suggestion of Fowler [7] and also Fetisov [8], that a resonance in the $^3He+^3He$ reaction near the threshold could explain the discrepancies between the measured neutrino flux and the predictions of the standard solar model. However, indirect measurements via other reactions could not confirm, the direct measurements did not rule out the existence of such a state. Therefore improved experimental data in the low energy range appeared desirable.

The experimental situation is more complicated, because the reaction leads to three particles, which results in continuous particle spectra. (From earlier measurements as well as from the present studies [9], it is known that the shape of the spectra at low energies are given predominantly by phase space arguments, favoring a direct three particle break up as the reaction mechanism.) This feature makes it difficult to

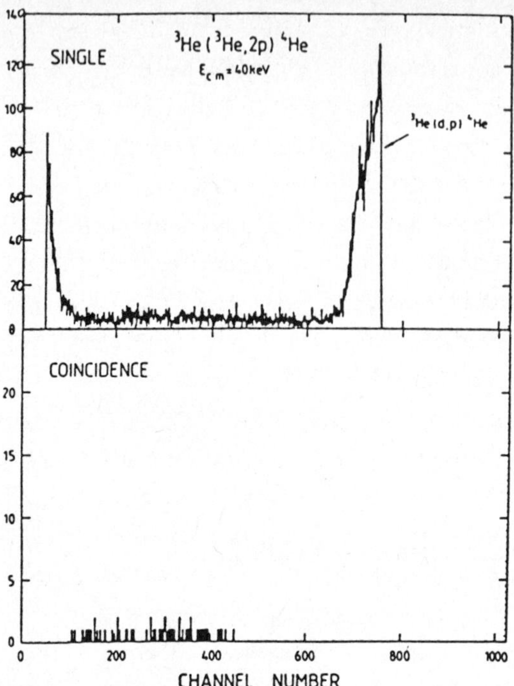

Fig.8 Single and coincidence proton spectrum of the $^3He(^3He,2p)^4He$ reaction. The beam and the residual ^4He nuclei are stopped in the detector foil. The single spectrum shows also the proton peak from the contaminant reaction $^3He(d,p)^4He$.

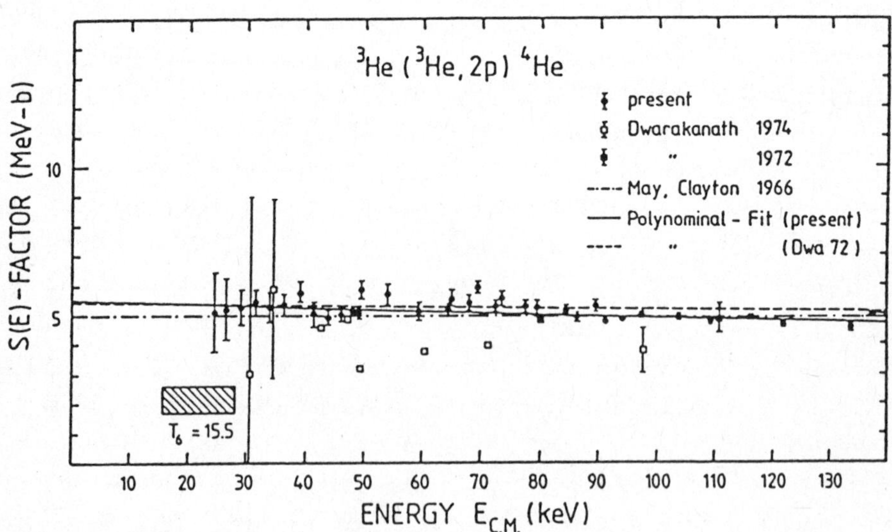

Fig.9 S(E)-factor data for the $^3He(^3He,2p)^4He$ reaction are compared with previous results ([9] and references therein)

distinguish the reaction products from any kind of background.

The major contribution to the background were induced by a HD^+-component in the beam, leading to the $^3He(d,p)^4He$ reaction and therefore to the same type of residual nuclei as in the $^3He+^3He$ reaction. Although beeing quite low ($\approx 10^{-7}$), this beam impurity causes due to the large cross-section of the ^3He+D reaction a violent background for energies below E_{cm} = 50 keV (fig.8). A unique signature for the $^3He+^3He$ reaction could be obtained by the coincident detection of the two protons in a pair of opposite mounted particle detectors (see fig.3). This solution however reduces of course the efficiency of the set-up.

For the consequent low count rates (at E_{cm}=25 keV $\approx$ 1 event/10 h) cosmic rays as well as high frequency electronic noise such as produced by high voltage discharge of the accelerator leads to coincidence rates comparable to or even larger than the reaction itself. To solve the problem of the cosmic background, the whole target chamber were surrounded by a large plastic scintilator, all coincidence events between this scintillator and one of the particle detectors were rejected. In order to exclude coincidences, triggered by the electronic noise, all simultaneous events in more than one pair of the detectors were also rececected. In addition, the target chamber with the scintillator as well as the electronics and the connecting cables were placed in Faraday cages. With this improvements no background event could be observed during one week of running.

In this way the measurements were extended into the thermal energy region of the sun down to E_{cm} = 25 keV (fig.9) and upper limits to the reaction yield have been obtained down to E_{cm} = 18 keV. The results show no evidence for the suggested resonance, which has to be narrow and below E_{cm} = 16 keV to solve the problem of the solar neutrinos. The present work could reduce the limits on such a state by about a factor of 30 compared to the previous conclusions.

REFERENCES

1. C.Rolfs, H.P.Trautvetter and W.S.Rodney, Rep.Progr.Phys. 501(1987)3
2. A.Krauss, H.W.Becker, H.P.Trautvetter, C.Rolfs and K.Brand, Nucl. Phys. A 465(1987)150
3. C.Rolfs, J.Görres, K.U.Kettner, H.Lorenz-Wirzba, P.Schmalbrock, H.P.Trautvetter and W.Verhoeven, Nucl. Instr. Meth. 157(1978)19
4. G.Steigmann, Ann. Rev. Nucl. Part. Sci. 29(1979)313
5. W. A.Fowler, G.R.Caughlan and B.A.Zimmerman, Ann. Rev. Astron. Astroph. 5(1967)525 and 13(1975)373

6. H.J.Assenbaum, K.Langanke and C.Rolfs, Zeitschr. f. Phys. A (in print)
7. W.A.Fowler, Nature 238(1972)24
8. V.N.Fetisov and Y.S.Kopysov, Phys. Lett. B40(1972)602; Nucl. Phys. A 239(1975)511
9. A.Krauss, H.W.Becker, H.P.Trautvetter and C.Rolfs, Nucl. Phys. A 467(1987)273

A NEW γ-RAY DETECTOR FOR STUDIES OF CAPTURE REACTIONS INVOLVING RADIOACTIVE NUCLEI

F. B. Waanders*

Institut für Kernphysik, Universität Münster, W. Germany

1. INTRODUCTION

Radiative capture reactions $A(x,\gamma)B$ are amongst the most important reactions for the formation of some elements and are usually studied by detecting the emitted γ-rays [1,2]. If the capture cross section is small, and/or if competing reactions produce high γ-ray background, these measurements require often high resolution Ge-detectors with absolute efficiencies in the order of 10^{-2} to 10^{-3} for E_γ = 1 to 10 MeV.

In hot and explosive astrophysical scenarios nuclear burning times can be measured in seconds or less. If the lifetimes of the radioactive nuclei are comparable to (or larger) than the burning times, they become involved in the nuclear burning processes [1,3]. The laboratory study of such reactions, which are predominantly of the (p,γ), (p,α) and (α,γ) type, requires new experimental techniques and new detectors.

The study of capture reactions, when a radioactive target is being investigated, is hampered by the presence of high γ-ray background due to the radioactivity. For example, 1 $\mu g/cm^2$ targets of ^{22}Na ($T_{1/2}$ = 2.6 y), ^{18}F ($T_{1/2}$ = 100 min) and ^{13}N ($T_{1/2}$ = 10 min) have 511- (and 1274 -) keV γ-ray activities of 7 x 10^8; 7 x 10^{12} and 1 x 10^{14} per second, respectively. Reduction of, say, the ^{13}N flux to a level of, say 10^4 per second would require lead shielding of about 15 cm thickness, reducing also the yield of typical 3 - 9 MeV γ-rays by a factor of about 2 x 10^4.

---

*Permanent address: Department of Physics, Potchefstroom University, Potchefstroom, South Africa

An ideal detector for the study of such 'hot' targets should have the following properties: (i) zero sensitivity for 511- (and 1274-) keV γ-rays, ii) high and nearly uniform efficiency for 4-8 MeV γ-rays, (iii) high γ-ray energy resolution, (iv) 4π geometry, (v) low background, and (vi) reasonable cost.

TEST SYSTEM

2.1. Construction

A detector fulfilling all of the above requirements, except for γ-ray energy resolution, has recently been developed by Rolfs and Kavanagh at the Kellogg Radiation Laboratory in California [5]. The detector is based on the photodisintegration of deuterium with subsequent neutron counting. Thus, the reaction D(γ,n)H with a Q-value of = -2.22 MeV has been used to detect γ-rays with energy $E_\gamma > Q$. The γ-ray source was surrounded by a cylindrical tank filled with heavy water. Neutrons from the photodisintegrated deuterium were moderated in the heavy water and the surrounding graphite. The emitted neutrons were counted in 4π geometry with 12 ^{3}He-filled proportional counters symmetrically embedded near the surface of a 1.4 m cube of high-purity graphite (fig. 1).

2.2. Total efficiency

The total efficiency of the detector system is given by the relation [5]

$$\varepsilon_\gamma = \sigma_d(E_\gamma) N_d \, l_{eff} \, \alpha_\gamma \, \varepsilon_n$$

where $\sigma_d(E_\alpha)$ is the cross section for photodisintegration of deuterium, N_d the number of deuterium nuclei per cm^3 (= 6.7×10^{22} D/cm^3), l_{eff} the effective path length of γ-rays in the heavy water, α_γ the correction for γ-ray absorption in the heavy water, the walls of the target holder and the tank, and ε_n the detection for the neutrons

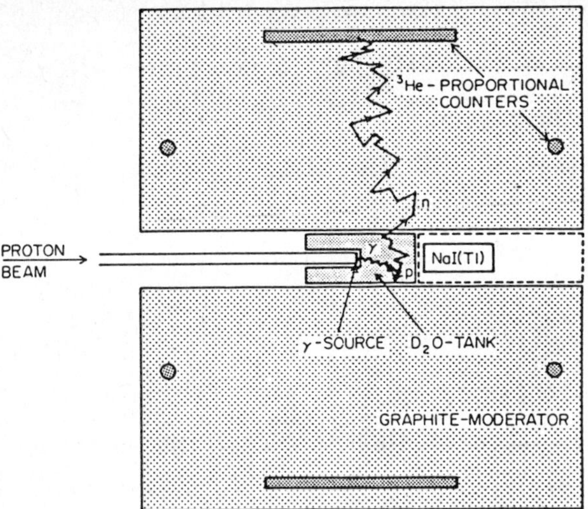

Fig. 1: Cross section of the test system (schematically)

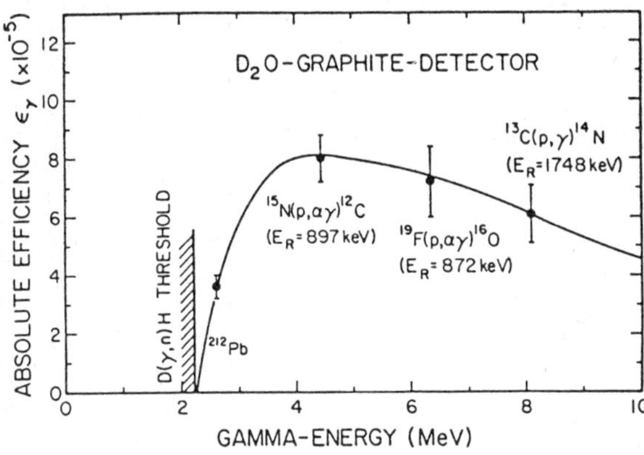

Fig. 2: The absolute efficiency of the D_2O graphite detector is shown as a function of γ-ray energy.The energy dependence reflects predominantly the photodesintegraion of deuterium [4].

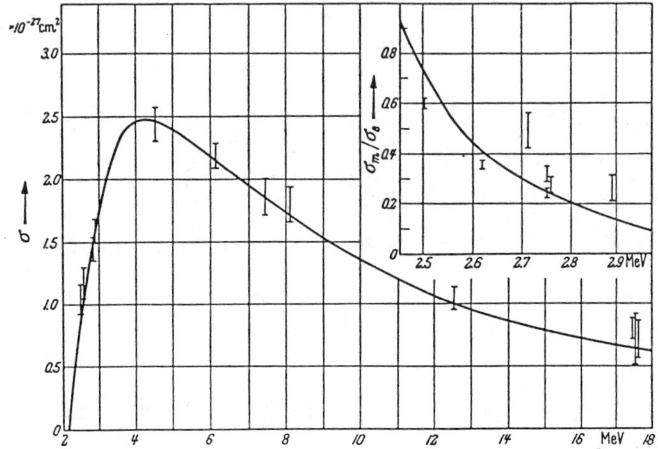

produced in D_2O. The absolute detection efficiency of the small test system has been determined for energies of E_γ = 2.6 to 8.1 MeV and found to agree well with calculations (fig. 2).

2.3. Control test

For a simulation of the study of capture reactions using radioactive targets, the E_p = 992 keV resonance of $^{27}Al(p,\gamma)^{28}Si$ was used with and without a 3 mCi ^{22}Na source placed near the ^{27}Al target. The results were found to be independent of the γ-ray flux from the ^{22}Na source (fig. 3).

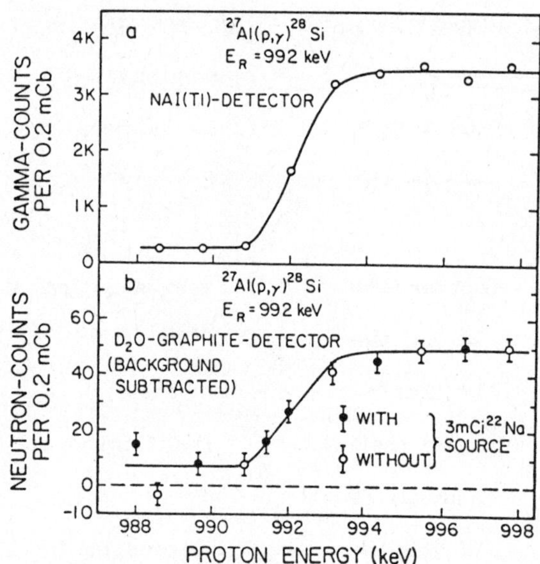

Fig. 3: Thick target yield curves for the E_p = 992 keV resonance of $^{27}Al(p,\gamma)^{28}Si$ as observed (a) with a NaI(Tl) crystal and (b) with the D_2O test detector. In the latter case the filled-in points were measured by placing a 3 mCi ^{22}Na source near the ^{27}Al target.

DETECTOR BEING CURRENTLY CONSTRUCTED

Encouraged by the results of the small test system a larger detector with an expected total efficiency of about 2 x 10^{-3} for the study of the capture reactions involving 'hot' targets is presently being constructed at the Universität Münster. A schematic diagram of the detector is shown in fig. 4 and some details regarding matters concerning electronics, construction and background contributions are given.

3.1. Construction

For a cylindrical tank of length L and radius L/2 the effective pathlength l_{eff} is related to L by $l_{eff} \approx 0.56$ L. Since the mean free pathlength of an 8 MeV γ-ray in D_2O is $l_{eff} \approx 42$ cm the length of the tank should be L $\approx$ 68 cm.

According to Fermis' formula, the root-mean square distance from a point source of a few MeV neutrons reached by neutrons of thermal age is about 30 cm in 99.8 % D_2O, thus giving another reason for the size of the tank. The D_2O serves the purpose of γ-ray conversion as well as of neutron moderation.

To enhance further the neutron thermalization the tank will be surrounded by a 10 cm thick layer of polyethylene. Although it was tested experimentally that a layer of thickness in the order of 5-7 cm would be sufficient for thermalization, the thicker polyethylene will be used due to its availabiltiy (fig. 5a).

The absorption of the thermalized neutrons in the 3 mm thick stainless steel walls of the D_2O tank will be in the order of only a few percent (fig. 5b).

3.2. Background contributions

Sources of background contributions are:

(i) intrinsic α-radioactivity in the walls of the ^{3}He proportional counter, (ii) α-radioactivity in the heavy water producing neutrons via $^{18}O(\alpha,n)^{21}Ne$ and $^{13}C(\alpha,n)^{16}O$, (iii) 'external' neutrons produced by

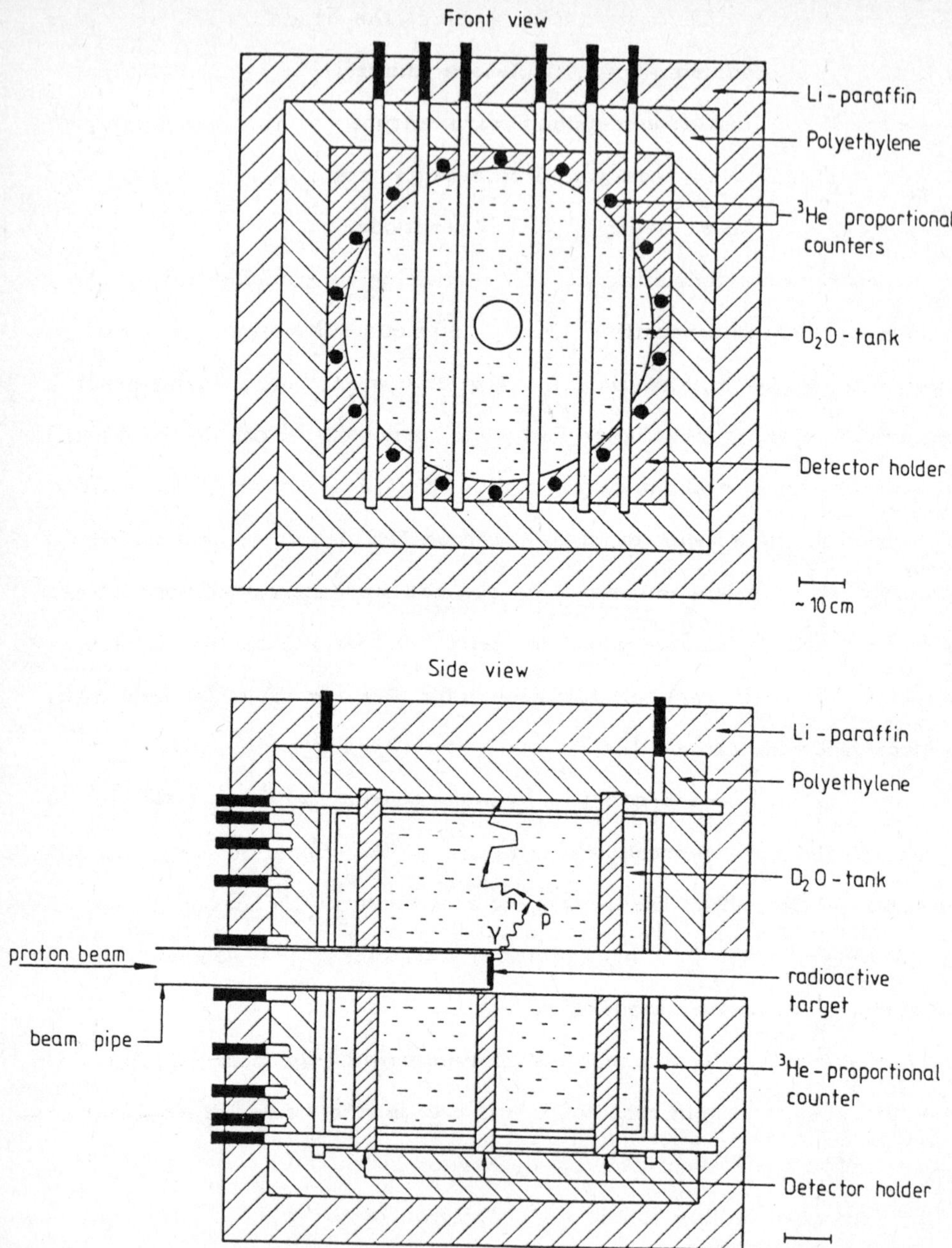

Fig. 4: Cross section of the larger D_2O γ-ray detector shown schematically.

cosmic rays and radioactivity in the walls of the target room, (iv) cosmic rays producing signals in the counters themselves, (v) the 2.61 MeV γ-ray from room background producing neutrons via photodesintegration of deuterium, and (vi) target contaminants may produce background γ-ray fluxes and/or neutrons via (p,n) reactions.

A test system consisting of a ^{3}He counter surrounded by a polyethylene cube and Li-paraffin blocks of various thicknesses was used to investigate these background contributions with the following results. The α-radioactivity (arising probably from the natural α-emitters Th and U) was found to be about 0.0003 c/sec. For the 30 counters the summed background will be about 0.008 c/sec. A 10 cm layer of Li-paraffin reduced the background resulting from 'external' neutrons by a factor of 20 leading to a count rate of about 0.004 c/sec (fig. 5c). This scales to a background rate of 0.1 c/sec for the 30 ^{3}He counters. The Li-paraffin will sorround the D_2O tank (fig. 4). The radioactivity for the 240 l of D_2O is about 5 mCi and if this radioactivity is entirely due to α-decays they will produce neutrons in the tank via $^{18}O(\alpha,n)^{21}Ne$ leading to a negligible background rate of about 0.005 c/sec.

3.3. Electronics and data handling

A high-voltage unit and a low-voltage unit supply power to all 30 counters and preamplifiers. After amplification the signals are analyzed by a differential discriminator associated with each counter with a window correspoonding to the energy region of the $^3He(n,p)^3H$ events (0.2 to 0.8 MeV). The logic signal of say the nth counter is tailored to an unique pulse height proportional to n and fed to a common junction for all counters. Pulses at the junction are then recorded with one pulse heigth analyzer, providing an easy way to monitor the proper operation of all counters.

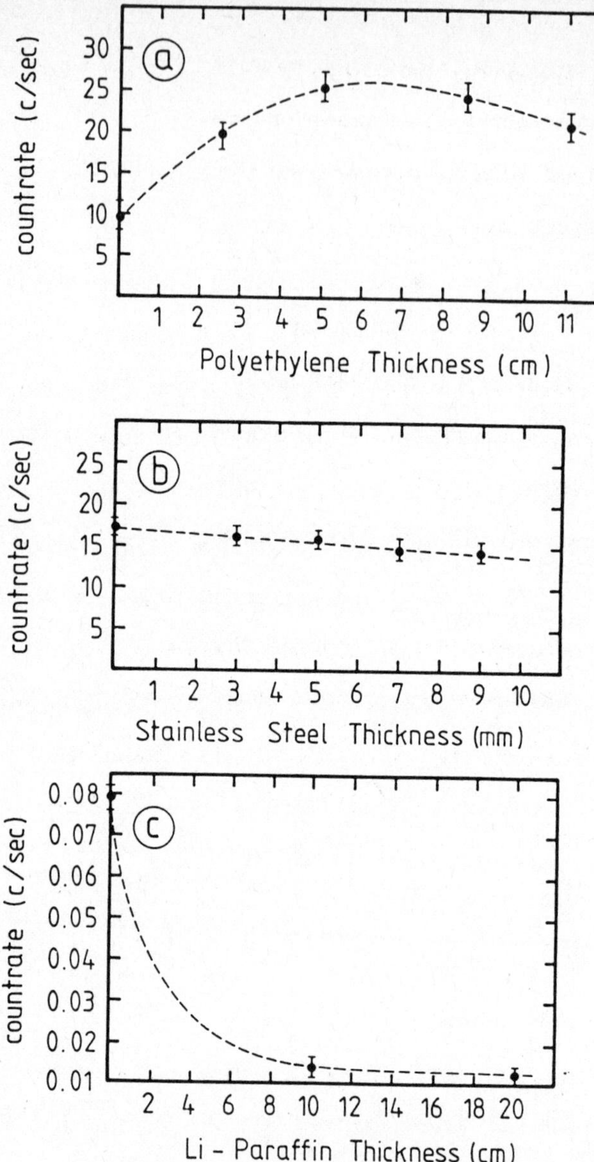

Fig. 5: Graphs of neutron count rate against: (a) polyethylene thermalization thickness, (b) absorption of thermal neutrons in stainless steel and (c) neutron absorption in various thicknesses of Li-paraffin.

CONCLUSION

It should be pointed out that this new γ-ray detection technique might be helpfull in general for studies of the wide field of reactions induced on short-lived nuclei mentioned in the introduction as well as the study of more exotic nuclei.

REFERENCES

1. W.A. Fowler, Rev. Mod. Phys. 56 (1984) 149
2. L. Buchmann and J. d'Auria, Proc. Workshop on Radioactive Ion Beams, Vancouver, BC, Canada (1985)
3. C. Rolfs, H.P. Trautvetter and W.S. Rodney, Rep. Progr. Phys. 50 (1987) 3
4. L. Hulthen and M. Sugawara, Handbook of Physics, vol. 39 (1957) 14
5. C. Rolfs and R.W. Kavanagh, Nucl. Instr. Meth. A244 (1986) 507

COULOMB DISSOCIATION AS A SOURCE OF INFORMATION ON RADIATIVE CAPTURE PROCESSES OF ASTROPHYSICAL INTEREST

H. Rebel

Kernforschungszentrum Karlsruhe GmbH
Institut für Kernphysik P.O.B. 3640,
D-7500 Karlsruhe, Federal Republic of Germany

Abstract

As an alternative to the direct experimental study of radiative capture processes at low relative energies we consider the inverse process, the photodisintegration of nuclear projectiles by means of the virtual photons provided by a nuclear Coulomb field of a target nucleus: $Z+a \rightarrow Z+b+c$.

The Coulomb dissociation cross section proves to be enhanced due to the large virtual photon number, seen by the passing projectile, and the kinematics of the process leads to particular advantages for studies of the interaction of the two break-up fragments at small relative energies E_{bc}. The theoretical implications and the conditions of dedicated experimental investigations are discussed. They are demonstrated by recent experimental and theoretical studies of the break-up of 156 MeV ^{6}Li projectiles, used as an access to the $\alpha+d \rightarrow {}^6$Li + γ reaction.

1. INTRODUCTION

Most of the laboratory approaches to experimental nuclear astrophysics, investigating charged-particle-induced reactions in stellar burning processes, involve the bombardement of rather thin targets by low-energy protons, ^{3}He, α-particles or other light ions [Fow 84, RT 78, Rol 86] . As a rule, the cross sections are almost always needed at energies far below those for which measurements can be performed in the laboratory. They must be obtained by extrapolation from the laboratory energy region, using procedures which are not free from theoretical bias and belief.

Tab. 1 presents some selected cases of interest at various astrophysical sites. The ^{3}He(^{4}He,γ)^{7}Be radiative capture reaction which at solar temperatures affects the solar neutrino flux and bears strongly on the longstanding solar neutrino problem [Kaj 86, OBK 84, NDA 69, ABL 84] is experimentally studied down to the c.m. energy $E_{c.m.}$ = 165 keV, while the cross section is actually needed at 15 keV. A similar situation is found for the

$^{12}C(\alpha,\gamma)^{16}O$ reaction [KBB 82] which plays an important role for the stellar helium-burning processes in red giant stars. To which extent ^7Li and ^6Li are synthesized in the expanding universe, in amounts comparable to the observed abundances, depends on the $(\alpha+t)$ and $(\alpha+d)$ radiative capture cross sections [Aus 81, Wag 73, SW 77, KTA 86] . The cross sections have been studied in the laboratory at c.m. energies $E_{c.m.} \geq 1$ MeV [RDW 81]. The present conclusion that ^7Li is produced in the primeval Big Bang, ^6Li, however, in the galactic cosmic rays, is based on purely theoretical estimates and extrapolations of the reaction rates.

EXAMPLE	$E_{measured}$	ASTROPHYSICAL INTEREST
Hydrogen Burning $\alpha + {}^3He \rightarrow {}^7Be + \gamma$ $E_0 \approx 10$ keV	$\gtrsim 165$ keV	Solar Neutrino Problem
Helium Burning $\alpha + {}^{12}C \rightarrow {}^{16}O + \gamma$ $E_0 \approx 300$ keV	$\gtrsim 1.34$ MeV	Ashes of Red Giant (C/O Ratio)
Big Bang Nucleosynthesis $\alpha + t \rightarrow {}^7Li + \gamma$ $\alpha + d \rightarrow {}^6Li + \gamma$ $E_0 \approx 100$ keV	$\gtrsim 1$ MeV	Li Be B Production Test of the Standard Big Bang Model

Tab. 1 Some examples of radiative nuclear capture reactions of actual astrophysical interest.

Direct (radiative) capture processes are electromagnetically induced transitions from continuum states, described by Coulomb distorted waves, to bound final states with particular angular moment a and with emission of γ-rays of corresponding multipolarities L.The capture cross sections

$$\sigma\left(E,L,J_i^\pi \rightarrow J_f^\pi\right) = \frac{e^2}{\hbar} \frac{8\pi\left(L+1\right)}{L\left[\left(2L+1\right)!!\right]^2} k_\gamma^{2L+1} B_{capt}\left(E,L,J_i^\pi \rightarrow J_f^\pi\right)$$

can be expressed in terms of reduced electromagnetic transition probabilities. The quantity B_{capt} depends on the energy of the entrance channel and is dominated by the Coulomb barrier penetration, which strongly suppresses the cross sections at low energies.

In view of possible uncertainties of astrophysical considerations, introduced by various experimental difficulties in investigations of radiative capture reactions, any alternative experimental access to the transition probabilities B_{capt} would be of obvious interest.In the following, we consider a recently proposed approach [Reb 85, BBR 86, SR 86] which suggests the use of the Coulomb field of a large-Z nucleus for inducing photodisintegration processes of fast projectiles.

2. COULOMB BREAK-UP OF PROJECTILES

Instead of studying directly the capture reaction

$$b + c \rightarrow a + \gamma$$

one may consider the time reversed process (with "a" being in the groundstate)

$$\gamma + a \rightarrow b + c\,.$$

The corresponding cross sections are related by the detailed balance theorem

$$\sigma(b+c \rightarrow a+\gamma) = \frac{(2j_a+1)\cdot 2}{(2j_b+1)(2j_c+1)} \; \frac{k_\gamma^2}{k^2} \; \sigma(a+\gamma \rightarrow b+c)\,.$$

The wave number in the (b + c) channel is

$$k^2 = \frac{2\mu_{bc}E}{\hbar^2}$$

with μ_{bc} the reduced mass while the photon wave number is given (neglecting a small recoil correction)

$$k_\gamma = \frac{E_\gamma}{\hbar c} = \frac{E+Q}{\hbar c}$$

in terms of the Q value of the capture reaction. Except for extreme cases very close to threshold ($k \rightarrow o$), the phase space would favour the photodisintegration cross section as compared to the radiative capture. However, direct measurements of the photodisintegration near the break-up threshold do hardly provide experimental advantages and seem presently impracticable (see Reb 85). On the other hand, the copious source of virtual photons acting on a fast charged nuclear projectile when passing the Coulomb field of a (large Z) nucleus might offer a more promising way to study the photodisintegration process as Coulomb dissociation. Fig. 1 indicates schematically the main features of the dissociation reaction.

At a sufficiently high projectile energy the two fragments b and c emerge with rather high energies (around the beam-velocity energies) which facilitates the detection of these particles. At the same time the choice of adequate kinematical conditions for coincidence measurements allows to study rather low *relative* energies of b and c and ensures that the target nucleus stays in the ground state (elastic break up). In addition, it turns out that the large number of virtual photons seen by the passing projectile leads to an enhancement of the cross section, promising an experimental access to the electromagnetic transition matrix elements of interest.

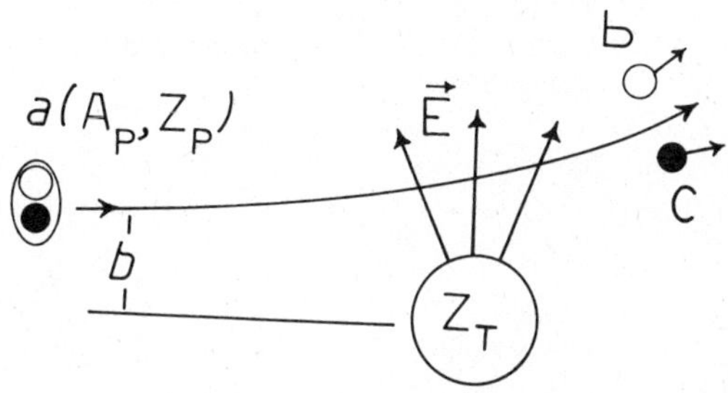

Fig. 1 Coulomb dissociation $a \rightarrow b + c$ in the field of a target nucleus (Z_T).

The double-differential cross section for Coulomb excitation of a projectile by an electric multipole transition of the order L as given by the first order theory of Alder and Winther [AW 75] can be rewritten in a factorized form [Fer 24, Weiz 34, Will 34]

$$\frac{d^2\sigma}{d\Omega dE_x} = \frac{1}{E_x}\frac{d\eta_{EL}}{d\Omega} \cdot \sigma^{photo}_{EL},$$

where

$$\sigma^{photo}_{EL} = \frac{(2\pi)^3 (L+1)}{L[(2L+1!!)]^2} k_\gamma^{2L-1} B(EL; I_i \rightarrow I_f)\, \rho_f(E_\gamma)$$

is related to the B_{capt} (EL)-value and the capture cross section, respectively.

The function $d\eta_{EL}/d\Omega$ only depends on the excitation energy E_x ($\approx E_\gamma = \hbar\omega$) and on the relative motion, but not on the internal structure of the projectile. We call $d\eta_{EL}/d\Omega$ the *virtual photon spectrum* per unit solid angle seen by the projectile when passing through the Coulomb field with the velocity v. It is actually a function of the incident energy E, of the target charge Z_T and of the Coulomb parameter

$$\eta = \frac{Z_P Z_T e^2}{\hbar v}$$

The impact parameter

$$b = a \, ctg\, \Theta$$

with

$$a = \frac{Z_P Z_T e^2}{m v^2}$$

being half the distance of closest approach in a head-on collision, has to be chosen large enough in order to avoid violent *nuclear* interactions. This requirement implies for larger projectile energies rather small scattering angles Θ. In this case ($\varepsilon = 1/(\sin \Theta/2) >> 1$) the shape of the virtual photon spectrum is approximately a function of an adiabaticity parameter

$$x = \omega b/v$$

(which appears to be adequate for nonrelativistic energies when $(1 - v^2/c^2) \approx 1$ with the condition $\Theta << 1$).

For the E1 case, we have

$$\frac{dn_{E1}}{d\omega} = \frac{Z_T^2}{4\pi^2} \alpha\varepsilon^2 \left(\frac{c}{v}\right)^2 x^2 \left\{K_o^2(x) + K_1^2(x)\right\} = \frac{Z_T^2}{4\pi^2} \alpha\,\varepsilon^2 \left(\frac{c}{v}\right)^2 \phi_1(x)$$

and for the E2 case

$$\frac{dn_{E2}}{d\Omega} = \frac{Z_T^2}{4\pi^2} \alpha \frac{1}{\xi} \left(\frac{c}{v}\right)^4 \cdot \phi_2(x)$$

where

$$\phi_2(x) = x^2 \left\{K_1^2(x) + x^2(K_1^2 + K_0^2) + x K_o K_1\right\}.$$

Here $K_i(x)$ are the modified Bessel functions.

As in low-energy Coulomb excitation, where the cross section drops exponentially with the adiabaticity parameter

$$\xi = \frac{\omega a}{v} = \frac{\textit{collision time}}{\textit{transition time}}$$

a requirement, obvious in Fig. 2 is

$$x = \xi \; ctg\, \Theta/2 \ll 1.$$

The structure of the virtual photon spectra has been explored in detail by several authors [BB 85, Gol 84]. It should be emphasized that the field generating the transitions is source-free and equivalent to a real photon field since the scattering charges do not penetrate the nuclear volumes for sufficiently large impact parameters [Gol 84].

It can be seen that the E2 virtual photon numbers are in many interesting cases much larger (depending on the value of b) than the corresponding E1 ones. Varying the experimental conditions with different relative E1 and E2 virtual photon numbers the quantity σ_{EL}^{photo} may be individually determined. For larger b the E1 component increases relatively. In coincidence studies interference effects between different multipoles will show up in general, which can in principle help to disentangle the various multipole contributions. A selective population of magnetic substates of the system b+c is expected. It can be directly

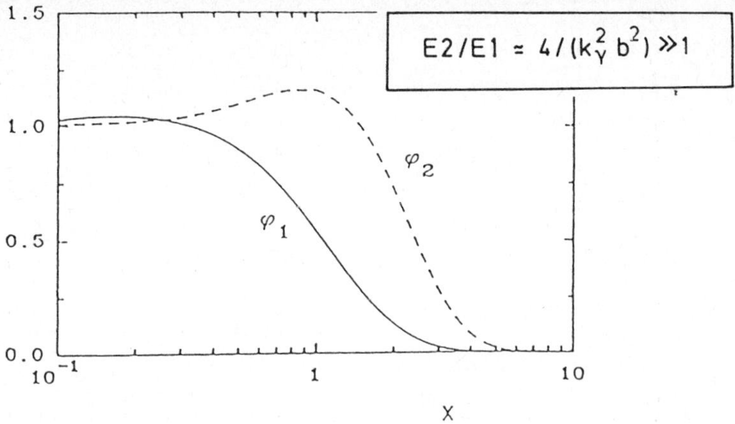

Fig.2 The shape of the virtual photon spectrum as a function of the adiabaticity parameter x for the multipolarities E1 and E2.

calculated from the theory of Coulomb excitation. A more detailed account of the theoretical basis and calculations of the Coulomb break-up cross sections is given elsewhere [BBR 86, Bau 85].

Figure 3 displays the electric dipole component relevant for the examples of the dissociation of ^{7}Be and ^{16}O when passing ^{208}Pb with an impact parameter b = 10 fm at two different projectile energies. The corresponding break-up thresholds are marked.The most interesting feature is the high intensity of the equivalent photon spectra which actually leads to an enormous enhancement of the photodissociation cross section. This is one of the main advantages of the proposed method. The examples given in Tab. 2 demonstrate the effect. The table gives the double-differential cross sections for the excitation of the projectile to the continuum energy E_{bc} of the emerging fragments when the projectile or the fragment center-of-mass, respectively, is scattered to dΩ. Assuming a specific detection geometry, this cross section can be transformed* into the triple differential cross section, which we are actually measuring in the laboratory. Obviously the resulting values appear to be experimentally accessible, in contrast to the corresponding σ_{capt}-values. With the assumed conditions the quadrupole component of the virtual photon field appears to be much stronger than E1 component at the particular values of the impact parameter and projectile energy, so that the ^{6}Li break-up cross section is enhanced.

Our simplified consideration, pointing out the idea, is only of first order, and similarily to Coulomb excitation of bound states, higher order contributions, which in our case involve continuum coupling effects, have most likely to be taken into account (AIK 87).

---

* For sake of simplicity isotropic decay of the excited projectile has been assumed for the example given in Tab. 2. However, there is no problem to take into account the angular correlation between the emitted fragments.

REACTION b + c ↔ a	E_{bc} [MeV]	σ_{capt} [n b]	$\frac{d^2\sigma^{Diss}}{dE_{bc}\, d\Omega}$ [μb MeV^{-1} sterad^{-1}]	$\frac{d^3\sigma^{Diss}}{dE_b d\Omega_b d\Omega_c}$ [μb MeV^{-1} sterad^{-2}]	E_{thr} [MeV]
E1 $\alpha + {}^3He \leftrightarrow {}^7Be$	0.1	≈ 0.5	11	52 $\theta_\alpha = 5°$ $\theta_{He} = 7°$	1.58
E1 $\alpha + {}^{12}C \leftrightarrow {}^{16}O$	1.0	≈ 0.1	2		7.162
E2 $\alpha + d \leftrightarrow {}^6Li$	0.5	≈ 1.0	10^4		1.47
			Elastic Coulomb break up with ^{208}Pb E_{Proj} = 30MeV/ amu - Impact parameter 10fm		

Tab.2 Numerical values of break-up cross sections for selected examples of astrophysical interest.

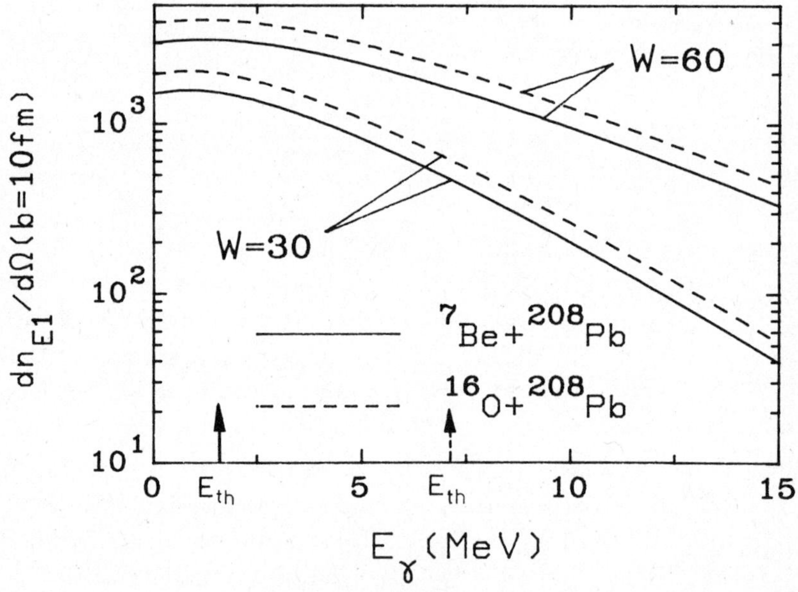

Fig. 3 E1 virtual photon spectra seen by the projectile with b = 10 fm at different projectile energies W (MeV/amu).

3. EXPERIMENTAL APPROACH OF ^{6}LI BREAK-UP.

In order to explore the feasibility of the proposed break-up approach we have recently concentrated our experimental efforts to the case of ^{6}Li. The D(α,γ)^{6}Li capture cross section has been experimentally investigated for energies $E_{\alpha d} \geq 1$ MeV and analyzed on the basis of a capture model [RDW 81]. The L = 2 resonance at $E_{\alpha d} = 0.71$ MeV corresponds to the first excited state at $E_{3_1^+} = 2.185$ MeV in ^{6}Li (Fig. 4). The resonance strength can be deduced from the electromagnetic transition probability B(E2; $1^+ \rightarrow 3_1^+$) known from inelastic scattering [Eig 69].

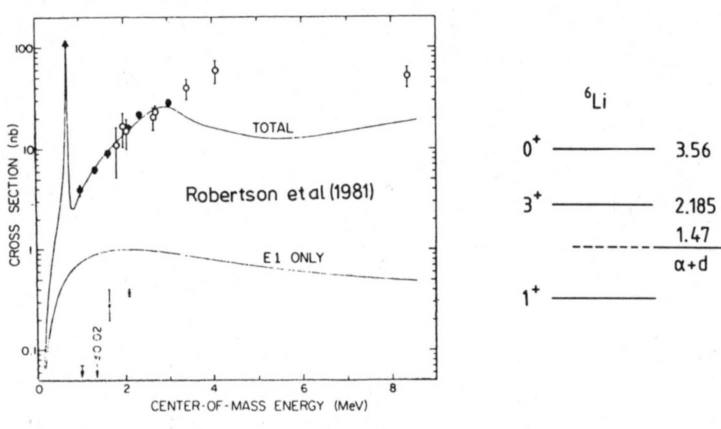

FIG. 1. Cross section for the reaction ^{2}H(α, γ)^{6}Li. Open circles, MSU data; closed circles, CRNL data; triangles, ^{6}Li(e,e'd) (Ref. 7); crosses, CRNL data for E1 component. The curves are a direct-capture calculation.

Fig. 4 Cross section for the D(α,γ)^{6}Li capture reaction [RDW 81]

The break-up experiments [Reb 86] are performed at the 156 MeV ^{6}Li beam of the Karlsruhe Isochronous Cyclotron, using the magnetic spectrometer "Little John" [Gil 80, GBZ 80]. This spectrometer is especially designed and equipped for the observation of the ejectile emission at extreme forward angles.Fig. 5 shows spectra of α-particles from collisions of 156 MeV ions with ^{208}Pb. At forward angles these spectra are dominated by a bump around the beam-velocity energy, indicating break-up processes as being the origin [NBR80]. However, the bump is mainly related to nonelastic break-up processes, where the nonobserved deuteron interacts nonelastically with the target, in particular by break-up fusion [PKB 86].

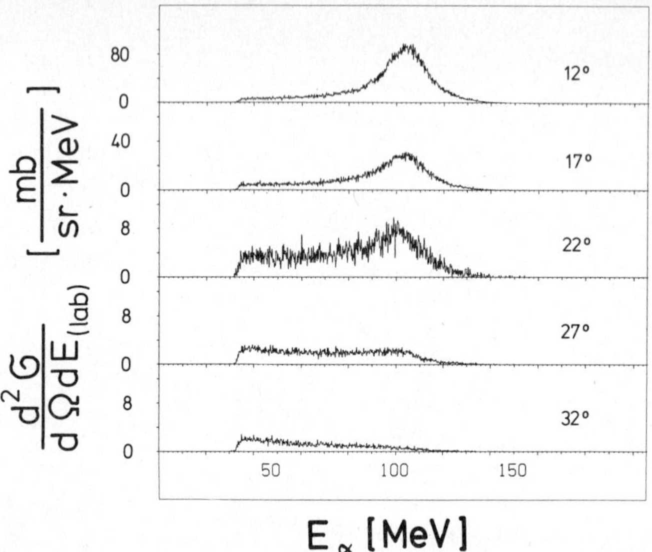

Fig. 5 Inclusive α-particle spectra of break-up of ^{6}Li from ^{208}Pb at 156 MeV, observed at various emission angles [NRB 80]

These inclusive measurements of the break-up yields have been extended to emission angles smaller than 5°, where Coulomb effects are expected to show up. The main experimental difficulties arise from the elastic scattering of ^{6}Li, especially as beam-velocity deuterons and α-particles are focussed onto the same position of the focal plane detector, due to the same magnetic stiffness. Fig. 6 shows the energy-integrated inclusive cross sections of the α-particle and deuteron components from collisions of 156 MeV ^{6}Li with ^{208}Pb.

In order to isolate the elastic component of the break-up bump, i.e. the mode of a correlated emission of deuterons and α-particles leaving the "catalyst" for break-up, the target nucleus, in the ground-state, we have to perform correlation measurements and take advantage of the three-body kinematics. The kinematical situation for a typical detector arrangement with a α-particle and a deuteron detector in fixed-angle-position is displayed in Fig. 7. The kinematics for three particles in the final state lead to a correlation of the α-particle and deuteron energy (for a particular value of target excitation). For a heavy target this is an approximately linear relation, as shown in Fig. 7 for the case of ^{208}Pb, which remains in the ground state. Along this kinematical line all events of *elastic* break-up are distributed. Fig. 7 shows additionally the relative energy $E_{\alpha d}$ plotted over the E_α^{Lab} axis, and one recognizes that a particular $E_{\alpha d}$ value appears twice (once the α-particle the slower fragment, once the deuteron). There is a remarkably slow variation of $E_{\alpha d}$ around the $E_{\alpha d}^{min}$-value ("magnifying glass effect") which allows a good resolution on the relative-energy scale. We have just to measure the coincidence cross section on the kinematical curve around the minimum region.

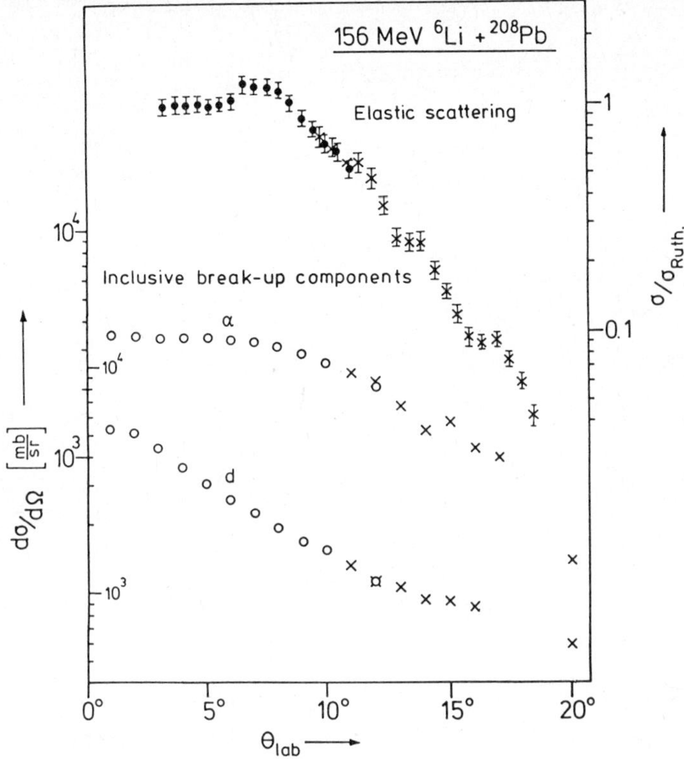

Fig. 6 Elastic scattering and inclusive break-up components from 156 MeV ^{6}Li collisions with ^{208}Pb [Jel 87].

Due to considerable cancellations of various contributions to the relative energy $E_{\alpha d}$, the energy resolution $dE_{\alpha d}$ on the $E_{\alpha d}$ scale is much better than on the scale of the laboratory energies. Since for the velocities $v_{\alpha d}$, v_{α}, v_d

$$v_{\alpha d}^2 = v_\alpha^2 + v_d^2 - 2\, v_\alpha v_d \cos\theta_{\alpha d}$$

then

$$v_{\alpha d}\, dv_{\alpha d} = (v_\alpha - v_d \cos\theta_{\alpha d})\, dv_\alpha + (v_d - v_\alpha \cos\theta_{\alpha d})\, dv_d$$

As for beam-velocity particles ($v_\alpha \approx v_d$) emerging within a narrow angle cone ($\cos\theta_{\alpha d} \approx 1$)

$$dE_{\alpha d} \ll dE_\alpha, dE_d \,.$$

However the resolution is affected by the accuracy of $\theta_{\alpha d}$

$$dE_{\alpha d} = \frac{2\sqrt{m_\alpha m_d E_\alpha \cdot E_d}}{m_\alpha + m_d} \sin\theta_{\alpha d}\, d\theta_{\alpha d}$$

and requires a good angular accuracy of the experimental set-up.

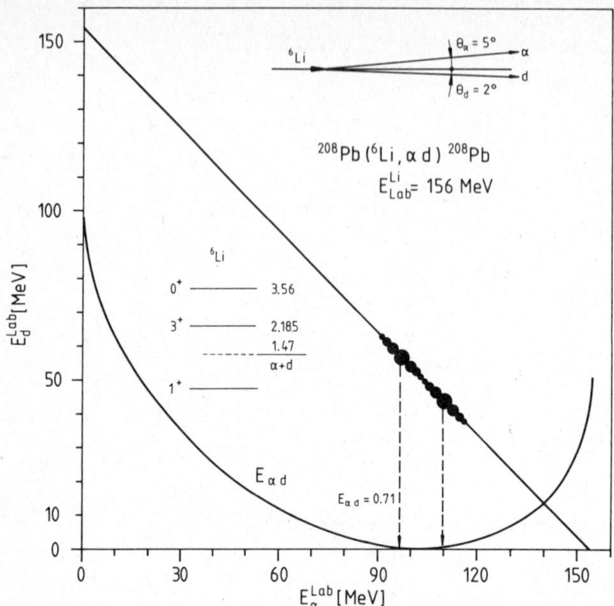

Fig. 7
Kinematical loci of the emerging deuteron and α-particle from a ^{6}Li dissociation on ^{208}Pb at E_{lab} = 26 MeV/amu.

Fig. 8 shows a first result which corresponds to the kinematical situation of Fig. 7. The two peaks of the cross section projected onto the E_α-axis represent the sequential (resonant) break-up mode via the first excited state of ^{6}Li. Due to unsufficient energy resolution of the solid-state-detector (used as second detector), there appears some deficiency; the inelastic break-up mode with excitation of the $3_1{}^+$ - state in ^{208}Pb (which may itself be of other interest) is not well separated. In any case the result of this test demonstrates that experiments are feasible under these conditions. What we had to expect under the experimental conditions is displayed by Monte-Carlo-simulations (Fig. 9).

In addition to the resonance peaks Fig. 9 shows (with an enlarged scale) a prediction based on a recent alternative theoretical consideration (SR 86) of the nonresonant Coulomb break-up by a DWBA approach in the Rybicki-Austern (RA 71) formulation of the break-up theory for L = 2 transitions.

Our interest is focused to the direct (nonresonant) Coulomb break-up of ^{6}Li, not yet discovered up to now and represented by cross-section in regions of the kinematic loci away from resonance peaks from sequential processes. With a narrow angular spacing of the detectors the region of very low relative energies can be considerably stretched (Fig. 9) in the laboratory energy scale.

In order to improve our experimental set-up and to reduce the problems with the elastic scattering and solid-state detectors, we have recently installed a split focal plane detector ("Multi-hit detector"), which enables the observation of α-d-deuteron coincidences at very

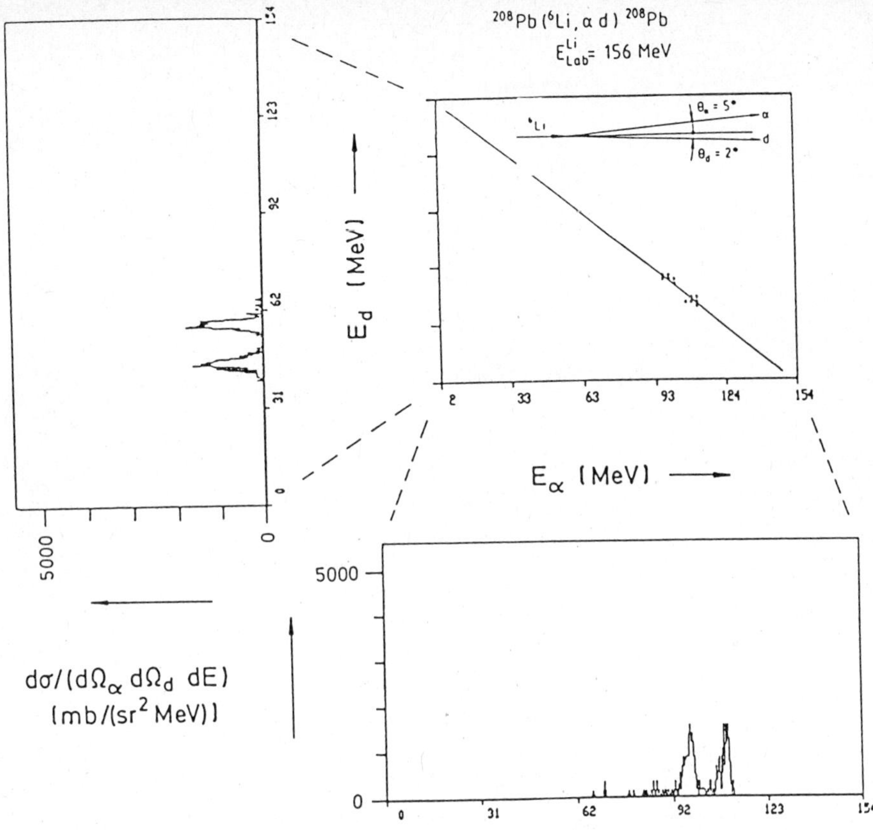

Fig. 8 Experimental α-d coincidence spectra at very forward angles from collisions of 156 MeV ^{6}Li-ions with ^{208}Pb [Jel 87].

forward angles with the spectrometer [GK 86]. A position sensitive trigger detector at the position of the entrance slit will be additionally used, so that we can cover several angle-combinations simultaneously in one run, at expense of a more complicated arrangement and data analysis. This improved set-up is just in the test-stage.

It is interesting to note that the features predicted by current theories of the nonresonant Coulomb break-up (SR 86, Bas 87) resemble very much experimental results of Coulomb break-up of ^{7}Li observed E_{Li} = 70 MeV [SRD 84]. In view of the present experimental uncertainty of the $\alpha + t \rightarrow {}^7$Li capture rate [GMR 61] ^{7}Li is certainly a further case of interest.

MONTE-CARLO-SIMULATION

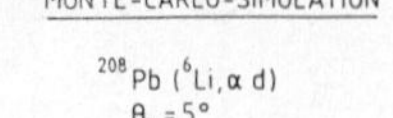

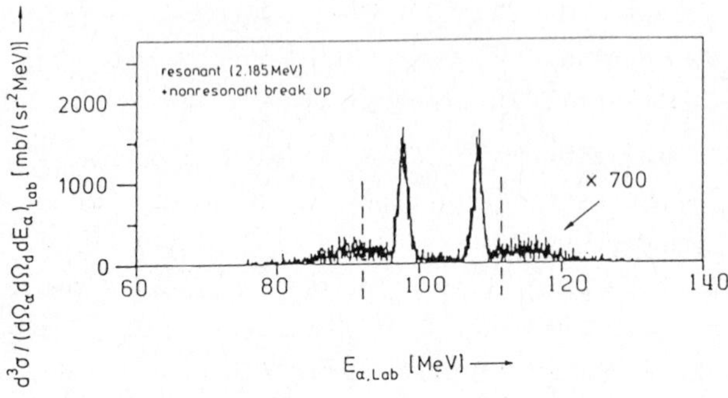

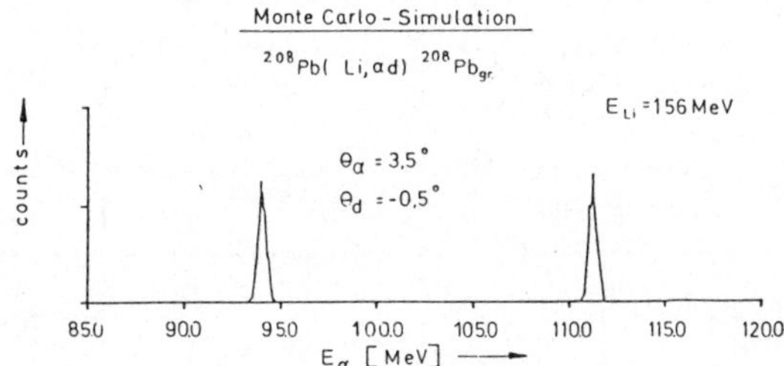

Fig. 9 Monte-Carlo-simulation of coincidence (α-d) coincidence spectra [Jel 87].

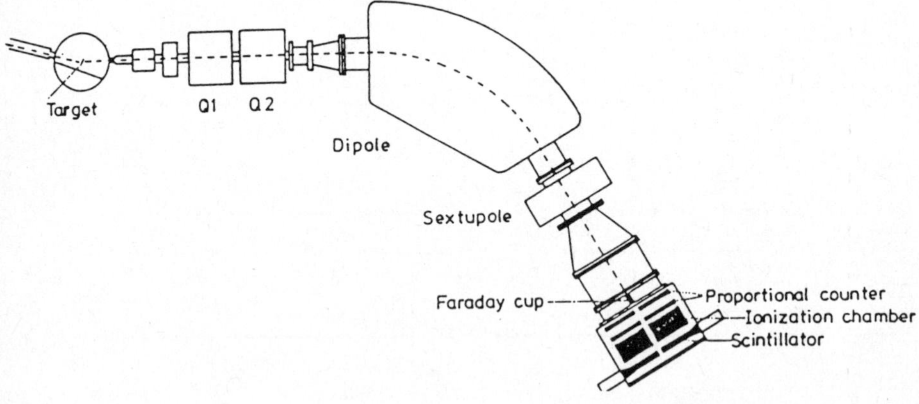

Fig. 10 Spectrometer "Little John" with split focal plane detector [Gil 86]

4. CONCLUSIONS

Experimentally the proposed approach requires measurements with moderately high incident projectile energies and at extreme forward angles, in order to exclude the influence of the nuclear field. The choice of the adequate projectile energy needs a careful analysis of the specific case under consideration. The values of the estimated coincidence cross sections are rather small, but they appear to be measurable by present days' experimental techniques.

The kinematical situation with three outgoing particles provides particular advantages and flexibilities for studies of the excitation function, i.e. with tuning of the relative energy of the emerging fragments, and of the angular distribution in the rest frame of the decaying subsystem. Studies of the latter aspect require a particularly good angular resolution. The cross sections can be interpreted in terms of electromagnetic matrix elements which just determine the radiative capture cross sections for the interaction of *naked* nuclei. However, we expect that finally a higher-order Coulomb excitation theory has to be invoked.

In detail, there is number of open problems of experimental and theoretical nature. The interpretation has to be refined with respect to the *orbital dispersion* and *external Coulomb distortion and polarisation effects*, i.e. effects on the ejectile movements due to the presence of the "catalyst".

Table 3 Radiative capture reactions of interest for light element nucleosynthesis [Reb 86].

Reaction	Half-life	Relevance
^{3}He(α,γ)^{7}Be ^{7}Be(p,γ)^{8}Be ^{7}Be(α,γ)^{11}C	(53.3 d) (770 ms) (20.4 m)	Solar neutrino problem ^{3}He abundancy
^{4}He(d,γ(^{6}Li ^{6}Li(p,γ)^{7}Be ^{6}Li(α,γ)^{10}B ^{4}He(t,γ)^{7}Li ^{7}Li(α,γ)^{11}B ^{11}B(p,γ)^{12}C ^{9}Be(p,γ)^{10}B ^{10}B(p,γ)^{11}C	(stab.) (53.3 d) (stab.) (stab.) (stab.) (stab.) (stab.) (20.4 m)	Primordial nucleosynthesis of Li Be B - isotopes
^{12}C(p,γ)^{13}N ^{16}O(p,γ)^{17}F ^{13}N(p,γ)^{14}O ^{20}Ne(p,γ)^{21}Na	(10 m) (65 s) (70.6s) (22.5s)	CNO - cycles
^{15}O(α,γ)^{19}Ne	(17.2s)	RP - Process
^{12}C(α,γ)^{16}O ^{16}C(α,γ)^{20}Ne ^{14}N(α,γ)^{18}F	(stab.) (stab.) (109.7 m)	Helium-burning

Very interesting an improved experimental possibilities would be provided by a dedicated set-up at a synchroton-cooler ring with suitable spectrometers [GHR 86], enabling particle coincidence studies at very forward emission directions. The use of a storage ring seems to be indispensable when working with radioactive beams like ^{7}Be. Tab. 3 presents examples of capture reactions, which are of importance at various astrophysical sites. The list gives an impression on the field and emphasizes the necessity of the radioactive beams. However, presently the proposed experimental approach requires still the successful demonstration of the experimental feasibility and of the theoretical analysis.

This lecture is based on the scientific discussion and the studies of a larger experimental research group. In particular, I'm grateful to G. Baur and D.K. Srivastava for their theoretical support, to H.J. Gils, H. Jelitto, J. Kiener, G. Schatz and S. Zagromski for conceptive and experimental contributions to the considered problems.

REFERENCES

ABL 84 T.K. Alexander, G.C. Ball, W.N. Lennard, H. Geissel, H.B. Mak, Nucl. Phys. A 427 (1984) 526

AIK 87 N. Austern, Y. Iseri, M. Kamimura, M. Kawai, G. Rawitscher and M. Yahiro, Phys. Rep. (to be published).

Aus 81 S.M. Austin, Prog. Part. Nucl. Phys. **7** (1981) 1.

AW 75 K. Alder and A. Winther "Electromagnetic Excitation", North Holland, Amsterdam, 1975.
L. Alder, A. Bohr, T. Huns, B. Mottelson and A. Winther, Rev. Mod. Phys. 28 (1956) 432

Bau 85 G. Baur, Lecture presented at the 1985 Varna International Summer School on Nuclear Physics, Sept. 22 - Oct. 1, 1985.

BB 85 C.A. Bertulani and G. Baur, Nucl. Phys. A 442 (1985) 739.

Bas 85 D.N. Basu, Workshop on Break-Up Phenomena in Nuclear Physics, Bhabha Atomic Research Centre, Calcutta (India) February 9-11, 1987

BBR 86 G. Baur, C.A. Bertulani and H. Rebel, Nucl. Phys. A 459 (1986) 188 - Proc. Int. Symp. on Weak and Electromagnetic Interactions in Nuclei, 1-5 July 1986, Heidelberg (Germany).

DBH 71 D.L. Disdier, G.C. Ball, O. Häuser and R.E. Warner, Phys. Rev. Lett. **27** (1971) 1391.

Eig 69 F. Eigenbrod, Z. Phys. 238 (1969) 337.

Fer 24 E. Fermi, Z. Phys. 29 (1924) 315

Fow 84 W.A. Fowler, Rev. Mod. Phys. 56 (1984) 149.

GBZ 80 H.J. Gils, J. Buschmann, S. Zagromski, H. Rebel, J. Krisch, M. Heinz, Internal Reports Kernforschungszentrum Karlsruhe

GHR 86 H.J. Gils, D. Heck, H. Rebel and G. Schatz, Karlsruhe Radioactive Ion Beam Instrumentation and Cooling: KARIBIC Internal Note, Kernforschungszentrum Karlsruhe 1986.

Gil 80 H.J. Gils, KfK-Report 2972 (1980)

GK 86 H.J. Gils and J. Kiener, Internal Report, Kernforschungszentrum Karlsruhe 1986

GMR 61 G.M. Griffiths, R.A. Morrow, P.J. Riley, J.B. Wassen, Canad. J. Phys. 39 (1961) 1397.

Gol 84 A. Goldberg, Nucl. Phys. A 420 (1984) 636

Jel 87 H. Jelitto PhD Thesis, Universiy of Heidelberg 1987 KfK-Report 4259 (May 1987)

Kaj 86 T. Kajino, Nucl. Phys. **A 460** (1986) 559

KBB 82 K.U. Kettner, H.W. Becker, L. Buschmann, J. Görres, H. Krähwinkel, C. Rolfs, P. Schmalbrock, H.P. Trautvetter, and A. Vlieks, Z. Phys. **A 308** (1982) 73.

KTA 86 T. Kajino, H. Toki and S.M. Austin, submitt to Astrophys. Journal - MSUCL-574 (1986).

NDA 69 K. Nagatani, M.R. Dwarakanath, and D. Ashery, Nucl. Phys. A 128 (1969) 325.

NBR 80 B. Neumann, H. Rebel, J. Buschmann, H.J. Gils, H. Klewe-Nebenius, and S. Zagromski, Z. Phys. A 296 (1980) 113.

OBK 84 J.L. Osborne, C.A. Barnes, R.W. Kavanagh, R.M. Kremer, G.J. Mathews, J.L. Zyskind, P.D. Parker, and A.J. Howard, Phys. Rev. Lett. 48 (1982) 1664 - Nucl. A 419 (1984) 115.

PKB 86 P. Planeta, H. Klewe-Nebenius, J. Buschmann, H.J. Gils, H. Rebel, and S. Zagromski, T. Kozik, L. Freindl, and K. Grotowski, Nucl. Phys. A 448 (1986) 110.

RA 72 F. Rybicki and N. Austern, Phys. Rev. C6 (1972) 1525.

RDW 81 R.G.H. Robertson, P. Dyer, R.A. Warner, R.C. Melin, T.J. Bowles, A.B. Mc Donald, G.C. Ball, W.G. Davies, and E.D. Earle, Phys. Rev.Lett. 47 (1981) 1867.

Reb 85 H. Rebel, Workshop "Nuclear Reaction Cross Sections of Astrophysical Interest" unpublished report, Kernforschungszentrum Karlsruhe, February 1985.

Reb 86 H. Rebel, Lectures presented at the International Summer School "Symmetries and Semiclassical Features of Nuclear Dynamics" 1.-13. Sept. 1986, Poiana Brasov (Romania) - KfK-Report 4158 (1986).

Rol 86 C. Rolfs, Rep. Progr. Particl. and Nucl. Physics 17 (1986) 365.

RT 78 C. Rolfs and H.P. Trautvetter, Ann. Rev. Nucl. Sci. 28 (1978) 115.

SR 86 D.K. Srivastava and H. Rebel, Journ. Phys. G: Nucl. Phys. 12 (1986) 717.

SRD 84 A.C. Shotter, V. Rapp, T. Davinson, D. Bradford, N.E. Sanderson and M.A. Nagarajan, Phys. Rev. Lett. 53 (1984) 1539.
A.C. Shotter, in Proc. 4th Int. Conference on Clustering Aspects of Nuclear Structure and Nuclear Reactions, Chester, U.K., 23-27 July, 1981, D. Reidel Publ. Company p. 199.

SW 77 D.N. Schramm and R.V. Wagoner, Ann. Rev. Nucl. Sci. 27 (1977) 37.

Wag 73 R.V. Wagoner, Astrophys. J. 179 (1973) 343.

Weiz 34 C.F. Weizsäcker, Z. Phys. 88 (1934) 612.

Will 34 E.J. Williams, Phys. Rev. 45 (1934) 729.

NEW EXPERIMENTAL RESULTS FOR NUCLEAR REACTIONS IN EXPLOSIVE HYDROGEN BURNING

M. Wiescher, J. Görres, L.O. Lamm, C.P. Browne
Department of Physics,
University of Notre Dame
B.W. Filippone, B. Vogelaar
W.K. Kellogg Radiation Lab.
California Institute of Technology

Proton capture reactions on proton rich radioactive nuclei play an important role in nucleosynthesis during explosive hydrogen burning via the r(apid) p(roton) process. Due to the usually short life times of the nuclei involved in such burning processes, little experimental information is available about the particular reaction cross sections and rates in such burning sequences. Model calculations for explosive nucleosynthesis processes in high temperature and density environments are currently based on estimates of these rates derived from nuclear structure information.

We will discuss two specific examples, $^{19}Ne(p,\gamma)^{20}Na$ and $^{22}Na(p,\gamma)^{23}Mg$, an indirect and a direct approach for experimentally determining reaction cross sections and rates for proton capture on unstable target nuclei. Furthermore, we will present the first results of these experiments and will discuss the possible impact on explosive nucleosynthesis.

The reaction $^{19}Ne(p,\gamma)^{20}Na$

Recent spectroscopic observations of nova ejecta in the ultraviolet and infrared range indicate strong enhancements compared to solar abundances for Ne, Na, Mg, Al in nova CrA 1981 [1] for Ne, Mg, Si, S in nova Aql 1982 [2] and for Ne, Mg in nova Vulpeculae 2 1984 [3]. This is in contradiction with abundance predictions from model calculations which interpret a nova event as thermal runaway in an accreting hydrogen shell on top of a degenerate C-O-white dwarf in a stellar binary system [4]. It is suggested [5,6] that some nova events might be described by a thermal runaway on O-Ne-Mg white dwarfs, which would explain the overabundance in these elements. Another explanation [7] suggests that at such temperature and density conditions breakout of the hot CNO cycles via $^{15}O(\alpha,\gamma)^{19}Ne(p,\gamma)^{20}Na$ may occur, which would transfer CNO material into the Ne-Na-Mg region.

The current estimate of the rate of $^{19}Ne(p,\gamma)^{20}Na$ [8] is largely based on the experimentally well known structure of the analog nucleus ^{20}F. The excitation energies of the proton unbound states in ^{20}Na were derived from Thomas Ehrman shift calculations between analog states. The results indicate several proton unbound states in the excitation range E_x = 2.86 - 3.05 MeV, well above the proton threshold at Q = 2.196 MeV. This suggests that the reaction rate for $^{19}Ne(p,\gamma)^{20}Na$ at nova conditions is mainly determined by the low energy tail of those resonance states as well as by nonresonant direct capture to bound states in ^{20}Na. However, the calculated level shifts depend critically on the single particle spectroscopic factor and the orbital momentum of the state. Since these were not known for all of the discussed levels, an experimental verification is desirable.

Little experimental information is available about the level structure of ^{20}Na above the proton threshold. Previous $^{20}Ne(^{3}He,t)^{20}Na$ charge exchange experiments [9,10] indicate a state just below the threshold at E_x = 1.92 MeV and a broad (or several unresolved) level at E_x = 2.89 MeV.

With improved experimental conditions we have investigated the excitation range of ^{20}Na between E_x = 1.30 and 3.20 MeV to determine the level density and level energies and to enable comparison with the theoretical predictions. The experiments were performed at the FN-Tandem-VdG accelerator at the University of Notre Dame using bombarding energies of $E(^3He^{++})$ = 24 - 27 MeV and

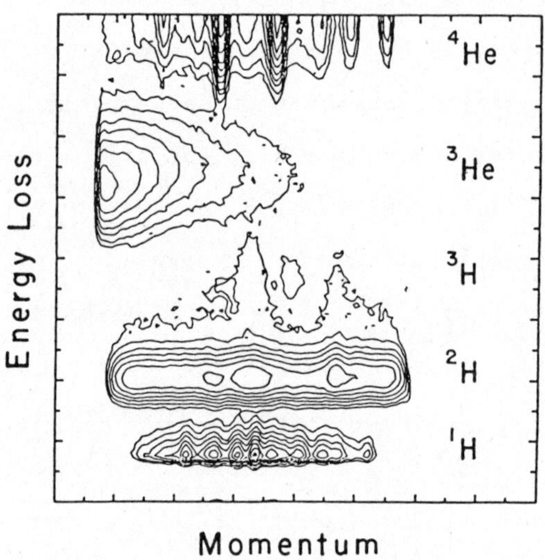

Figure 1. Two dimensional data display of energy loss vs momentum showing distinct particle bands.

beam currents of 0.5 - 3.5 μA . The target consisted of an extended gascell (2cm diameter), filled with 99.5% enriched ^{20}Ne gas. As entrance and exit windows for beam and reaction products we used thin (2 μm) Ni-foils, which were able to withstand a gas pressure of 0.2 atm in the gas cell. The particle spectra were measured at various angles between $\theta = 10^o - 40^o$ using the 100cm broad range magnetic spectrograph [11]. With a two wire position sensitive proportional counter in the focal surface of the spectrograph we measured in two parameter event mode the momentum and the energy loss of the reaction products. Figure 1 shows the contour plot of a two dimensional spectrum measured at a bombarding energy of 25 MeV and an angle of 15°. Strong groups from the background reactions ^{20}Ne(^{3}He,p), (^{3}He,d), (^{3}He,^{3}He) and (^{3}He,α) with cross sections in the mbarn to barn range can be identified, while the observed triton groups are considerably weaker (μbarn). In figure 2 two triton spectra are shown, measured at different energies, angles and field settings, which display triton groups populating states in ^{20}Na below and above the proton threshold. The observed energy resolution of the triton peaks of 80 keV, significantly larger than the known resolution of the spectrograph of 15 - 20 keV, is mainly determined by straggling effects of beam particles

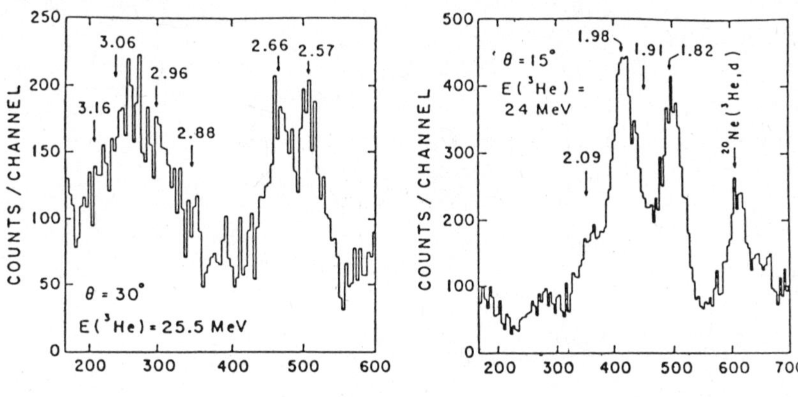

Figure 2. Triton spectra populating levels in ^{20}Na above (left) and below (right) the proton threshold.

and reaction products in the Ni windows and in the gas itself. The spectra indicate a cluster of four levels below the proton threshold at excitation energies between 1.8 and 2.1 MeV, a doublet just above the threshold between 2.5 and 2.7 MeV, and a broad structure around 3 MeV. The exact peak positions and excitation energies were derived by standard peakfitting procedures; an example is given in figure 3. The results indicate four bound levels at E_x = 1.82, 1.91, 1.98 and 2.09 MeV and two unbound states at E_x = 2.57 and 2.66 MeV. The broad structure can be well fitted in all obtained spectra by four overlapping triton peaks corresponding to level energies of E_x = 2.88, 2.96, 3.06 and 3.16 MeV.

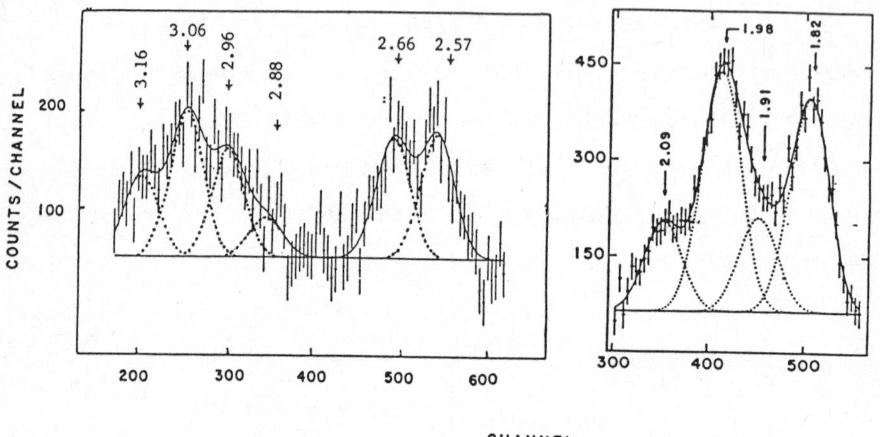

Figure 3. Gaussian fit to the data shown in fig.2

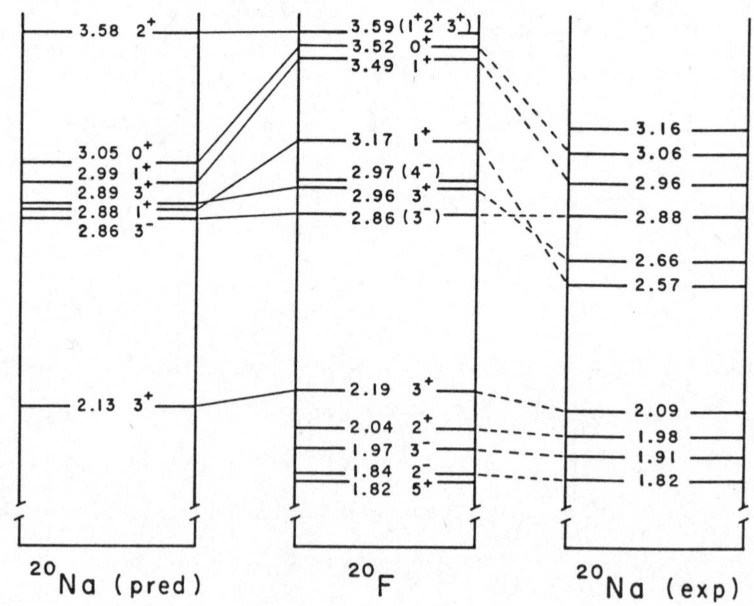

Figure 4. Observed levels in ^{20}Na and their tentative assignment to known states in ^{20}F. Also shown are the predicted level energies from ref. [8]

The measured angular distributions do not yet allow a unique spin and parity assignment for the observed states. Currently, only a tentative assignment to the known analog states in the mirror nucleus ^{20}F is possible on the basis of the observed energies. This is shown in figure 4 together with the predicted assignments of reference [8]. A fairly straight forward assignment is indicated for the bound states, where no large level shifts are expected. The observed states at 2.88, 2.96 and 3.06 MeV agree well with the predicted levels at 2.86 MeV (3^-), 2.99 MeV (1^+) and 3.05 MeV (0^+) calculated from the Thomas Ehrman shifts between bound and unbound analog states. The shifts calculated for the 1^+ state at 3.17 MeV

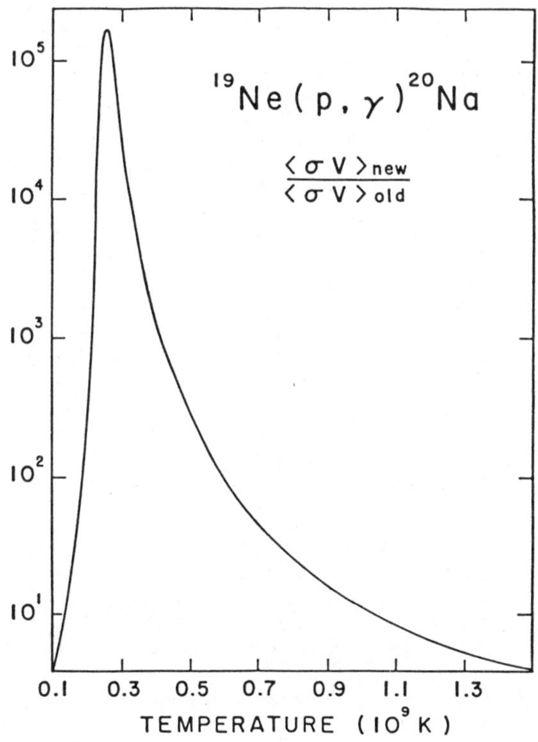

Figure 5. The estimated rate of $^{19}Ne(p,\gamma)^{20}Na$ based on the assignment shown in fig. 4 compared to the previous rate of ref. [8]

and the 3^+ state at 2.96 MeV in ^{20}F are relatively small due to the small single particle factors used in the calculations. The shifts tentatively suggested here may be explained by larger spectroscopic factors than assumed in reference [8].

Using the spectroscopic factors, derived from the level shifts, to calculate the proton widths Γ_p of the states and adopting the gamma-partial widths Γ_γ from the proposed analog states in ^{20}F, the strengths of the two resonance states in the ^{19}Ne(p,γ)^{20}Na can be calculated using the standard expression:

$$\omega\gamma = \frac{2J+1}{12} \cdot \frac{\Gamma_p \cdot \Gamma_\gamma}{\Gamma} .$$

Mainly due to the low resonance energies, E_r = 0.37 MeV (1^+) and 0.46 MeV (3^+), the resulting reaction rate increases significantly in the temperature range T_9 = 0.1 - 0.8 compared to the previous predictions (Figure 5).

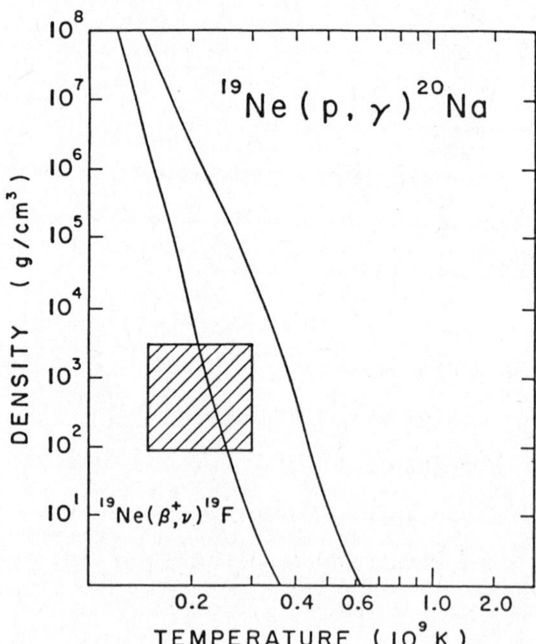

Figure 6. Breakout conditions of the hot CNO cycle depending on stellar temperature and density. The right hand curve is based on the previous [8], the left hand curve on the present rate.

This modifies considerably the breakout conditions of the hot CNO cycle by the $^{19}Ne(p,\gamma)^{20}Na$ reaction ,providing the temperature and density conditions in the particular environment allow a sufficiently high production of ^{19}Ne by either $^{15}O(\alpha,\gamma)$ or $^{18}F(p,\gamma)$. The temperature density plane in figure 6 displays the conditions for the two possible depletion reactions of ^{19}Ne, $^{19}Ne(\beta^{+}\nu)^{19}F$ and $^{19}Ne(p,\gamma)^{20}Na$. The two solid lines indicate the boundary conditions where both reactions have the same strength, calculated on the basis of reference [8] and on the basis of the new enhanced (p,γ) reaction rate. The dashed square marks approximately the region of typical density and temperature conditions for explosive hydrogen burning in novae. While the old boundary line is well above this area, indicating predominantly the β^{+}-decay of the produced ^{19}Ne, the present rate allows a significant depletion of ^{19}Ne toward heavier masses suggesting that a significant transfer of CNO material towards the NeNaMg region might be possible.

Further measurments to verify the proposed structure of the observed levels are currently in progress.

The reaction $^{22}Na(p,\gamma)^{23}Mg$

The reaction $^{22}Na(p,\gamma)^{23}Mg$ is of considerable interest for the understanding of nucleosynthesis in the hot NeNa-cycle and in the rp-process in explosive stellar environments [12]. Of particular importance is the knowledge of the rate for determining the abundances of ^{22}Na in such burning events, because the decay of ^{22}Na ($T_{1/2}$ = 2.6 y) after the freeze out provides an extremely attractive mechanism for the formation of ^{22}Ne-enriched neon isotopic anomalies (NeE) observed in meteoritic inclusions [13].

Little was known about the structure of the compound nucleus ^{23}Mg above the proton threshold at Q = 7.58 MeV when the first estimate of the rate was published [12]. Therefore considerable effort was raised to study the proton unbound levels in ^{23}Mg by ^{25}Mg(p,t)^{23}Mg [13] and ^{24}Mg(^{3}He,α)^{23}Mg [14,15]. The experimental results indicate a fairly high level density in this excitation range, the excitation energies of the observed states are reported within an uncertainty of 10 - 20 keV. Hovewer, the results do not

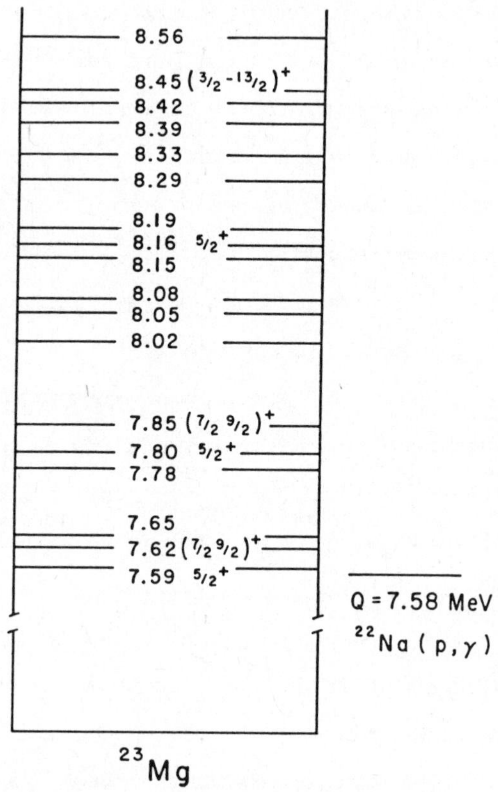

Figure 6. Proton unbound states in ^{23}Mg observed in ^{25}Mg(p,t) and ^{24}Mg(^{3}He,α) reactions

allow a unique spin and parity assignment for the levels, making it difficult to derive a reasonable estimate for the ^{22}Na(p,γ) - reaction rate. Despite the lack of sufficient input information such an attempt to estimate the rate was made [7], but due to the discussed uncertainties [16] an experimental verification seems to be highly desirable.

The difficulty in measuring the $^{22}Na(p,\gamma)^{23}Mg$ reaction is caused by the extremly large γ-background due to the β^+-decay of ^{22}Na to the first excited state (98%) in ^{22}Ne. This creates γ-radiation from the annihilation of the positrons (0.511 MeV) and from the subsequent γ-decay of the first excited state in ^{22}Ne to the ground state (1.275 MeV).

The target material was commercially obtained in the form of a 1.3 mCi/ml solution of $^{22}NaCl$. As target backing a 0.5 mm thick sheet of Ta was used coated with a layer of 50 $\mu g/cm^2$ Ni to reduce the typical drifting of Na-atoms into the backing. The $^{22}NaCl$ was uniformly distributed over the target spot area of 0.3 cm^2 by slowly evaporating small drops of the solution. The total target activity was 60 μCi, which corresponds to $1 \cdot 10^{15}$ ^{22}Na-atoms/cm^2. Scans of the target activity verified the homogeneous distribution of the ^{22}Na over the target area. Test experiments showed that the target could withstand beam currents of 10 μA (E_p = 700 keV) without noticable deterioration (<1%); higher beam currents however did result in loss of target material due to sputtering effects ($\lesssim$ 20% in a period of one week).

The experiments were performed with the 2MV Pelletron at the California Institute of Technology. We investigated the reaction in the bombarding energy range E_p = 0.55 - 0.9 MeV with beam currents of 8 - 30 μA. The γ-radiation was measured at 0° with a large volume Ge-detector shielded with 2.2cm of lead against the target activity. A low frequency ramping voltage of 8 kV on the target allowed us to cover an excitation range of 8 keV per run. Significant impurities in the target material (^{23}Na, Cl, Si) could be observed by measuring $^{23}Na(p,\gamma)$-, $^{35}Cl(p,\gamma)$-, $^{29}Si(p,\gamma)$- and $^{30}Si(p,\gamma)$-resonances. From the known resonance strengths the number of atoms for the particular impurities could be determined: $1 \cdot 10^{15}$ ^{23}Na, $1.5 \cdot 10^{15}$ Cl and $1 \cdot 10^{17}$ Si atoms/cm^2. Figure 8 shows the yield curve and the γ-spectrum of the $^{23}Na(p,\gamma)^{24}Mg$ resonance at

E_r = 0.677 MeV ($\omega\gamma$ = 1 eV). This allows the derivation of the target thickness and the stopping power since ^{23}Na in the target material behaves chemically as ^{22}Na.

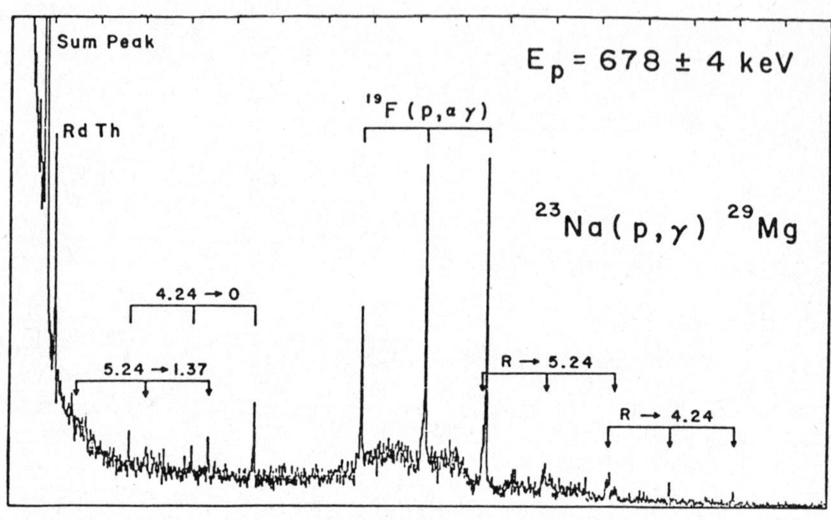

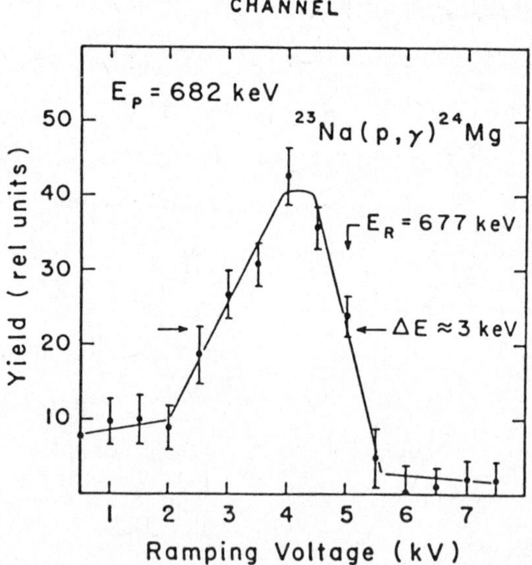

Figure 8. The lower part shows the yield curve over the E_r= 677 keV $^{23}Na(p,\gamma)^{24}Mg$ resonance. The associated γ-spectrum is shown in the upper part.

The excitation curve was measured in energy steps of 8 keV collecting a charge of 0.1 - 0.2 Cb per run. The obtained γ-spectra were analyzed in terms of all possible γ-transitions to bound states in ^{23}Mg below E_x = 4.4 MeV. Figure 9 shows two typical spectra, obtained at the energy E_p = 0.614 MeV, where a strong 1 eV resonance was expected ,and at E_p = 0.796 MeV as background run.

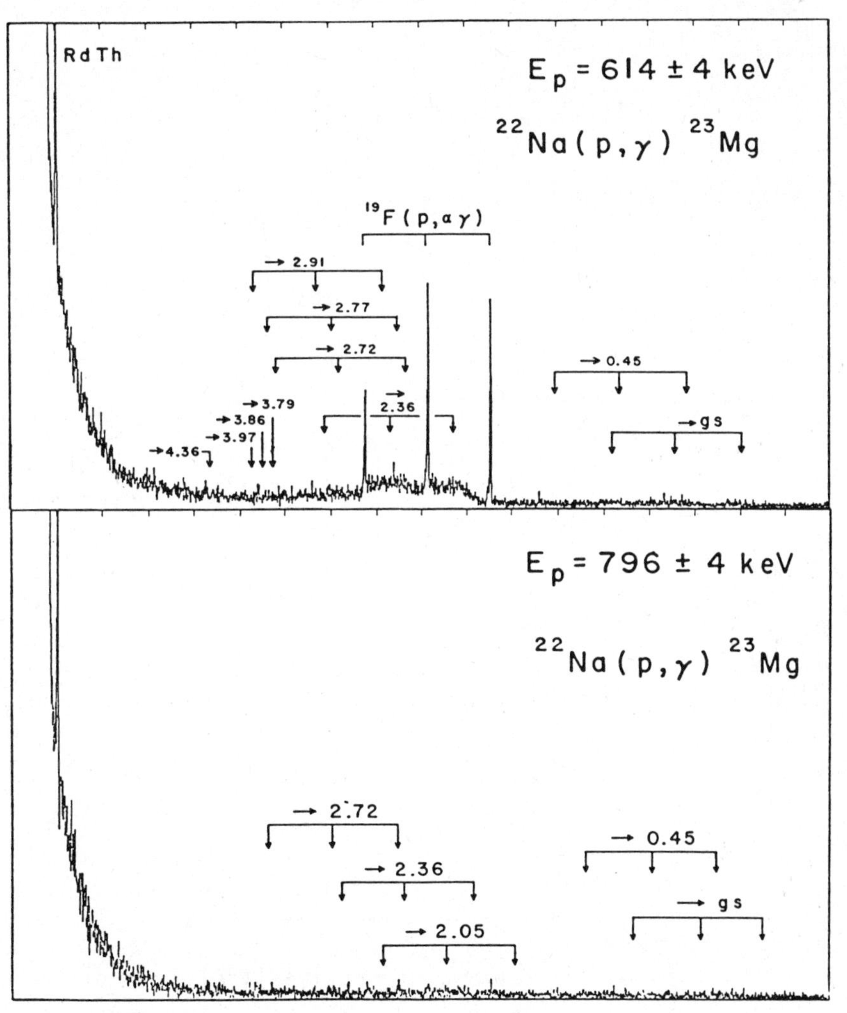

Figure 9. Typical γ-spectra obtained at E_p=614, 796 keV in search of resonances in ^{22}Na(p,γ)^{23}Mg

Indicated are the positions of possible γ-transitions in $^{22}Na(p,\gamma)^{23}Mg$. No obvious transition could be observed. From the derived yield an upper limit for the strength of possible resonances can be determined by assuming the same target composition for ^{22}Na as for ^{23}Na. Figure 10 shows the resulting upper limits (1σ) as a function of the proton energy for the expected transitions to the ground state ($3/2^+$) and the first excited state ($5/2^+$) in ^{23}Mg. The derived upper limits of the resonance strengths are between 30 meV and 200 meV, depending on the reaction background and are therefore significantly lower than the predicted values in reference [16].

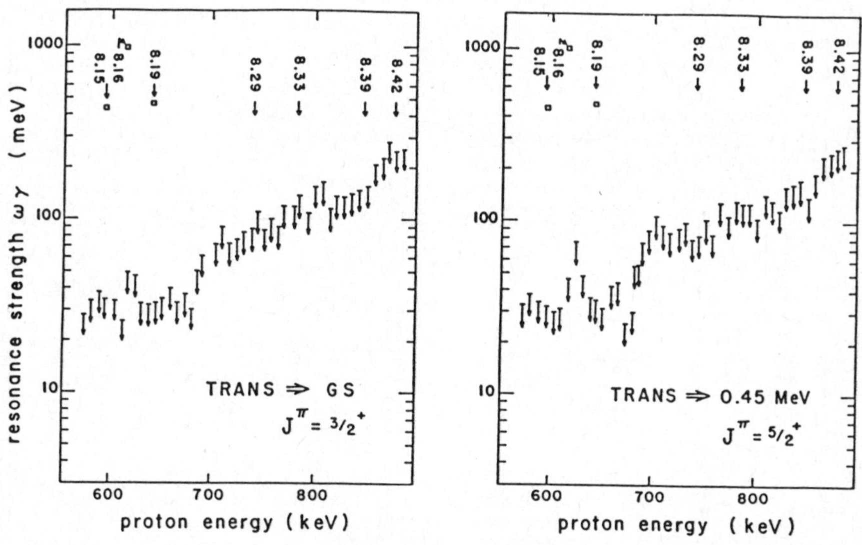

Figure 10. Upper limits for the resonance strengths in the transitions to the groundstate and the first excited state in $^{22}Na(p,\gamma)^{23}Mg$. The squares indicate the estimated resonance strength from reference [16]

Measurements with improved targets are in progress to search for strong $^{22}Na(p,\gamma)^{23}Mg$ resonances expected at higher bombarding energies. Furthermore it is planned to extend the investigation to lower bombarding energies (E_p < 500 keV) which corresponds to the energy range of explosive hydrogen burning scenarios.

References

[1] Williams, R.E., Ney, E.P., Sparks, W.M., Starrfield, S.G., Truran, J.W., Wyckoff, S.: 1985, Monthly Notices Roy. Astron. Soc.
[2] Snijders, M.A.J., Batt, T.J., Seaton, M.J., Blades, J.C., Morton, D.C., 1984, Monthly Notices Roy. Astron. Soc. 211, 7
[3] Gehrz, R.D., Grasdalen, G.L., Greenhouse, M., Hackwell, J.A., Hayward, T., Bentley, A.F., 1986, Ap.J. 308, L63
[4] Starrfield, S., Truran, J.W., Sparks, W.M., 1978, Ap. J. 226, 186
[5] Law, W.Y., Ritter, H., 1983, Astron. Astrophys. 123, 33
[6] Truran, J.W., 1986, in Audouze, J., Mathieu, N. (eds), Nucleosynthesis and Its Implications on Nuclear and Particle Physics, p.97, D. Reidel Publishing Co.
[7] Wiescher,.M., Görres, J., Thielemann, F.-K., Ritter, H., 1986, Astron. Astrophys. 160, 56
[8] Langanke, K., Wiescher, M., Fowler, W.A., Görres, J., 1986, Ap.J. 301, 629
[9] Pehl, R.H., Cerny, J., 1965, Phys. Lett. 14, 137
[10] Donovan, P.F., Parker, P.D., 1965, Phys. Rev. Lett. 14, 147
[11] Goss, J.D., Rollefson, A.A., Browne, C.P., 1973, NIM 109, 13
[12] Wallace, R.K., Woosley, S.E., 1981, Ap. J. Supp. 45, 389
[13] Eberhardt, P., Jungck, M.H.A., Meier, I.O.,Niederer, F., 1979, Ap. J. Lett., 234, L169
[14] Nann, H., Saha, A., Wildenthal, B.H., 1981, Phys. Rev. C23, 606
[15] Schmalbrock, P.M.F., Donoghue, T.R., Hausman, H.J., Wiescher, M., Wijekumar, V., Browne, C.P., Rollefson, A.A., 1985, AIP Proc. 125, 785, and to be published
[16] Wiescher, M., Langanke, K., 1986, Z. Phys. A325, 309

BETA-DECAY HALF-LIVES OF VERY NEUTRON-RICH NUCLEI AND THEIR CONSEQUENCES FOR THE ASTROPHYSICAL R-PROCESS

Karl-Ludwig Kratz
Institut für Kernchemie, Universität Mainz
Fritz-Strassmann-Weg 2, D-6500 Mainz

Abstract

Half-lives for Gamow-Teller β-decay of very neutron-rich nuclei have been calculated using the RPA shell-model code of Krumlinde and Möller. For the examples of the isotope sequences of $_{27}$Co and $_{37}$Rb, and of nuclei around ^{132}Sn it is demonstrated that close agreement between experiment and theory can be obtained, provided an appropriate choice of model parameters is made for each mass region. On the basis of this agreement, $T_{1/2}$ predictions for isotopes up to the r-process path are made and compared to earlier model calculations. Possible implications of the results on the site and the cycle time of the r-process are discussed.

Introduction

Despite years of effort, the basic astrophysical site (or sites) for the r(apid neutron-capture)-process is not yet known or at best vaguely defined (for review, see e.g. Refs. [1-4]). The problem of identifying the site is complicated, because it requires a detailed knowledge of both the nuclear physics to form neutron-rich nuclei and the astrophysical conditions of neutron density, stellar temperature and exposure time in evolved stars.

Following the essential ideas of Burbidge, Burbidge, Fowler and Hoyle (B^2FH) [1] for stellar neutron capture in the 'classical' r-process, in many models a great simplification is introduced by assuming that the stellar temperature and neutron density are sufficiently high so that an (n,γ)-(γ,n) equilibrium can be achieved. Then, r-abundance calculations can be performed without using reaction cross sections, and the most important nuclear data needed are neutron-separation energies (S_n) and β-decay half-lives ($T_{1/2}$) along the r-process path. The length of time required for such nucleosynthetic processes from Fe to the actinides is mainly defined by the generally assumed longer-than-average $T_{1/2}$ of the so-called 'waiting-point' nuclei at the neutron shell closures N = 50, 82 and 126. At these points the neutron-capture flow is halted, and the nuclei must 'wait' until ß-decay to a new element allows it again to capture neutrons. As a result, distinct r-abundance peaks near A = 130 and 195 and a 'pygmy' peak near A=80 occur (see, e.g., Fig. 1 of Ref. [5]). With two exceptions - the very recently identified 'waiting-point' nuclei ${}^{80}_{30}Zn_{50}$ [6] and ${}^{130}_{48}Cd_{82}$ [7] - for the nuclei in the r-process path the important ß-decay properties ($T_{1/2}$ and the ß-delayed neutron-emission probability P_n) are still unknown and had - and still have - to be predicted from nuclear models. As reviewed in [8] (see also the more recent Ref. [4]), with the uncertainty in the calculated $T_{1/2}$ far from stability, the estimated ß-decay cycle time (τ_β) was - and still is believed - to lie within the rather broad range of

$$0.1 \leqslant \tau_\beta \leqslant 30 \text{ s}.$$

With this, the resultant uncertainties in the time scale for the r-process and thus the required neutron-exposure time did not allow to put substantial constraints on astrophysical scenarios.

A few years ago, considerable improvement over past attempts has been obtained with regard to reproduction of the r-process abundances [9], in particular the location of the abundance peaks at A ≃ 130 and 195 and their relative heights. This improvement was mainly attributed to new $T_{1/2}$ based on a more detailed model for the microscopic nuclear structure [10] than the earlier used Gross Theory [11]. The astrophysical scenario was a supernova shock front passing through the helium-burning shell of a massive star. In the calculations of [9] it was necessary, however, to assume s-process enhanced seed abundances. Furthermore, when starting from the iron-group nuclei, not even the N = 50 neutron shell (corresponding to the 'pygmy' r-abundance peak at A ≃ 80) could be reached because of the short duration of the neutron flux (a few tenths of a second) in explosive He-burning. Therefore, the calculations to reproduce the solar r-process abundances started from ${}^{78}_{28}Ni_{50}$. Moreover, in this astrophysical model, the calculated r-abundances are extremely sensitive to $T_{1/2}$. As was demonstrated in [12], a change in the TDA shell model $T_{1/2}$ of [10] by (only) a factor of three in either direction would already destroy the good agreement with the observed r-abundances. This supports the conclusion of [4] that there still seems to be too much uncertainty in the input astrophysics and model assumptions to claim that explosive He-burning in massive stars has solved the r-process puzzle [10].

However, the above calculations highlight the need to further reduce the uncertainties of the nuclear physics properties before the r-process can be understood. Recent $T_{1/2}$ measurements for very neutron-rich nuclei (see, for example, Refs. [13,14]) have suggested that the $T_{1/2}$ predictions of [10] may not be entirely reliable far from stability. Therefore, the RPA shell model of Krumlinde and Möller

[15], which provides a considerable improvement over the TDA shell model approach of [10], has been used to calculate the $T_{1/2}$ of a number of isotope sequences from β-stability up to the neutron drip line. This paper reports on the improvements (and limitations) of $T_{1/2}$ predictions in the $_{26}Fe$ - $_{30}Zn$, the $_{35}Br$ - $_{39}Y$ and the $_{46}Pd$ - $_{49}In$ regions against the background of recent measurements. By relating the $T_{1/2}$ and P_n-values of ^{130}Cd [7] and $^{131,132}In$ [16] to the observed r-abundances of ^{130}Te and $^{131,132}Xe$ [5], a new constraint on the mode of operation of the r-process is obtained which relies for the first time on experimental data only. The viability of the proposed steady-flow solution is supported by a new estimate of the seed-to-fission cycle time on the basis of experimental and theoretical $T_{1/2}$ of N=50, 82 and 126 'waiting-point' nuclei.

The RPA Shell Model

The RPA shell model of Krumlinde and Möller [15] - including the modifications described in [17] - has been used to calculate Gamow-Teller (GT) β-decay. This model uses calculated Nilsson model wave functions, spherical or deformed, as the starting point for determining the wave functions of mother and daughter nucleus in β-decay. Pairing is treated in the BCS approximation, and the pairing strength parameter Δ is chosen according to [18]. To account for the retardation of low-energy GT decay rates, a simple residual GT interaction is added which is treated in the RPA. Appropriate χ- and μ-parameters for the proton and neutron single-particle potentials for different mass regions are taken from Bengtsson and Ragnarsson [19]. Deformation parameters ε_2 and ε_4 from [20] are used in the calculations, with the simplified assumption of equal deformation for mother and daughter nucleus. From the GT strength functions theoretical $T_{1/2}$ (and P_n-values) were derived as described in detail in [13]. Nuclear masses were taken from [20-22],and the integral Fermi function according to [23] was used.

The $T_{1/2}$ Puzzle in the Iron-Group Region

During the last few years, decay properties of a number of new neutron-rich isotopes in the region of $_{25}Mn$ to $_{30}Zn$ have been measured [6,14,24], among them with $^{80}_{30}Zn_{50}$ the identification of the first r-process 'waiting-point' nucleus. The most interesting feature of $_{25}Mn$ to $_{28}Ni$ isotopes with $N \gtrsim 36$ is the fact that their $T_{1/2}$ are up to an order of magnitude shorter than the TDA shell model predictions of [10], whereas the $T_{1/2}$ of neutron-rich $_{29}Cu$ and $_{30}Zn$ isotopes lie between the predictions of the statistical [11] and the microscopic model [10], thus following the trend observed for medium-heavy and heavy nuclei far from stability (see, e.g., Refs. [13,14,25]). Apart from considerable nuclear-structure interest, a general tendency towards short $T_{1/2}$ in the Fe-Co-Ni region up to the N = 50 isotones would be of importance in calculations of the r-process. As is discussed in [14], in explosive He-burning the neutron-capture times in this area are comparable to the theoretical $T_{1/2}$ of [10], so that the time scale for the r-process starting from Fe-group seed abundances and proceeding to the N = 50 magic shell (with the build-up of the 'pygmy' A ≃ 80 abundance peak) would no longer be given by the sum of the $T_{1/2}$ at the N=50 'waiting point' [8].

In order to understand the occurence of short $T_{1/2}$ in this limited mass region, the RPA shell model was used to calculate theoretical $T_{1/2}$ for GT decay. These values were then compared to the predictions from the earlier models [10,11]. For all three $T_{1/2}$ predictions the same mass formulae were used, so that the results do not depend on different assumptions on Q_β, respectively $f(E_\beta)$.

However, it was clear from the beginning that reliable RPA calculations could not be made in a straightforward way. Sudden changes of $T_{1/2}$ within isotope or isotone sequences in other mass regions were found to be related to spherical shell closures or nuclear shape changes (see, for example, Refs. [13,26]). Such effects are also expected to occur in the $_{25}$Mn to $_{30}$Zn region: spherical shell closures at Z = 28 and N = 38,40 and 50, as well as predicted local prolate quadrupole deformation [20] up to $\varepsilon_2 = 0.29$ (e.g. ^{61,63}Cr, $^{62-64}$Mn) with even some oblate deformed isotopes (e.g. ^{62}Cr, ^{64}Fe) in between. Therefore, already the reproduction of known decay properties such as $T_{1/2}$, level schemes and log(ft)-values with the present RPA shell model requires considerable fine-tuning of the nuclear physics input parameters χ, μ, ε_2 and Δ. Moreover, RPA calculations for mother-daughter pairs with different deformations - in the extreme a 'prolate mother' (e.g. ^{64}Mn, $\varepsilon_2 = 0.27$) ß-decaying into an 'oblate daughter' (e.g. ^{64}Fe, $\varepsilon_2 = -0.08$) - are not possible with the present shell model code [15]. With the model-inherent simplification of equal deformation for mother and daughter nucleus, in such cases one may well be able to reproduce the low-energy level scheme, but with a $T_{1/2}$ easily being off by an order of magnitude. On the other hand, a 'correct' $T_{1/2}$ can be obtained, however with a completely wrong ß-strength function, respectively level scheme. Fortunately, such problems seem to occur in the $_{24}$Cr to $_{27}$Co region only for a few isotopes with $34 \lesssim N \lesssim 40$. Nevertheless, this makes reliable $T_{1/2}$ predictions for 'neighbouring', still deformed isotopes, such as ^{64,65}Mn, ^{65,66}Fe and ^{68,69}Co, difficult.

As an example for the fine-tuning of model parameters entering the RPA code of [15] to reproduce simultaneously a known level scheme, the corresponding quasi-particle (QP) configurations, the log(ft)-values for GT-decay and the measured $T_{1/2}$, Fig.1 shows a comparison of the relevant experimental features of $^{63}_{27}Co_{36}$ ß-decay to $^{63}_{28}Ni_{35}$ [24] with RPA shell model calculations for different assumptions on quadrupole deformation. In these calculations, χ- and μ-parameters optimized for this mass region [19] were used and pairing reduction by 25% [18] was taken into account. As is clearly seen from the figure, for a deformation of $\varepsilon_2 = 0.10$, as predicted by Möller and Nix [20], apart from the ground-state (g.s.) configuration of the mother nucleus there is no agreement between experiment and theory. Best

overall agreement is obtained for a considerably stronger deformation of $\varepsilon_2 \simeq 0.2$. The same trend towards larger quadrupole deformation than predicted in [20] was also observed for the heavier Co as well as for odd-neutron Fe isotopes [27]. Coming back to ^{63}Co, it is also evident from Fig.1 that another dramatic change of the GT-decay pattern occurs around $\varepsilon_2 = 0.25$ which involves different 1QP-configurations for the g.s. of both mother and daughter nucleus, but - fortuitously - yields the 'correct' number for the $T_{1/2}$.
This kind of comparison between experiment and RPA prediction has been made for a number of known neutron-rich isotopes of each nuclear type (odd-Z, odd-N, odd-odd, even-even) in the $_{24}$Cr to $_{30}$Zn region in order to obtain a consistent picture on deformation and 1-3QP structure in N $\geqslant$ 34 isotones of these elements [27]. On the basis of this investigation one can conclude that the most important parameter in this mass region is deformation which seems to change rapidly and differently from the Möller-Nix predictions [20]. As mentioned above, this makes reliable $T_{1/2}$ predictions for the next heavier isotopes difficult. However, the calculation of $T_{1/2}$ for more exotic nuclei with 40 $\lesssim$ N $\lesssim$ 54 may become more reliable again because of the increasing influence of the spherical N = 50 magic shell.

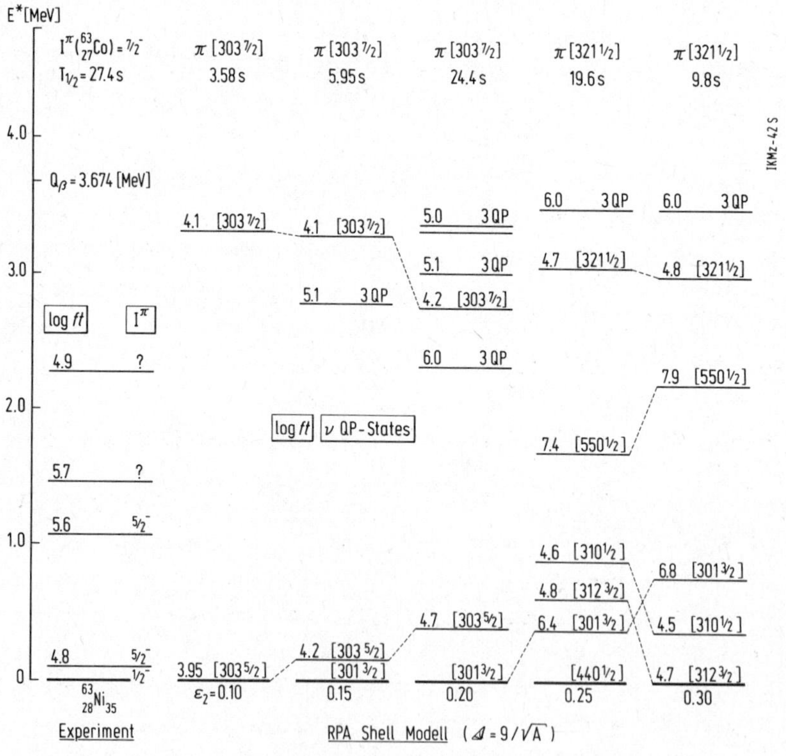

Fig. 1: Comparison of the relevant experimental features of ^{63}Co GT-decay to ^{63}Ni [24] with RPA shell model calculations for different assumptions on quadrupole deformation. For discussion, see text.

In any case, the present RPA shell model calculations support the general tendency towards shorter $T_{1/2}$ in the Fe-group region up to the neutron drip line. As a typical example, in Fig.2 the $T_{1/2}$-ratios of Exp/RPA, Exp/Gr.Th., TDA/Gr.Th. and RPA/Gr.Th. are shown for the Co isotope sequence. It is seen that for the range of astrophysical interest up to the r-process path at N = 50 the RPA-$T_{1/2}$ are shorter than the TDA-$T_{1/2}$ [10] by roughly an order of magnitude. Surprisingly, even the

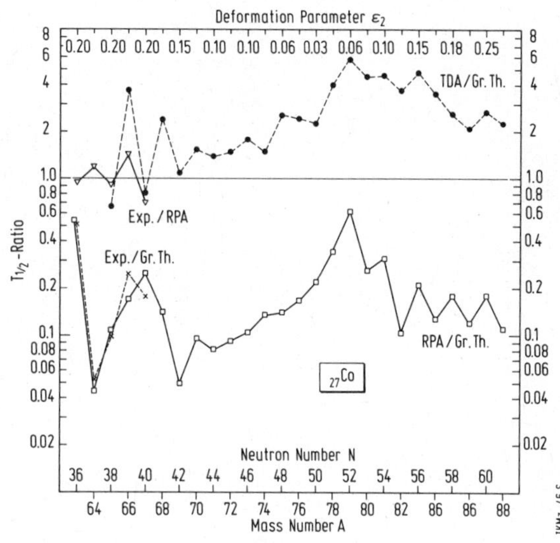

Fig. 2: Comparison of $T_{1/2}$-ratios for neutron-rich Co isotopes:
▽ Exp/RPA,
x Exp/Gr.Th.,
• TDA/Gr.Th. and
□ RPA/Gr.Th.

statistical Gross Theory [10], which is known to systematically overestimate $T_{1/2}$ far from stability, predicts shorter $T_{1/2}$ in the mass region than the microscopic model of [10]. Another interesting result of the present $T_{1/2}$ calculations is the indication that besides ^{80}Zn with $T_{1/2}$ = (530 ± 20) ms [6] the only other N = 50 'waiting-point' nucleus with a longer-than-average $T_{1/2}$ is ^{79}Cu, whereas the RPA-$T_{1/2}$ of ^{76}Fe, ^{77}Co and ^{78}Ni are only in the range of 10 - 30 ms [27].
With respect to astrophysical implications, the present $T_{1/2}$ predictions justify the assumption in the r-process calculations discussed in [14], to apply the experimental 'short' $T_{1/2}$ for extrapolation in the Fe-group region where the r-process starts. With the even shorter RPA-$T_{1/2}$ of the present work, it might be worth to repeat such calculations as a function of neutron exposure in order to resolve the so far existing puzzle of the abundance distribution around A = 80.

The Z ≃ 40 Region

The region of neutron-rich nuclei around Z = 40 and A = 100 is characterized by the interplay between a number of spherical and deformed (sub-)shells and a strong $[\nu g_{7/2}, \pi g_{9/2}]$ interaction. As a result, for $Z < 37$ smooth and for $Z \geqslant 37$ sudden transitions from spherical to deformed g.s. shapes occur, which bring along dramatic changes in the β-decay pattern and, hence, in $T_{1/2}$ (see, e.g., Refs.

[13,15,26]. Thus, $T_{1/2}$ predictions are again difficult. To check the validity of the microscopic model of [10], in Fig.3 the $T_{1/2}$ of the heaviest known isotopes of $_{35}Br$ - $_{42}Mo$ are compared to the TDA predictions. On the average, for these nuclei which lie about half-way between ß-stability and the r-process path, the ratio

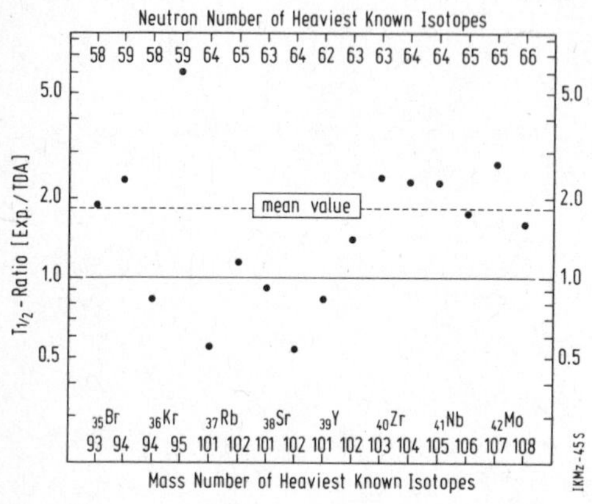

Fig. 3:
$T_{1/2}$-ratios between experimental and predicted [10] values for the two heaviest known isotopes of $_{35}Br$-$_{42}Mo$, each.

between observed and predicted values is 1.85. When extrapolating from the last known isotopes over another 10 - 15 mass units to the r-process path, the deviation from the 'true' (but at present time unmeasurable) $T_{1/2}$ may well reach the factor of three which was regarded as the critical deviation from the TDA-$T_{1/2}$ for the viability of the explosive He-burning scenario [12].

After extensive spectroscopic studies during the past years, together with considerable fine-tuning of the nuclear-physics input parameters for our RPA shell model, we now believe to have understood the main features of the GT strength functions and the gross ß-decay properties of exotic nuclei in this mass region.

Fig. 4:
Comparison of experimental $T_{1/2}$ of very neutron-rich Rb isotopes with predictions from different nuclear models.

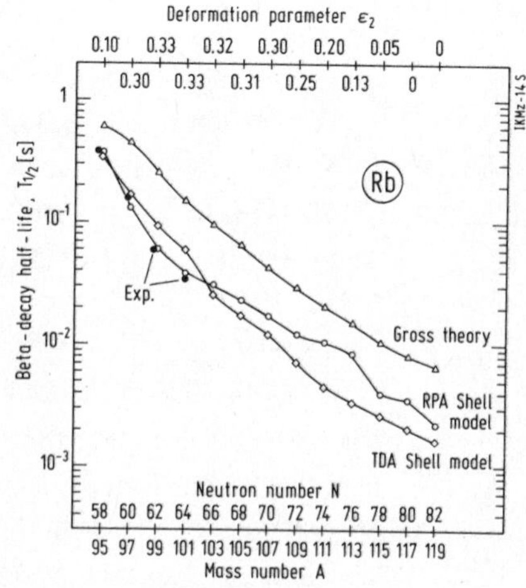

Based on these results, we have calculated so far the $T_{1/2}$ of neutron-rich $_{35}Br$ - $_{39}Y$ isotopes. As can be seen from the example shown in Fig.4, the RPA-$T_{1/2}$ are systematically shorter than the predictions based on the Gross Theory [11], but longer (near the r-process path by factors of 2-3) than the TDA-$T_{1/2}$ [10].

The N = 82 'Waiting-Point' Region

Besides N = 50, the other important 'waiting-point' region, which lies just at the present limit of experimental accessibility, is that at N = 82. With respect to the observed strong A ≃ 130 r-abundance peak compared to the 'pygmy' A ≃ 80 peak [5], however, the N = 82 region is even more important for the understanding of the r-process.

When assuming, for example, that the equilibrium steady-state solution is realized in nature, one can directly relate the observed r-abundances in the A ≃ 130 peak, i.e. $^{128,130}_{52}Te$ and $^{129,131,132}_{54}Xe$, to the abundances and - within the 'waiting-point' concept - the $T_{1/2}$ (and P_n-values) of their N ≃ 82 progenitors in the r-process path [2], i.e. $^{128}_{46}Pd$, $^{129}_{47}Ag$, $^{130}_{48}Cd$ and $^{131,132}_{49}In$. With this, the measurement of the $T_{1/2}$ of even a single N = 82 'waiting-point' nucleus may already be sufficient to determine the scenario in which the heavy elements in nature were formed. For example, with the above steady-state assumption, the $T_{1/2}$ of ^{130}Cd was predicted by Hillebrandt in Ref. [7] from the known r-abundances of ^{130}Te and $^{131,132}Xe$ [5] and the known $T_{1/2}$ and P_n-values of $^{131,132}In$ [16] to be (205 ± 20) ms, in good agreement with the recently measured value of (195 ± 35) ms [7]. Unless the above agreement is accidental, this would favour the steady-flow solution for the r-process, which was originally suggested by B^2FH in 1957 [1], and would rule out short-duration scenarios. Fortuitous agreement could, however, be excluded by an independent check, e.g. via the (in the near future possible) measurement of the $T_{1/2}$ of $^{129}_{47}Ag_{82}$. From the same arguments given before, a $T_{1/2}$ ≃ 150 ms can be predicted.

In order to obtain at least a rough idea, whether this astrophysical $T_{1/2}$ estimate for ^{129}Ag is plausible also from the nuclear physics point o view, RPA shell model calculations were performed for a number of known isotopes around $^{132}_{50}Sn_{82}$. With the appropriate Nilsson parameters [19], their spherical single-particle level structures could be reasonably well reproduced. Examples for the level of agreement between measured and predicted gross β-decay properties are given in Tab.1. The $T_{1/2}$ roughly indicate the amount of GT strength at low excitation energy, whereas the P_n-values reflect the ratio of high-lying to low-energy β-strength. Within the uncertainties to be seen from Tab.1, also the RPA predictions of $T_{1/2}$ ≃ 120 ms and P_n ≃ 20% for $^{129}_{47}Ag_{82}$ should not be too far from reality. This result, which is close to the above astrophysical $T_{1/2}$ estimate, is quite promising with respect to a possible consistent picture of $T_{1/2}$ of 'waiting-point' nuclei and of r-abundances for the whole A ≃ 130 peak within the 'classical' r-process model.

Tab. 1: Comparison of measured $T_{1/2}$ and P_n-values of isotopes around $^{132}_{50}Sn_{82}$ with RPA shell model predictions

Isotope	$T_{1/2}$ Exp.	$T_{1/2}$ RPA	P_n [%] Exp.	P_n [%] RPA
$^{130}_{48}Cd_{82}$	0.2	0.3	4	2.5
$^{131}_{49}In_{82}$	0.28	0.24	1.7	1.8
$^{132}_{49}In_{83}$	0.20	0.31	5.5	2.2
$^{133}_{49}In_{84}$	0.18	0.12	85	100
$^{132}_{50}Sn_{82}$	40	44	-	-

The N ≃ 126 'Waiting-Point' Region

In contrast to the N = 50 and 82 areas, 'waiting-point' nuclei at N = 126, such as ^{195}Tm and ^{196}Yb, will not become accessible to experimental investigation in the near future. Therefore, it will remain necessary in this latter region to rely entirely on predicted β-decay properties of the relevant nuclides. Moreover, since there are no nearby neutron-rich N = 126 isotones known which can be used to optimize the model parameters and to check the validity of the predictions, the uncertainties of any model-$T_{1/2}$ are particularly large. Recent measurements of $T_{1/2}$ values of neutron-rich $_{59}$Pr - $_{63}$Eu isotopes [25] and $_{69}$Tm - $_{71}$Lu nuclei [24] have revealed the same overall discrepancies with model predictions as already observed for the A ≃ 100 region. The Gross Theory [11] systematically overestimates and the TDA model [10] generally underestimates the experimental $T_{1/2}$ by roughly a factor of two. Based on the experience with medium-heavy nuclei, this pattern will presumably extend up to the inaccessible N = 126 'waiting-point' nuclei and will thus influence the A ≃ 195 r-abundance peak.

The R-Process Time Scale

A quantity which plays an important role for the evaluation of different astrophysical models is the length of time required for the r-process. A lower limit for this number may be obtained by simply adding up the time (τ_β) spent at the N = 50, 82 and 126 'waiting points' (for review, see e.g. Ref. [8]).
With the measured $T_{1/2}$ of two of the most critical 'waiting-point' nuclei, ^{80}Zn [6] and ^{130}Cd [7], together with the present RPA shell model calculations and a renormalization of the TDA-$T_{1/2}$ for the N ≃ 126 region, one now can derive a new

estimate for τ_β. By adding up the $T_{1/2}$ of the 'waiting-point' nuclei for the three neutron magic shells

a) N = 50	$T_{1/2}$ (^{76}Fe - ^{80}Zn)	≃	790 ms
b) N = 82	$T_{1/2}$ (^{128}Pd - ^{130}Cd)	≃	440 ms
c) N = 126	$T_{1/2}$ (^{183}Ho - ^{196}Yb)	≃	440 ms

one obtains an r-process waiting time limit of about 1.7 s. This is actually an underestimate of τ_β because the also longer-than-average $T_{1/2}$ of the $_{31}$Ga-, $_{49}$In- and $_{50}$Sn-isotopes in the r-process path have not been considered. Their (partly known) $T_{1/2}$ add up to further 400 ms. Together with the remaining about 50 nuclei in the r-process path between Ge and U, for which an average $T_{1/2} \simeq 10$ ms can be estimated, one obtains a total ß-decay cycle time for the r-process of about 2.6 s. The actual r-process would, however, last even longer because of the time required for neutron capture and for fission cycling if it occurs.

Conclusions

Recent experimental $T_{1/2}$ for very neutron-rich nuclei together with improved $T_{1/2}$ predictions have revealed new constraints on the models of stellar environments appropriate to the r-process. It seems clear, that a short-duration scenario, such as explosive He-burning which was favoured in recent years [10] as the site of the r-process, cannot be the main environment for the nucleosynthesis of heavy elements. Instead, on the basis of the present results a long-duration, steady-flow solution for the r-process seems to be more likely, although an astrophysical site for such a scenario is not known.

Acknowledgement

The nuclear-structure studies of this work were financially supported by the German BMFT, and their astrophysical applications by the DFG.

References

[1] G.R. Burbidge et al., Rev. Mod. Phys. 29, 547 (1957).
[2] W. Hillebrandt, Sp. Sci. Rev. 21, 639 (1978).
[3] A.G.W. Cameron et al., in: Nucleosynthesis, Challenges and New Developments, Univ. of Chicago Press, p.190 (1985).
[4] G.J. Mathews and R.A. Ward, Rep. Prog. Phys. 48, 1371 (1985).
[5] F. Käppeler et al., Ap. J. 257, 821 (1982).

[6] E. Lund et al., Proc. AMCO-7, THD Schriftenreihe Wissenschaft und Technik 26, 102 (1984), and Physica Scripta 34, 614 (1986).
[7] K.-L. Kratz et al., Z. Physik A325, 489 (1986).
[8] D.N. Schramm, in: Essays in Nuclear Astrophysics, Cambridge Univ. Press, p.325 (1982).
[9] W. Hillebrandt et al., Astron. Astrophys. 99, 195 (1981).
[10] H.V. Klapdor et al., Z. Physik A299, 213 (1981), and At. Data Nucl. Data Tables 31, 81 (1984).
[11] K. Takahashi and M. Yamada, Prog. Theor. Phys. 41, 1470 (1970); and At. Data Nucl. Data Tables 12, 101 (1973).
[12] F.-K. Thielemann, Proc. Int. Workshop on Stellar Nucleosynthesis, Erice, Sicily (1983); and private communication.
[13] K.-L. Kratz, Nucl. Phys. A417, 447 (1984).
[14] U. Bosch et al., Phys. Lett. 164B, 22 (1985).
[15] J. Krumlinde and P. Möller, Nucl. Phys. A417, 419 (1984).
[16] E. Lund et al. Z. Physik A234, 233 (1980).
[17] K.-L. Kratz et al., ACS Symp. Ser. 324, 165 (1986).
[18] D.G. Madland and J.R. Nix, to be published.
[19] T. Bengtsson and I. Ragnarsson, Physica Scripta T5, 165 (1983).
[20] P. Möller and J.R. Nix, At. Data Nucl. Data Tables 26, 165 (1981).
[21] A.H. Wapstra and G. Audi, Nucl. Phys. A432, 1 (1985).
[22] E.R. Hilf et al., CERN Report 76-13 (1976).
[23] D.H. Wilkinson and B.E.F. Macefield, Nucl. Phys. A232, 58 (1974).
[24] E. Runte et al., Nucl. Phys. A399, 163 (1983); and A441, 237 (1985).
[25] R.C. Greenwood et al. Phys. Rev. C35, 1965 (1987).
[26] K.-L. Kratz et al., ACS Symp. Ser. 324, 159 (1986).
[27] K.-L. Kratz et al., to be published.

Experimental Studies of Thermal Effects During s-Process Nucleosynthesis

F. Käppeler
Kernforschungszentrum Karlsruhe, Institut für Kernphysik
P.O. Box 3640, D-7500 Karlsruhe
Federal Republic of Germany

Abstract

Two recent experimental studies are reported which contribute to our understanding of the temperature during the **s**-process as well as to possible thermal effects on beta decay life-times.

(i) The determination of the temperature characteristic for the weak **s**-process component depends critically on the beta decay half-life of the isomeric state in ^{79}Se. A first measurement of this half-life is presented and the implications for the **s**-process are discussed.

(ii) In a second experiment we investigate the mechanism by which the ground state and the isomeric state in ^{176}Lu might be thermally equilibrated. This problem is the key to a possible interpretation of ^{176}Lu as a cosmic clock. As direct transitions between the long-lived ground state ($t_{1/2}$= 3.6 10^{10}y) and the 3.7h isomer are strongly forbidden, equilibration may occur through thermally populated states at higher excitation. Branchings in the decay of these excited states to the ground state and to the isomer as well are searched for by high resolution gamma-ray spectroscopy at the ILL high flux reactor in Grenoble.

What Does the Classical Model Tell Us About the s-Process Temperature ?

With the availability of rather precise information on abundances and stellar neutron capture rates the classical model turned out to be an excellent approximation for the observed features of **s**-process nucleosyn-

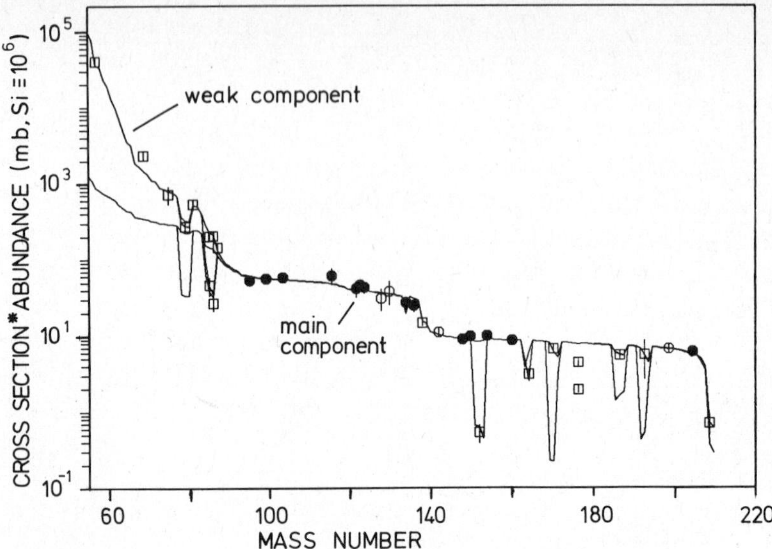

Fig. 1 The characteristic **s**-process quantity σN(A). The empirical products (symbols) can be described by the classical model (solid lines); **s**-process branchings are considered.

thesis. This is the more surprising as this approach is purely phenomenological and independent of any stellar model. With the assumption of two exponential distributions of neutron exposure, $\rho(\tau)$, it allows for reproducing the characteristic quantity of the **s**-process, the product abundance times cross section, σN_s, over the entire mass range from the Fe seed to Bi by means of only four parameters (figure 1). These stand for the seed abundances, f, and mean neutron exposures, τ_o, of the two components indicated in figure 1. While the main component accounts for the abundances in the mass range A>100, the weak component needs to be added for describing the steep increase towards the Fe group isotopes. As follows from figure 1, the product σN_s varies smoothly with mass number, the model calculations (solid lines) showing perfect agreement with the empirical σN_s values (symbols) of the **s**-only isotopes (Käppeler et al. 1982, Beer et al. 1984). The only discontinuities occur at the **s**-process branchings where radioactive nuclei on the **s**-process path cause the neutron capture chain to split due to competition between neutron capture and beta decay.

Analyses of these branchings yield estimates for the neutron density and /or the temperature during the **s**-process. The branching ratio

$$B_n = \lambda_n / (\lambda_n + \lambda_\beta) \tag{1}$$

can be expressed by the neutron capture rate $\lambda_n = n_n \sigma v_T$ and the beta decay rate $\lambda_\beta = \ln 2 / t_{1/2}(T)$ of the branching point isotope, and is determined by the deviation of the involved **s**-only isotope from the smooth $\sigma N_s(A)$ curve. The **s**-process temperature can be inferred from some branchings through the temperature dependence of the beta decay rate (Takahashi and Yokoi 1987, Yokoi and Takahashi 1985). The present situation for the main component is summarized in figure 2. The solid line marks the allowed region for neutron density and thermal energy as derived from the combined analysis of the branchings at ^{151}Sm, ^{170}Yb, and ^{185}W (Beer et al. 1984). This region could be reduced by inclusion of the branchings at A=147,148 (Winters et al. 1986). However, the influence of temperature is still ambiguous in this case, depending on the question whether the ground state and the 137 keV isomer in ^{148}Pm - which are about equally populated by neutron capture in ^{147}Pm - are thermally equilibrated on the beta decay time scale of a few days. This question cannot be solved at present because the level scheme of ^{148}Pm is almost unknown: beside the ground state ($t_{1/2}$= 5.4 d) and the isomer ($t_{1/2}$=41 d) only one additional level is reported at 76 keV. As direct transitions between the isomer and the other two states are too slow, equilibrium could only be achieved via mediating states at higher

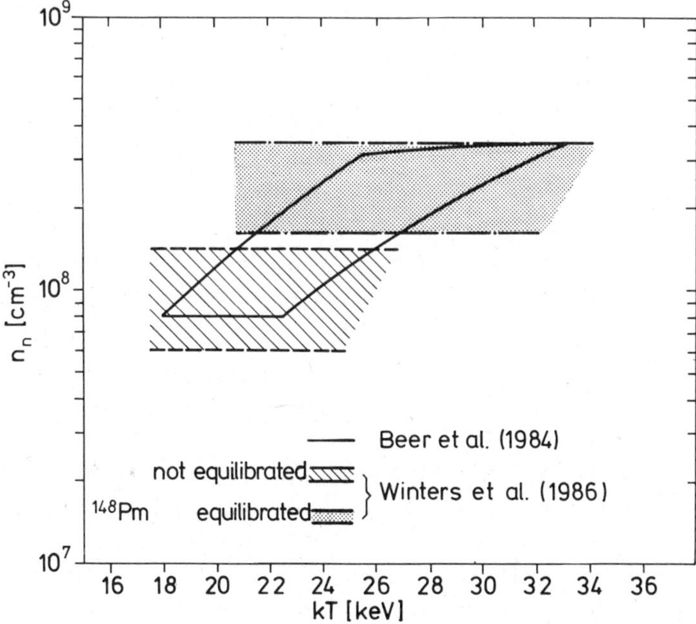

Fig. 2 Allowed regions of neutron density and temperature from analyses of various branchings. Note the ambiguity due to the open problem of thermal equilibration in 148Pm.

excitation. Evidence for such states was found by Norman et al. (1985) in $^{149}Sm(d,{}^{3}He)$ reactions but spin and parity assignments are still missing. In this situation, the two extremes (no equilibration and complete equilibrium) have to be treated separately as indicated in figure 2. Accordingly, one finds two different regions for the **s**-process temperature. If one of the two regions could be excluded by improved branching analyses, this would have consequences for the neutron source of the **s**-process: a low temperature would favor the $^{13}C(\alpha,n)$ reaction, whereas the $^{22}Ne(\alpha,n)$ reaction requires higher temperatures to operate.

Further efforts are therefore necessary to remove this ambiguity, in particular

- the investigation of the level scheme of ^{148}Pm;
- an improved analysis of the various branchings;
- the analysis of additional branchings to obtain more constraints on the temperature (e.g. at ^{134}Cs, ^{154}Eu).

^{79}Se: Stellar Decay Rate and the ^{79}Se Branching

The only branchings to characterize the weak **s**-process component are those at ^{79}Se and ^{85}Kr. Figure 3 shows the **s**-process path in the mass region 75 < A < 90 with the terrestrial half-lives indicated at the branch points. The fact that both branching ratios are about equal (Walter et al. 1986a, 1986b) implies that the beta decay rate of ^{79}Se must be greatly enhanced at **s**-process temperatures.

The responsible mechanism is sketched in figure 4. The isomer at 96 keV may beta decay by allowed transitions to ^{79}Br, whereas the ground state decay is forbidden. Due to the 3.9 m electromagnetic decay of the isomer to the ground state, complete thermal equilibrium among all excited states is achieved leading to a population probability of ~1 % for the isomer. With this value and an estimate for the beta decay half-life of the isomer it is possible to derive the temperature dependence of the total decay rate of ^{79}Se. Such estimates were performed by several authors (Conrad 1976, Cosner and Truran 1981, Yokoi and Takahashi 1985) using different estimates of the log ft value for the beta decay of the isomer. It turned out that the contributions of all other levels were very small, but that bound state decays had to be considered. Figure 5 shows the various results, which differ by a factor 3 at the standard **s**-process temperature

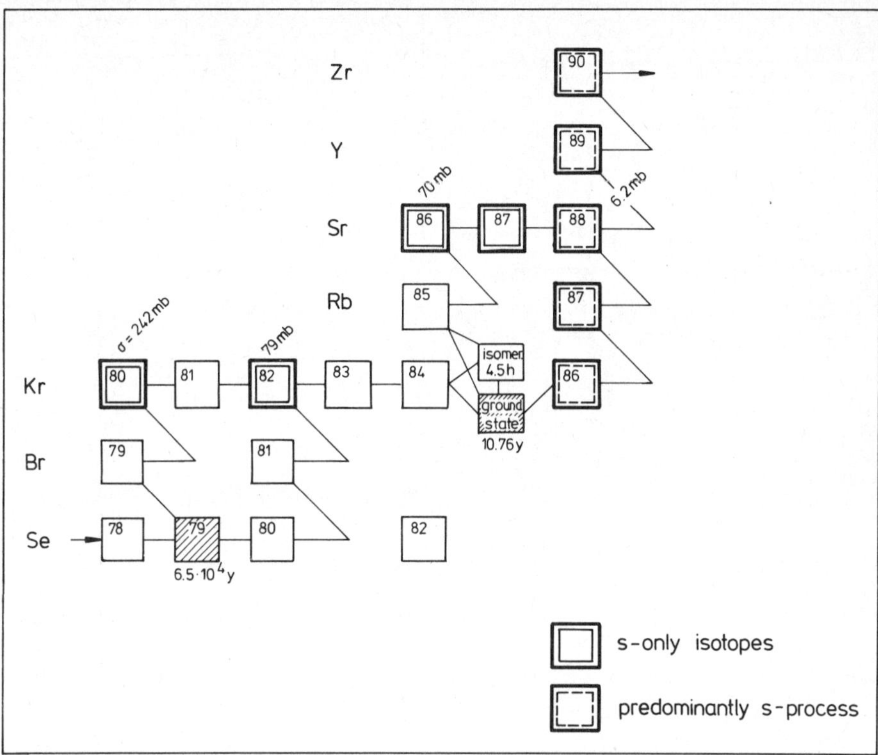

Fig. 3 The **s**-process capture chain in the mass region 78 < A < 90.

of 3.5 10^8 K. At the same time, this factor 3 reflects the individual uncertainties, owing to an uncertainty of 10% in the estimated log ft values for the beta decay of the isomer.

Therefore, an experimental determination of this quantity was called for. In principle, ^{79}Se nuclei in the isomeric state can be produced by neutron capture on ^{78}Se and the electrons from the subsequent beta decay could be detected. Such an experiment had to face two problems:

- The competing ground state decay is much faster leading to a branching ratio of 10000 in favor of the ground state decay. As this transition is converted to 90% , the conversion electrons had to be suppressed efficiently.
- To achieve reasonable statistics, activation in a thermal reactor was required for production of ^{79}Se^m. In order not to obscure the the expected effect, any traces of Al, Cu, Ti, V, and Cr had to be eliminated from the sample as activation of these elements leads to similar electron spectra and half-lives.

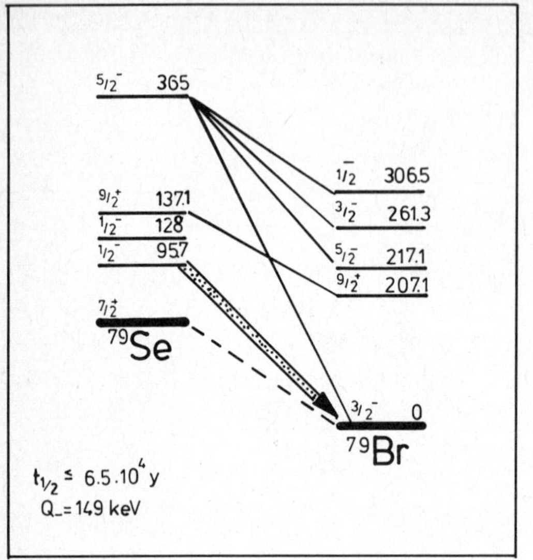

Fig. 4 Level scheme of 79Se and possible transitions at high temperature. The effective decay rate is determined by the isomeric state at 96 keV.

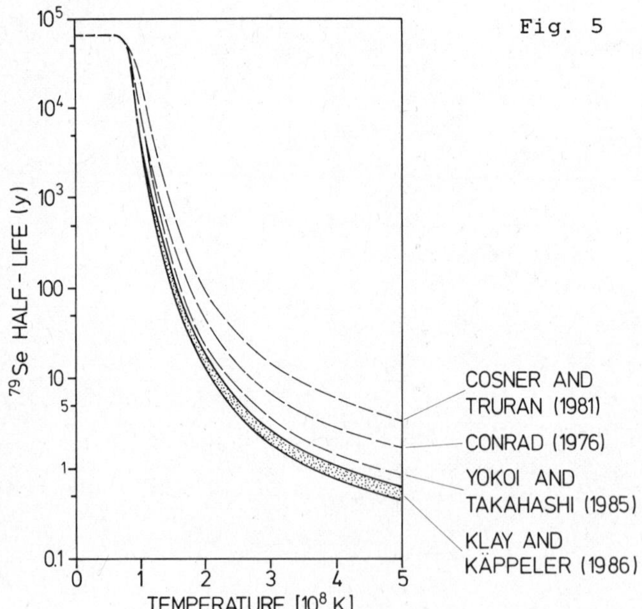

Fig. 5 The half life of 79Se as a function of temperature. Various theoretical estimates (dashed lines) are compared with the result of a recent experiment.

The experiment was carried out at the high flux reactor of the Institut Laue-Langevin in Grenoble (Klay and Käppeler 1986). The above points were solved by using a Mini-Orange beta spectrometer with a low energy cut-off near 100 keV for suppression of conversion electrons and by using an ultra pure graphite disk implanted with highly enriched ^{78}Se. After activating the sample for 4 min, it took another 4 min before the measurement could be started, including the transfer to a hot cell, unloading from the irradiation capsule, mounting in the spectrometer and evacuation. Then, the electron spectrum was followed for 16 min in 1 min intervals before the Mini-Orange was removed for subsequent detection of the conversion electrons. By repeated activations a branching ratio R = 1970 ± 350 was finally obtained, significantly smaller than previously estimated. The resulting log ft value of 4.8 ± 0.08 was then used to recalculate the temperature dependence of the ^{79}Se half-life with the formalism of Takahashi and Yokoi(1983) as it is shown by the shaded band in figure 5.

Now, the improved decay rate does no longer determine the uncertainty of the temperature estimate from the ^{79}Se branching. Figure 6 illustrates how the possible region for neutron density and temperature was reduced with the experimental result compared to a previous estimate, the remaining un-

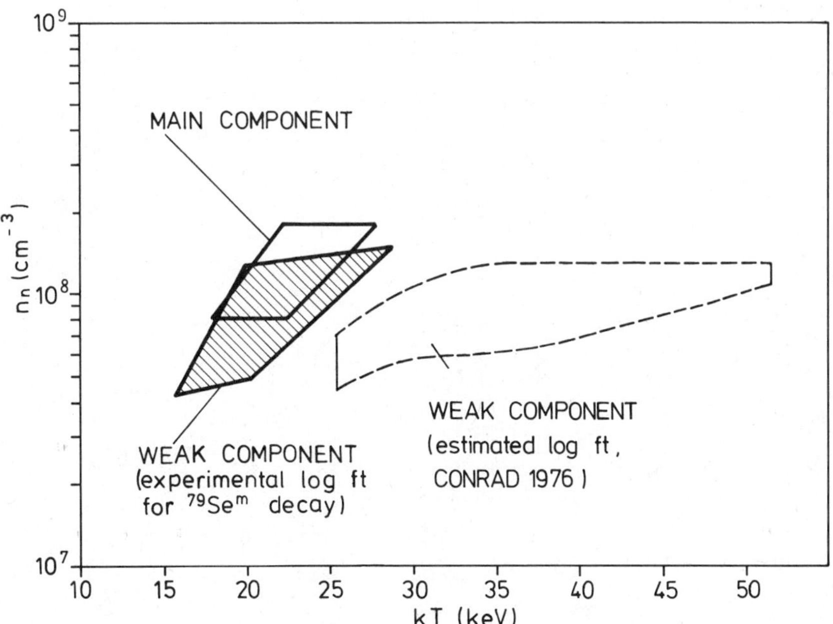

Fig. 6 Allowed range of neutron density and temperature for the two **s**-process components. (The ambiguity of figure 2 was not considered.)

certainty being mainly due to cross section uncertainties. Note, that the region for the main component does not consider the ambiguity outlined above. Clearly, this ambiguity must be resolved before the results for the weak component can be interpreted.

^{176}Lu: Cosmic Clock with Thermal Defects?

The suited half-life of 3.6 10^{10}y and the fact that it is shielded against the **r**-process (figure 7) made ^{176}Lu a good candidate for studying the age of the **s**-process elements (Audouze et al. 1972, Arnould 1973, McCulloch et al. 1976). The only difficulty seemed to be the proper treatment of the branching at ^{176}Lu as it is indicated in figure 7. This branching is due to the isomeric state at 127 keV which is significantly populated by neutron capture in ^{175}Lu and which decays with a half-life of 3.68 h to ^{176}Hf. Because of the large difference in the spin and - even more important - in the K quantum number, direct transitions are highly forbidden by selection rules, and ground state and isomer could be considered as independent of each other. Therefore, the **s**-process branching could be followed according to the relative population of ground state and isomer in the (n,γ) reaction, neglecting any transitions between them.

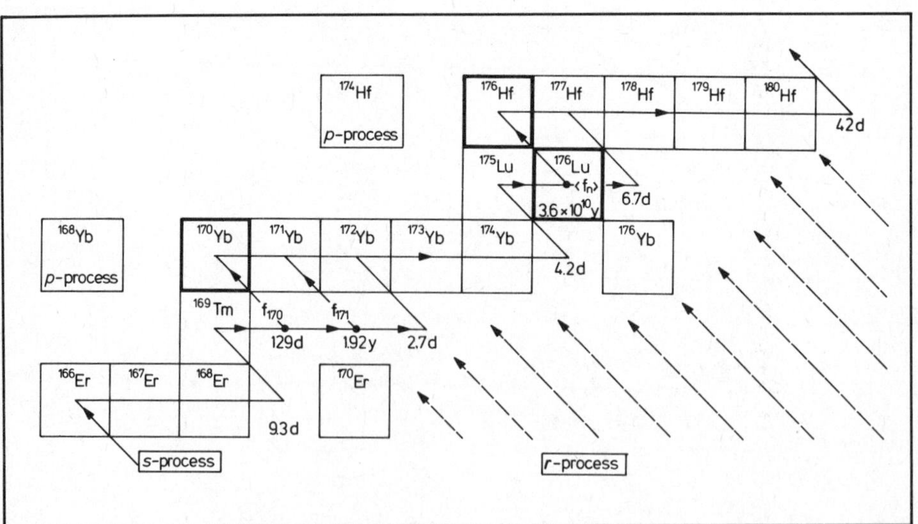

Fig. 7 The **s**-process capture chain in the mass region 170 < A < 180.

Soon after the first quantitative study of ^{176}Lu as a cosmic clock (Beer and Käppeler 1980), it was realized that the decay of ^{176}Lu might nevertheless be accelerated at high temperatures (Ward 1980, Beer et al. 1981). Though direct transitions are forbidden, ground state and isomer could well be connected via transitions through higher lying excited states. Any quantitative treatment requires a detailed knowledge of excited states; but in 1980 Ward was left with two completely different families of excited states: those with high K quantum numbers related to the ground state and the others with small K connected with the isomer. The only possible links between these groups were K-forbidden transitions, e.g. starting from the high spin members of the rotational bands with low K, for which the hindrance factor for interband transitions

$$\delta = 10^{-2||\Delta K|-\lambda|} \tag{2}$$

was reduced because of the higher multipolarity λ. By combination of the estimated intra- and interband transition rates with the population probabilities of the excitedd states it was found that the levels at 239, 376, and 537 keV with $J^{\pi}= 3^{-}, 4^{+}$ provided for the most efficient links between ground state and isomer. For the 376 keV state the respective transitions are indicated in figure 8. As a result, thermal equilibration in ^{176}Lu was suspected to appear already above ~1.5 10^{8} K, right in the temperature range of the **s**-process. However, this result was considered to be rather uncertain because of the qualitative estimates for transition rates and hindrance factors.

First experimental evidence for an electromagnetic link between ground state and isomer was obtained by photoactivation of Lu samples in intense gamma-ray fields. By subsequent spectroscopy the decay of the isomer was clearly identified, showing that it had been formed in this way (Veres et al. 1970, Norman and Kellogg 1985). These measurements found a positive effect only for gamma-ray energies higher than 1.1 MeV, a clear hint that excitation energies well above 500 keV might be important. This was supported when Coulomb excitation studies by Elze et al.(1984) revealed a rotational band with $K^{\pi}= 5^{-}$ at 872 keV (figure 8). For this state there is no K-hindrance according to equation (2), neither for the ground state transition nor for transitions to the 4^{-}, 3^{-} bands which ultimately lead to the isomer. Consequently, the important branching ratio for transitions towards ground state and isomer is rather large and hence accessible to experiment, contrary to the strongly hindered cases considered earlier. Figure 8 presents the updated level scheme of ^{176}Lu, which is no longer characterized by a separation in two families of very different K.

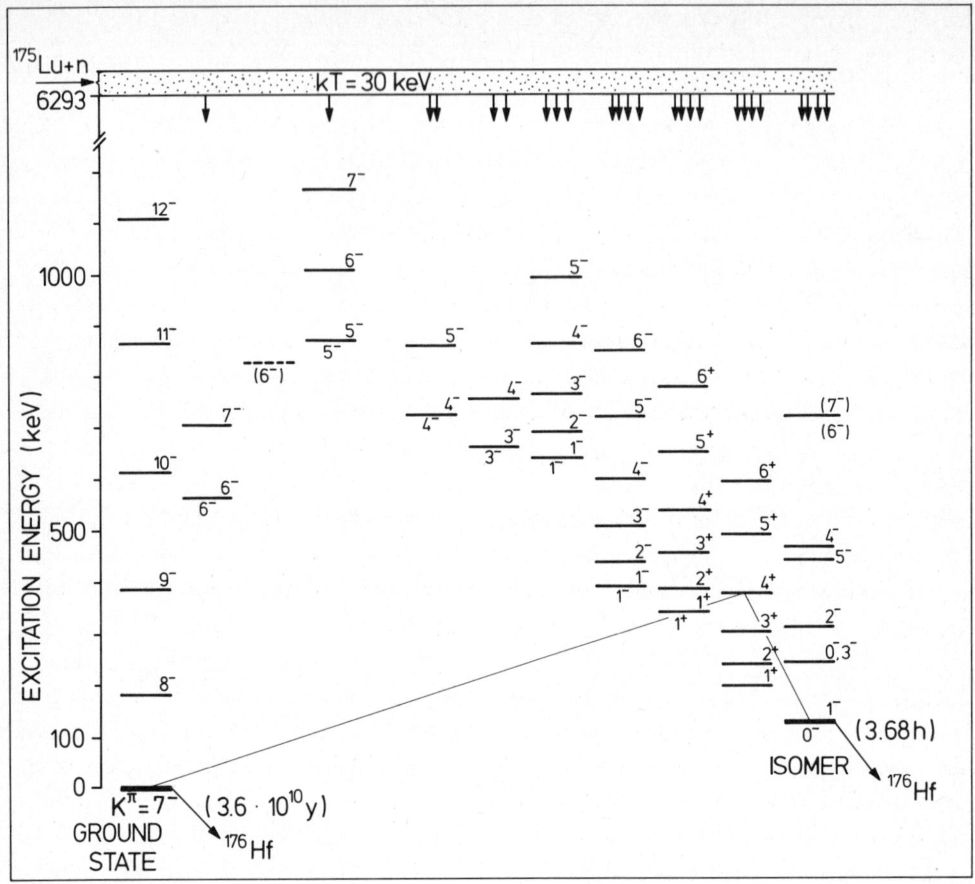

Fig. 8 The level scheme of ^{176}Lu with the various rotational bands plotted separately. Neutron capture populates excited states in the shaded area. The solid lines indicate the most efficient transitions in the equilibration mechanism considered by Ward (1980). Meanwhile higher lying bands with intermediate K quantum numbers were found (K,π = 5,- ; Elze 1984) which can provide for an equilibration without K-forbidden transitions.

An improved theoretical approach was reported by Gardner et al. (1986) who complemented the experimentally determined levels in ^{176}Lu by postulated states inferred from systematic trends in neighboring nuclei. With this 'complete' level scheme, which extended up to 1.5 MeV, they also found a strong temperature dependence of the ^{176}Lu half-life, but now using only unhindered transitions. The existence of a thermal enhancement of the ^{176}Lu decay rate was also reported by Beer et al. (1984) in an extensive s-process analysis in the mass range $160 < A < 180$. These authors showed that even the assumption of complete equilibrium was compatible with the observed abundances.

Given the situation of figure 8, a quantitative understanding of the fate of ^{176}Lu at high temperatures definitely requires an experimental assessment of level energies and information on branching ratios for interband transitions, e.g. from the 5^- band head. Therefore, we started an effort to extend the level scheme of ^{176}Lu up to ~1 MeV with improved precision and better sensitivity, using the high resolution spectrometers GAMS and BILL at the high flux reactor of the Institut Laue-Langevin in Grenoble (Klay et al. 1987). Meanwhile, the measurements have been completed and the spectra are under analysis. So far, we have identified about 500 gamma transitions, 200 of which have been placed in a level scheme comprising 58 levels. These include the important band heads with $K^\pi = 3^-$, 4^-, 5^-, and 6^-. The analysis is now being continued with the determination of transition multipolarities, which are derived from the conversion electron spectra. An additional ^{175}Lu(d,p) measurement is also envisaged in order to fix the energy of the 127 keV isomer.

Finally, it should be mentioned that positron annihilation might act as an additional excitation mechanism under stellar conditions (Watanabe et al. 1981, Norman et al. 1985). Though the available data do not allow for a quantitative treatment yet, it seems as if this process (when a positron annihilates with a K-shell electron) can strongly contribute to excitations around 1 MeV, especially at temperatures greater than 4 10^8 K.

Summary

The determination of the **s**-process temperature is not yet settled. A final answer requires the solution of a variety of interesting phenomena, of which only few have been investigated quantitatively. Therefore, the temperature effects relevant to **s**-process nucleosynthesis represent a wide field for further experimental and theoretical work.

References

Arnould, M. 1973, Astron. Astrophys. **22**, 311

Audouze, J., Fowler, W.A., Schramm, D.N. 1972, Nature Phys. Sci. **238**, 8

Beer, H., Käppeler, F. 1980, Phys. Rev. **C21**, 534

Beer, H., Käppeler, F., Wisshak, K., Ward, R.A. 1981, Ap. J. Suppl. **46**, 295

Beer, H., Walter, G., Macklin, R.L., Patchett, P.J. 1984, Phys. Rev. **C30**, 464

Conrad, J.H. 1976, PhD thesis, University of Heidelberg

Cosner, K., Truran, J.W. 1981, Ap. Space Sci. **78**, 85

Gardner, D.G., Gardner, M.A., Hoff, R.W. 1986, LLNL FY86 Nucl. Chem. Div. Annual Report, Lawrence Livermore National Laboratory

Elze, Th.W. 1984, priv. communication

Käppeler, F., Beer, H., Wisshak, K., Clayton, D.D., Macklin, R.L., Ward, R.A. 1982, Ap. J. **257**, 821

Klay, N., Käppeler, F. 1986, Int. Conf. on Weak and Electromagnetic Interactions in Nuclei, Heidelberg, July 4-7

Klay, N., Käppeler, F., Beer, H., Schatz, G., Börner, H., Hoyler, F., Schreckenbach, K., Krusche, B. 1987, in preparation

McCulloch, M.T., de Laeter, J.R., Rosman, K.J.R. 1976, Earth Plan. Sci. Lett. **28**, 308

Norman, E.B., Kellogg, S. 1985, in Capture Gamma-Ray Spectroscopy and Related Topics-1984, ed. S. Raman (American Inst. Phys., New York) p.753

Norman, E.B., Lesko, K.T., Crane, S.G., Larimer, R.M., Champagne, A.E. 1985, report LBL-20298, Lawrence Berkeley Laboratory, University of California

Takahashi, K., Yokoi, K. 1983, Nucl. Phys. **A404**, 578

Takahashi, K., Yokoi, K. 1987, At. Data and Nucl. Data Tables (in print)

Veres, A., Pavlicsek, I. 1970, Acta Phys. Acad. Sci. Hung. **28**, 419

Ward, R.A. 1980, priv. communication

Walter, G., Beer, H., Käppeler, F., Penzhorn, R.-D. 1986a, Astron. Astrophys. **155**, 247

Walter, G., Beer, H., Käppeler, F., Reffo, G., Fabbri, F. 1986b, Astron. Astrophys. **167**, 186

Watanabe, Y., Mukoyama, T., Katano, R. 1981, Phys. Rev. **C23**, 695

Winters, R.R., Käppeler, F., Wisshak, K., Mengoni, A., Reffo, G. 1986, Ap. J. **300**, 41

Yokoi, K., Takahashi, K. 1985, report KfK-3849, Kernforschungszentrum Karlsruhe

THERMONUCLEAR REACTIONS AT HIGH TEMPERATURES AND DENSITIES

James W. Truran
Max-Planck-Institut für Astrophysik, 8046 Garching b. München, FRG
and
Dept. of Astronomy, University of Illinois, Urbana, Ill., USA

Friedrich-Karl Thielemann
Department of Astronomy, Harvard University
Cambridge, Mass. 02138, USA

Marcel Arnould
Institut d'Astronomie et d'Astrophysique
Université Libre de Bruxelles, Belgium

Abstract

Statistical model calculations of thermonuclear reaction rates are described. Significant improvements over earlier work have been achieved with the use of: (i) realistic optical model potentials for interactions involving neutrons, protons and alpha particles; (ii) a revised procedure for the calculation of the total photon transmission function for electric dipole radiation; and (iii) a revised procedure for the determination of nuclear level densities as a function of energy. Our theoretical cross section predictions are found to be consistent with experimental values to within a factor of 2. General expressions are also provided for corrections due to the effects of electron screening on thermonuclear reaction rates, including both photodisintegration reactions and reactions involving charged particles in both the incoming and outgoing channels. These results provide a unified and consistent procedure for the determination of thermonuclear reaction rates as a function of both temperature and density.

I. Introduction

Investigations of stellar evolution (9), nova (29) and supernova explosions (34), thermonuclear outbursts on neutron stars (20), and nucleosynthesis (30) require as input a broad spectrum of nuclear properties. These include, specifically, the rates of thermonuclear reactions for a large number of nuclear species for temperature and density conditions compatible with these astrophysical environments. Calculations of nucleosynthesis associated with the late stages of evolution of massive stars and supernova explosions, for example, demand the use of complex networks of nuclear transformations involving tens to hundreds of nuclear species and hundreds of specific nuclear transformations. Similarly, studies of the neutron-capture processes responsible for the production of most of the nuclear species present in nature which are more massive than iron involve the use of vast networks of nuclear reactions.

The large number of reactions involved here suggests that it is clearly impractical to seek to obtain direct experimental cross section information in every case. It is rather necessary to develop some model of nuclear reactions to the point at which reliable theoretical predictions become possible. Substantial success has been achieved with the adoption of a Hauser-Feshbach (7) or equivalent expression for the energy-averaged cross section. Statistical model calculations for astrophysical purposes have previously been performed by Truran (27), Truran et al. (31), Michaud et al. (17), Michaud and Fowler (16), Truran (28), Arnould (1), Thielemann (21), Holmes et al. (8), and Woosley et al. (33). Our own recent calculations (23,24) are reviewed and discussed in the next section.

The treatment of thermonuclear reactions in high density astrophysical environments requires modification due to the effects of electron screening (10,11,18,19). Such conditions are typically encountered, for example, in studies of supernova core collapse leading to Type II supernovae, of central carbon ignition at high densities in accreting carbon oxygen white dwarfs, and in calculations of thermonuclear runaways on neutron stars or white dwarfs. Screening enhancements by substantial factors are possible for some reactions in very dense plasmas. In general, these plasma effects can be described by a screening potential which must be added to the nuclear and pure Coulomb potential. A general procedure (25,26) for the treatment of screening for particle-capture and photodisintegration reactions and for reactions involving charged particles in both the incoming and outgoing channels is outlined in section III.

II. Thermonuclear Reaction Rate Calculations

We consider reactions involving nuclei in the mass range A > 20 for which it can generally be assumed that there exists a sufficiently high density of excited states in the compound nucleus at the optimum bombarding energy to allow cross section calculations based upon statistical model (Hauser-Feshbach) assumptions. This model assumes that, for each angular momentum of the projectile, there exists a state in the compound nucleus at the bombarding energy with appropriate spin and parity. Assuming (for convenience in notation only) the target or final nucleus to exist or be produced only in its ground state, the cross section for a particle-induced reaction $i(j,k)\ell$ can be written

$$\sigma(E) = \frac{\pi\hbar^2/2\mu_{ij}E}{(2J_i+1)(2J_j+1)} \sum_{J,\pi} (2J+1) \frac{T_j(E,J,\pi)T_k(Q_k+E,J,\pi)}{\sum_n T_n(Q_n+E,J,\pi)} \tag{1}$$

where the $T_i(E,J,\pi)$ are the appropriate transmission coefficients for the particle and photon channels. The astrophysical thermonuclear reaction rate is given by the integral over a Maxwellian distribution

$$<\sigma v> = \left(\frac{8}{\mu_{ij}\pi}\right)^{\frac{1}{2}} (kT)^{-3/2} \int_0^\infty E\sigma(E) \exp(-E/kT)\, dE. \tag{2}$$

The general expression for the thermonuclear rate can then be written in the form

$$<\sigma v> = c\,(kT)^{-3/2} \int_0^\infty \sum_{J,\pi} (2J+1) \frac{T_j(E,J,\pi)\, T_k(Q_k+E,J,\pi)}{\sum_n T_n(Q_n+E,J,\pi)} \exp(-E/kT)dE \tag{3}$$

The challenge here is then to provide accurate estimates of the required neutron, proton, alpha particle, and photon transmission coefficients from statistical model assumptions. A knowledge of the nuclear level density as a function of energy represents critical input in this regard. Additionally, the transmission coefficients must of course be appropriately modified in the presence of electron screening.

Previous calculations of thermonuclear reaction rates for astrophysical studies used optical square well potentials (black nucleus model) for the particle channels, for all but perhaps s-wave and p-wave neutrons (28). We have achieved a significant improvement over earlier work by solving the Schrödinger equation with a more realistic optical potential for the appropriate particle-nucleus combination. For protons and neutrons, we employ the optical potential derived by Jeukenne et al. (12), with the correction for the imaginary part as discussed in Fantoni et al. (5) and Mahaux (13). For alpha particles, we used the phenomenological Woods-Saxon potential derived by Mann (14), based on the extensive data of McFadden and Satchler (15).

For the gamma ray transmission coefficients, both electric and magnetic dipole transitions (E1 and M1) are included in the calculation of the total photon width. M1 transitions were treated as discussed in Holmes et al. (8). The more important E1 contributions were calculated on the basis of the Lorentzian representation of the Giant Dipole Resonance, utilizing the phenomenological model for the resonance width proposed by Thielemann and Arnould (22).

Calculations of these transmission coefficients require as well a knowledge of the nuclear level densities. Our treatment of nuclear level densities closely parallels that of Gilbert and Cameron (6) and combines the use of the Bethe (3) formula, based on the Fermi gas model, at high energies

$$\rho(U) = \frac{1}{12\sqrt{2}\ \sigma\ a^{1/4}} \ \frac{\exp\ (2\sqrt{aU})}{U^{5/4}} \tag{4a}$$

and the empirically based Ericson (4) formula at low energies

$$\rho(E) = T^{-1} \exp\left[(E-E_0)/T\right] \tag{4b}$$

The parameter σ is the spin cutoff factor. The energy shift of the first excited state due to the necessity of breaking up a proton or neutron pair before exciting nucleons to higher states, not included in the independent particle Fermi gas model, is accounted for in the "back-shifted" Fermi gas formalism by adopting $U = E-\delta$ in Eq. (4), where δ is the pairing correction. We also identify $E_0 \equiv \delta$ in the Ericson formula. The level density at high densities is thus determined by the single parameter a, while fitting of the observed levels at low energies gives T. Further details of these procedures are presented elsewhere (23,24).

A measure of the success of our procedures in reproducing experimental trends is provided by comparisons of our radiation width and cross section predictions with the data. The ratios of our theoretical estimates of $<\Gamma_\gamma>$ to the measured average radiation widths determined from studies of thermal s-wave neutron capture (32) are shown in Fig. 1. The overall agreement between theory and experiment can be seen to be quite satisfactory. Similarly, a comparison of our calculated neutron capture cross sections with the 30 keV neutron capture cross sections compiled by Bao and Käppeler (2), shown in Fig. 2, indicates general agreement to a factor ~ 2. The absence of clear systematic trends gives us greater confidence in the extrapolation of these procedures to calculations of (n,γ) rates for very neutron-rich unstable nuclei. Finally, comparisons of our calculated energy-dependent cross sections for charged particle reactions with experiment, for representative intermediate mass nuclei which participate in stellar silicon burning, are shown in Fig. 3. We generally find that the energy-dependences and absolute values of charged particle cross sections are quite well reproduced, including specifically the magnitudes of cusps at the threshold energies of channel openings.

III. Electron Screening Corrections

In order to be able to provide compilations of thermonuclear reaction rates as a function of both temperature and density, we have sought to provide a simple but quite general expression for transmission coefficients in an environment where screening effects are important (25,26). We first note that, given the nuclear and pure Coulomb potential and the orbital quantum number ℓ, the transmission coefficients are a function only of the kinetic energy at infinity T(E). The Schrödinger equation in the presence of screening is

$$\frac{d^2u_\ell(r)}{dr^2} + [k^2 - \frac{\ell(\ell+1)}{r^2} - \frac{2\mu}{\hbar^2}(V_{nucl}(r) + V_{coul}(r) + V_{scr}(r))]\, u_\ell(r) = 0 \quad (5)$$

where we generally take

$$V_{scr}(r) = -U_0 + f(r) \quad (6)$$

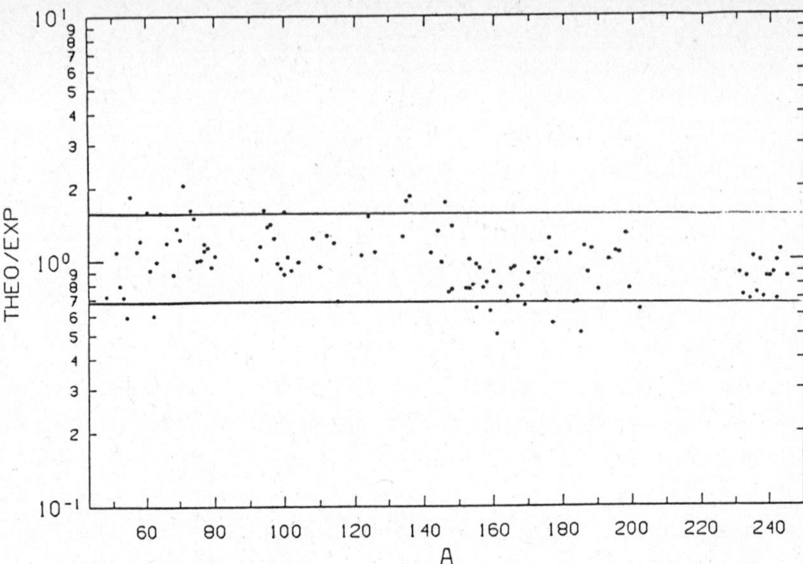

Fig. 1: Ratios of theoretical to experimental $<\Gamma_\gamma>$ versus A. The experimental data are from Weigmann and Rohr (32). The solid lines indicate that agreement for most cases is achieved to within a factor 1.5.

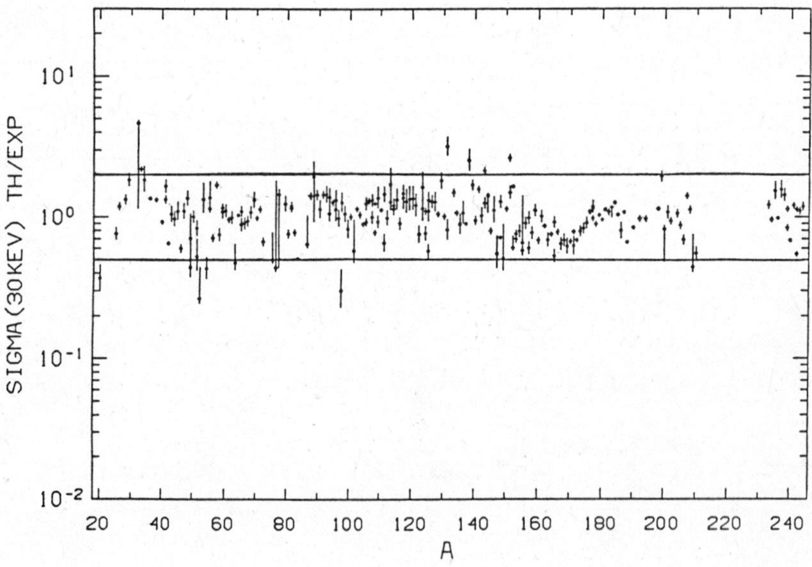

Fig. 2: Ratios of theoretical to experimental 30 keV neutron capture cross sections versus A. The experimental data are taken from Bao and Käppeler (2).

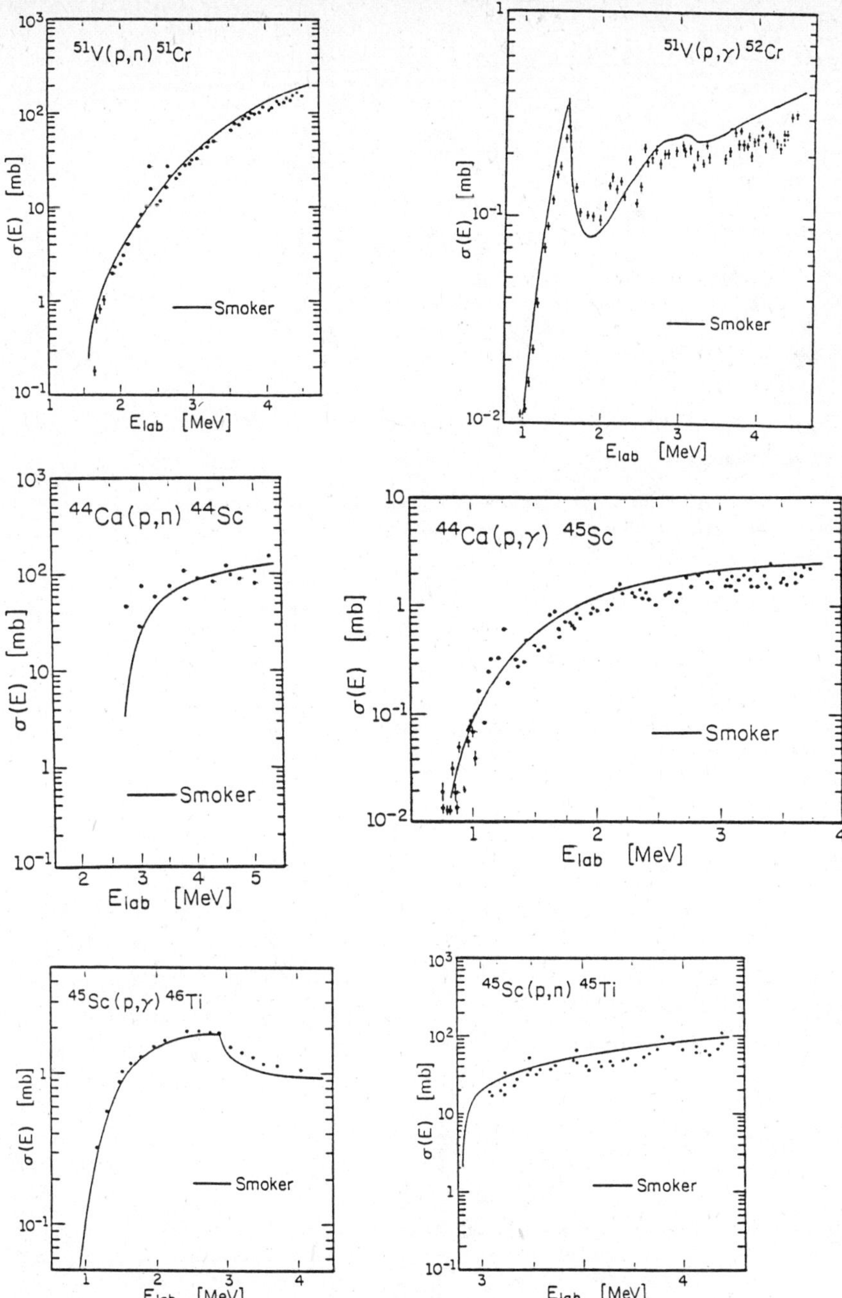

Fig. 3: Representative comparisons of theoretical and experimental cross sections for intermediate mass nuclei.

We consider barrier penetration in the WKB approximation, which yields the equation for the transmission coefficient

$$T(E) = \exp\{-2/\hbar \int_{x_1}^{x_2} [2\mu(V(r)-E)]^{1/2} dr\} \qquad (7)$$

where x_1 and x_2 are the classical turning points. In the region $r > R$, and for the case $\ell = 0$, we have $V(r) = V_{coul}(r) + V_{scr}(r)$ and, for our assumed form of $V_{scr}(r)$, it follows that

$$T_{scr}(E) = T(E+U_0)\ \xi\ (E+U_0) \qquad (8)$$

where $T_{scr}(E)$ is the screened transmission coefficient, $T(E+U_0)$ is the usual transmission coefficient evaluated at the elevated energy $(E+U_0)$, and $\xi(E+U_0)$ is a correction factor which arises due to the radial dependence of the screening potential $V_{scr}(r)$.

In the presence of screening, our integral expression for $<\sigma v>$ in Eq. (3) involves a sum of integrals of the form

$$\int_0^\infty \frac{T_{scr,j}(E,J,\pi)\ T_{scr,k}(Q_{scr,k}+E,J,\pi)}{\sum_n T_{scr,n}(Q_{scr,n}+E,J,\pi)} \exp(-E/kT)\, dE. \qquad (9)$$

where account is taken of the modifications of the Q-values for individual channels in the presence of screening. In particular, the nuclear potential in the incoming channel is lowered by an amount $U_{o,j}$ at $r = 0$ and thus a state with higher intrinsic excitation energy is produced. However, the nuclear potential in the outgoing channel is lowered by an amount $U_{o,n}$ at $r = 0$ and thus the available kinetic energy at infinity must be reduced accordingly. The Q-values of individual channels in the presence of screening may thus be written

$$Q_{scr,n} = Q_n + U_{o,j} - U_{o,n}\ . \qquad (10)$$

Our integral expression Eq. (9) may thus be written as

$$\int_0^\infty \frac{\xi_j(E+U_{o,j})T_j(E+U_{o,j})\ \xi_k(Q_k+E+U_{o,j})T_k(Q_k+E+U_{o,j})}{\sum_n \xi_n(Q_n+E+U_{o,j})T_n(Q_n+E+U_{o,j})} \exp(-E/kT)\,dE. \qquad (11)$$

We now make the usual variable transformation $E' = E+U_{o,j}$, which yields a factor $\exp(U_{o,j}/kT)$ in front of the integral similar to the screening factor in the absence of a radial dependence of the screening potential. We also approximate the integrand by evaluating the factors $\varepsilon_n(Q_n+E+U_{o,j})$ at the maximum of the integrand E'_{max} (Gamow peak energy) and arrive at the expression.

$$\exp(U_{o,j}/kT)\ \frac{\varepsilon_j(E'_{max})\,\varepsilon_k(Q_k+E'_{max})}{\varepsilon_{n,dom}(Q_n+E'_{max})}\int_{U_{o,j}}^{\infty}\frac{T_j(E')T_k(Q_k+E')}{\sum_n T_n\ (Q_n+E')}\exp\,(-E'/kT)dE'. \quad (12)$$

Here, $\varepsilon_{n,dom}(Q_n+E'_{max})$ denotes the correction factor in the dominant reaction channel at the maximum of the integrand. This expression holds for each integral in the sum and, assuming a weak J dependence of E'_{max}, for the screened thermonuclear rate in general. The screening factor for the reaction i(j,k)ℓ can thus be written

$$\exp(H_{ij,k}) = \exp(U_{o,j}/kT)\ \frac{\varepsilon(E'_{max})\,\varepsilon_k(Q_k+E'_{max})}{\varepsilon_{n,dom}(Q_n+E'_{max})}\ . \quad (13)$$

The analogous expression for the screening factor for the case of a photodisintegration reaction $(\lambda_{\gamma,j})$ is given by (25,26)

$$\exp(H_{\ell\gamma,j}) = \frac{\varepsilon(Q_j+E'_{max})}{\varepsilon_{n,dom}(Q_n+E'_{max})}\ . \quad (14)$$

In both instances, the determination of the screening correction due to the existence of a radial dependence of the screening potential $V_{scr}(r)$ requires a knowledge of which channel is dominant at the energy corresponding to the peak of the integrand $(Q_n+E'_{max})$. This must be provided either from experiment of from the predictions of theoretical calculations of thermonuclear rates. Future compilations of theoretical rates (26) will provide this information and E'_{max}.

IV. Summary and Conclusions

Our use of realistic optical model potentials in the determination of particle transmission coefficients and of revised and improved prescriptions for the determination of nuclear level densities and of electric dipole photon transmission coefficients has given rise to significantly improved estimates of thermonuclear reaction rates. The statistical model calculations described in this paper have been found specifically to predict thermonuclear reaction rates for both charged particle and neutron induced reactions which are generally in agreement with experimental determinations to within a factor of two.

The establishment of a general relation between particle transmission coefficients in screened and unscreened environments has also allowed us to derive a general expression for the screening enhancements of thermonuclear reaction rates, including screening of reactions involving charged particles in both the incoming and outgoing channels. The screening enhancement factors derived in the previous section can also be shown to be entirely consistent with the distribution of abundances in a nuclear statistical equilibrium, when proper account is taken of the changes of the chemical potentials for all species due to Coulomb interactions (26).

We note in conclusion that the combined studies outlined in this paper (23,24,25,26) allow for the first time a consistent and straightforward determination of reaction rates as a function of both temperature and density. The thermonuclear reaction rate dependence on the temperature follows from the appropriate integral over a Maxwellian distribution. Additionally, the approximate expressions for the screening enhancements derived in the preceding section may be applied if one knows only both the energy E'_{max} of the maximum of the integrand and the dominant reaction channel at that energy. The forthcoming compilation of thermonuclear reaction rates (24) performed for nuclei in the mass range $20 \leq A \leq 70$ with the SMOKER program will include this information.

Acknowledgements

This research was supported in part by the United States National Science Foundation under grant AST 86-11500 at the University of Illinois. J.W.T. wishes to express his thanks to the Alexander von Humboldt Foundation for support by a U.S. Senior Scientist Award and to Professor R. Kippenhahn for the hospitality of the Max-Planck-Institut für Astrophysik, Garching bei München.

References

(1) Arnould, M., 1972, Astron. Astrophys. 19, 82
(2) Bao, Z.Y. and Käppeler, F., 1986, KFK preprint
(3) Bethe, H., 1936, Phys. Rev. 50, 352
(4) Ericson, T., 1959, Nucl. Phys. 11, 481
(5) Fantomi, S., Friman, B.L., and Pandharipande, V.R., 1981, Phys. Lett. 104B, 89
(6) Gilbert, A. and Cameron, A.G.W., 1965, Can. J. Phys. 43, 1446
(7) Hauser, W. and Feshbach, H., 1952, Phys. Rev. 87, 366
(8) Holmes, J.A., Woosley, S.E., Fowler, W.A., and Zimmerman, B.A., 1976, At. Data Nucl. Data Tables 18, 306
(9) Iben, I., Jr. and Renzini, A., 1983, Ann. Rev. Astron. Astrophys. 21, 271
(10) Ichimaru, S. and Utsumi, K., 1983, Ap. J. (Letters) 269, L51
(11) Ichimaru, S. and Utsumi, K., 1984, Ap. J. 286, 363
(12) Jeukenne, J.P., Lejeune, A., and Mahaux, C., 1977, Phys. Rev. C16, 80
(13) Mahaux, C., 1982, Phys. Rev. C82, 1848
(14) Mann, F.M., 1978, Hauser 5, A. Computer Code to Calculate Nuclear Cross Sections, Hanford Engineering (HEDL-TME 78-83)
(15) McFadden, L. and Satchler, G.R., 1966, Nucl. Phys. 84, 177
(16) Michaud, G. and Fowler, W.A., 1970, Phys. Rev. C2, 2041
(17) Michaud, G., Scherk, L., and Vogt, E., 1970, Phys. Rev. C1, 864
(18) Mochkovitch, R. and Nomoto, K., 1986, Astron. Astrophys. 154, 115
(19) Salpeter, E.E., 1954, Austral. J. Phys. 7, 373
(20) Taam, R.E., 1985, Ann. Rev. Nucl. Part. Sci. 35, 1
(21) Thielemann, F.-K., 1980, Thesis, Technische Hochschule Darmstadt.
(22) Thielemann, F.-K., and Arnould, M., 1983, Proc. Int. Conf. on Nuclear Data for Science and Technology, Antwerpen, ed. K. Böckhoff, p. 762
(23) Thielemann, F.-K., Arnould, M., and Truran, J.W., 1987a, in Advances in Nuclear Astrophysics, eds. E. Vangioni-Flam, J. Audouze, M. Casse, J.-P. Chieze, and J. Tran Thanh Van (Gif-sur-Yvette: Editions Frontieres), p. 525
(24) Thielemann, F.-K., Arnould, M., and Truran, J.W., 1987b, in preparation
(25) Thielemann, F.-K., Truran, J.W., 1987a, in Advances in Nuclear Astrophysics, eds. E. Vangioni-Flam, J. Audouze, M. Casse, J.-P. Chieze, and J. Tran Thanh Van (Gif-sur-Yvette: Editions Frontieres), p. 541
(26) Thielemann, F.-K. and Truran, J.W. 1987b, in preparation
(27) Truran, J.W., 1966, Ph.D. thesis, Yale University
(28) Truran, J.W., 1972, Astrophys. Space Sci. 18, 306
(29) Truran, J.W., 1982, in Essays in Nuclear Astrophysics, eds. C.A. Barnes, D.D. Clayton, and D.N. Schramm (Cambridge: Cambridge Univ. Press), p. 467
(30) Truran, J.W., 1984, Ann. Rev. Nucl. Part. Sci. 34, 53
(31) Truran, J.W., Hansen, C.J., Cameron, A.G.W., and Gilbert, A., 1966, Can. J. Phys. 44, 151
(32) Weigmann, H. and Rohr, G., 1973, Reactor Centrum Nederland, Report 203, p. 194
(33) Woosley, S.E., Fowler, W.A., Holmes, J.A., and Zimmermann, B.A., 1979, At. Data Nucl. Data Tables 22, 371
(34) Woosley, S.E. and Weaver, T.A., 1986, Ann. Rev. Astron. Astrophys. 24, 205

THERMONUCLEAR FUNCTIONS

H.J. Haubold
Zentralinstitut fuer Astrophysik
Akademie der Wissenschaften der DDR
Potsdam, DDR - 1591

A.M. Mathai, W.J. Anderson
Department of Mathematics and Statistics
McGill University
Montreal, Canada H3A 2K6

1. Introduction

The generally accepted framework of nuclear astrophysics and its development have recently been reviewed by Fowler (1984). The experimental approach to nuclear astrophysics, that is concerned with the element-synthesizing nuclear reactions studied to a large extent in nuclear laboratories, is presented, e.g., in Rolfs and Trautvetter (1978). The aim of the present paper is to contribute to the problem of the parametrization of nuclear reaction rates as described in Fowler's paper (1984, Table I on p.154, and Table II on p.155; cp. also Bethe 1967). Thus, in the following sections we will go step by step through the abovementioned Tables and try to find in every step closed-form representations of the respective astrophysical quantity for constructing a reaction rate systematics as originally suggested by Fowler (1974). At least, we come up with analytic procedures for converting laboratory cross sections into astrophysical nuclear reaction rates (Haubold and Mathai 1985).

2. Solution of the Schroedinger Equation for the Coulomb Wave Function

The nuclear reaction of a particle mainly including neutrons, protons, and alphas with nuclei having atomic mass numbers $A \leq 30$ can be considered as a quantum mechanical scattering problem. Then, the problem will be described by the regular Coulomb wave function F_l satisfying the radial Schroedinger equation for two particles with electrical charges $Z_1 e$,

Z_2e moving in their Coulomb field with angular momentum $\hbar l$,

$$F_l = C_l(\eta)\rho^{l+1} e^{-i\rho}\, {}_1F_1(l+1-i\eta, 2l+2; 2i\rho) \tag{1}$$

In eq. (1), ${}_1F_1(.)$ denotes the usual confluent hypergeometric function, $k^2=2\mu E/\hbar^2=\mu^2 v^2/\hbar^2$, $\rho=kr$. An elaborate discussion of the Coulomb wave function in connection with nuclear reaction rates is given by Humblet (1985) who discussed extensively the analytic structure of Coulomb wave functions for the computation of nuclear reaction rates. Since the Coulomb wave function of angular momentum states other than $l=0$ vanish at the origin $r=0$, the value of F_l in (1) at $r=0$ is the value of the $l=0$ partial wave function at $r=0$. Therefore, from elementary properties of the Γ function, the Coulomb factor $C_l(\eta)$ depending on the Sommerfeld parameter $\eta=(\mu/2)^{\frac{1}{2}} Z_1 Z_2 e^2/\hbar E^{\frac{1}{2}}$,

$$C_l(\eta) = \frac{2^l e^{-\pi\eta/2}}{\Gamma(2l+2)} \left[\Gamma(l+1-i\eta)\,\Gamma(l+1+i\eta)\right]^{\frac{1}{2}}, \tag{2}$$

reduces for $l=0$ to

$$C_0(\eta) = \left[\frac{2\pi\eta}{e^{2\pi\eta}-1}\right]^{\frac{1}{2}} \tag{3}$$

(Humblet 1985).

3. Gamow's Penetration Factor

The penetration factor for the Coulomb potential barrier for zero angular momentum, P_0, is given by the asymptotic behaviour, $r\rightarrow 0$, of the square of the Coulomb wave function (1). Thus, we get

$$P_0 = \lim_{r\to 0} |F_l(kr)|^2 = C_0^2(\eta) = \frac{2\pi\eta}{e^{2\pi\eta}-1}. \tag{4}$$

In the typical astrophysical situation it holds $\eta>1$ and we introduce a negligible error in writing

$$P_0 \approx 2\pi\eta\, e^{-2\pi\eta}. \tag{5}$$

As far as there is no energy level near the incoming energy, $E=\mu v^2/2$, the cross section takes the simple form for quantum-mechanical reasons (Haubold and Mathai 1985),

$$\sigma \approx \frac{P_0}{v} = \frac{1}{v}\, 2\pi\, \frac{Z_1 Z_2 e^2}{\hbar v} \exp\left\{-\frac{Z_1 Z_2 e^2}{\hbar v}\right\}. \tag{6}$$

In the nuclear cross section (6) the rapidly varying term in the Gamow penetration factor governing transmission through the Coulomb barrier is eliminated. The factor in front of the exponential term in (6) is a geometrical factor because the quantum-mechanical interaction between two particles is always proportional to this factor ($\pi \lambda\!\!\!^{-} \sim E^{-1}$, $\lambda\!\!\!^{-}$: reduced De Broglie wave length). As we can see from eq. (6) the penetration factor approaches unity as v increases or as the electrical charges Ze decrease.

4. Salpeter's Astrophysical Cross Section Factor

Starting from (6) the following equation can be thought of as defining the astrophysical cross section factor S(E):

$$\sigma(E) = \frac{S(E)}{E} \exp\left\{-2\pi\left(\frac{\mu}{2}\right)^{\frac{1}{2}} \frac{Z_1 Z_2 e^2}{\hbar E^{1/2}}\right\} \tag{7}$$

(Salpeter 1952, cp. also Bethe 1967). The advantage of writing the nuclear cross section in this way is that two of the strongly energy-dependent factors of nonnuclear origin appearing in the nuclear cross section are factored out explicitely, leaving the problem of determination of S(E). It is mainly experimentally confirmed that the cross section factor S(E) is often found to be constant, or at least a slowly varying function of energy over a limited energy range (Rolfs and Trautvetter 1978). In the light of the weak energy dependence the cross section factor S(E) can be expanded into a power series up to the second order in energy as suggested by Salpeter (1952),

$$S(E) = S(0) + S'(0)E + \frac{1}{2} S''(0)E^2. \tag{8}$$

However, for the following analytic considerations the energy dependence of S(E) can even be more complicated than Salpeter's power series expansion, for example, as given by Fowler (1974).

5. The Average Over the Maxwell-Boltzmann Distribution

Under normal nucleosynthetic conditions the reacting nuclei are nondegenerate and nonrelativistic. In the hot and moderate dense interior of a star the reacting particles will be quickly moderated by elastic collisions with nuclei and the generally accepted assumption of a Maxwell-Boltzmann distribution of the relative velocities of the nuclei can be taken into account:

$$f(v) = \left(\frac{\mu}{2\pi kT}\right)^{\frac{3}{2}} \exp\left\{-\frac{\mu v^2}{2kT}\right\}. \tag{9}$$

However, it is quite interesting to note that already in the original paper of Atkinson and Houtermans (1929) physical reasons for a modification of the Maxwell-Boltzmann distribution are discussed. Particularly referring to the solar neutrino problem and to the controlled thermonuclear fusion in the laboratory, Vasil'ev et al. (1974) and Eder and Motz (1958), respectively, suggested that possibly a modification of the Maxwell-Boltzmann distribution must be taken into account for the computation of nuclear reaction rates (cp. Haubold and Mathai 1985, 1986a,d). Although there are definite hints from nuclear fusion experiments that deviations from the Maxwell-Boltzmann distribution, particularly in the high energy tail, can occur, no solution of a Boltzmann equation is known explaining physically the deviations in the Maxwell-Boltzmann tail.

6. Definition of the Nuclear Reaction Rate by Atkinson and Houtermans

Having the nuclear cross section in eqs. (7) and (8) as well as the distribution function (9), the reaction rate between two types of nuclei can be written as

$$r_{12} = \left(1 - \frac{1}{2}\delta_{12}\right) n_1 n_2 \langle\sigma v\rangle_{12}, \tag{10}$$

where n_1 and n_2 are the number densities of nuclei of type 1 and type 2, and where $\langle\sigma v\rangle_{12}$ is the reaction probability which is related to the product of the reaction cross section and the flux of particles, averaged over the Maxwell-Boltzmann velocity distribution. We have,

$$\begin{aligned}
\langle\sigma v\rangle &= \int_0^\infty dv\, \sigma(v)\, v f(v) \\
&= \left(\frac{8}{\pi\mu (kT)^3}\right)^{\frac{1}{2}} \int_0^\infty dE\, \sigma(E) \exp\left\{-\frac{E}{kT}\right\} \\
&= \left(\frac{8}{\pi\mu}\right)^{\frac{1}{2}} \sum_{\nu=0}^{2} \frac{1}{(kT)^{-\nu+(1/2)}} \frac{S^{(\nu)}(0)}{\nu!} \int_0^\infty dy\, e^{-y} y^{\nu} e^{-\frac{z}{y^{1/2}}},
\end{aligned} \tag{11}$$

where $y=E/kT$, and $z=2\pi(\mu/2kT)^{\frac{1}{2}}Z_1Z_2e^2/\hbar$. From eq. (11) one can see that the product of the steeply falling Maxwell-Boltzmann distribution, and the rapidly rising cross section in the kernel of the integral, gives a not quite symmetrical peak, which is called the Gamow peak. Additionally, one can imagine, that in charged particle reactions, the two dominant factors in the reaction probability are the Gamow factor, which inhibits the reaction probability at low energies, and the presence or absence of resonances. Regarding the latter point, the reactions of the pp-chain are nonresonant, the reactions of the CNO-cycles are dominated by one or a few resonances, and in the reactions of the C-, O-, and Si-burning the comound nucleus energy levels are close together and overlapping strongly.

7. Fowler's Basic Reaction Rates

Indeed, one can learn from the astrophysical literature many attempts to derive analytic expressions for nuclear reaction rates starting with the paper of Atkinson and Houtermans (1929, Bahcall 1966, Critchfield 1972, Haubold and John 1978, Haubold and Mathai 1985). In certain cases it is convenient to have analytic representations which can be used, along with their derivatives, extensively and with great confidence. Here we refer to the analytic results of Critchfield (1972) who gave the first elaborate discussion of what is called thermonuclear functions.

For the case of a nonresonant nuclear reaction rate the closed-form representation of the reaction rate integral in eq. (11) is given by

$$Ni_\nu(z) = \int_0^\infty dy\, e^{-y}\, y^\nu\, e^{-z/y^{1/2}} = \frac{1}{\pi^{1/2}}\, G^{3,0}_{0,3}\left(\left[\frac{z}{2}\right]^2 \middle|\, 0, \tfrac{1}{2}, 1+\nu\right), \tag{12}$$

where $G^{m,n}_{p,q}(.)$ is a generalized hypergeometric function known as Meijer's G-function (Mathai and Saxena 1973, Haubold and Mathai 1985). It is quite simple to find the asymptotic representation of this type of G-functions for large values of the characteristic parameter z, which is the well-known approximate representation of the nonresonant nuclear reaction rate integral:

$$Ni_\nu(z) \approx 2\left(\frac{\pi}{3}\right)^{1/2} \left(\frac{z}{2}\right)^{\frac{2\nu+1}{3}} e^{-3\left(\frac{z}{2}\right)^{2/3}}. \tag{13}$$

For the case of a resonant reaction rate one has to take into account the full energy dependent Breit-Wigner single level formula for a resonance,

$$\langle \sigma v \rangle = \left(\frac{(2\pi)^5}{\mu}\right)^{\frac{1}{2}} \frac{Z_1 Z_2 e^2}{(kT)^{3/2}} \frac{R_0 \omega \Gamma_{34} D}{1+(\frac{1}{2}\Gamma_1)^2} \int_0^\infty dx \, \frac{e^{-ax} \, e^{-z/(ax)^{1/2}}}{(b-x)^2 + g^2} \, . \quad (14)$$

In eq. (14) the following quantities are used:

$$x = E(1+[\Gamma_1/2]^2) \, , \qquad g = (\Gamma_0 + E_r\Gamma_1)/2 \, ,$$

$$a = \{kT(1+[\Gamma_1/2]^2)\} \, , \qquad b = E_r - \Gamma_0\Gamma_1/4 \, .$$

For a detailed discussion of the parametrization of eq. (14) we refer to Fetisov and Kopysov (1975) and Haubold and Mathai (1986b). The closed-form representation of the resonant reaction rate integral in eq. (14) can be given by

$$R_1(z,a,b,g) = \frac{1}{g^2 a} \sum_{k=0}^{\infty} \frac{(-1)^k}{(g^2)^k} \sum_{k_1=0}^{2k} \binom{2k}{k_1} \frac{(-1)^k}{a^{k_1}} b^{2k-k_1} \quad (15)$$

$$\times \frac{1}{\pi^{1/2}} G^{3,0}_{0,3}\left(\left[\frac{z}{2}\right]^2 \frac{1}{a} \,\middle|\, 0, \tfrac{1}{2}, 1+k_1\right).$$

8. A General Reaction Rate Systematics

In the nonresonant case we built up a reaction rate systematics generalizing eq. (12) in the following form:

$$Ni_\nu(z,\delta,a,t,d) = \int_0^d dy \, e^{-ay} \, e^{-by^\delta} \, y^\nu \, e^{-z/(y+t)^{1/2}} \, . \quad (16)$$

In the standard case of the nonresonant reaction rate ($d\to\infty$, $b=0$, $t=0$) the closed-form representation is given by,

$$N_1(z,a,\nu) = a^{-(\nu+1)} \frac{1}{\pi^{1/2}} G^{3,0}_{0,3}\left(\left[\frac{z}{2}\right]^2 a \,\middle|\, 0, \tfrac{1}{2}, 1+\nu\right). \quad (17)$$

For the asymptotic representation of eq. (17) we have

$$N_1(z\to\infty,1,\nu) \approx 2\left(\frac{\pi}{3}\right)^{1/2} \left(\frac{z}{2}\right)^{\frac{2\nu+1}{3}} e^{-3\left(\frac{z}{2}\right)^{2/3}} \, . \quad (18)$$

If due to plasma effects a depletion of the Maxwell-Boltzmann distribution has to be taken into account ($d\to\infty$, $b=1$, $t=0$) we get,

$$N_2(z,\delta,a,\nu) = \sum_{k=0}^{\infty} \frac{(-1)^k}{k!} a^{-(\nu+1+k\delta)} \frac{1}{\pi^{1/2}} G_{0,3}^{3,0}\left(\left[\frac{z}{2}\right] a \,\middle|\, 0, \tfrac{1}{2}, 1+\nu+k\nu\right), \qquad (19)$$

$$N_2(z\to\infty,\delta,1,\nu) \approx 2\left(\frac{\pi}{3}\right)^{1/2}\left(\frac{z}{2}\right)^{\frac{2\nu+1}{3}} e^{-3\left(\frac{z}{2}\right)^{2/3}} e^{-\left(\frac{z}{2}\right)^{2\delta/3}} \qquad (20)$$

(Eder and Motz 1958, Haubold and Mathai 1986d). Considering dissipative collision processes in a thermonuclear plasma a cut off of the high energy tail of the Maxwell-Boltzmann distribution could occur, thus we obtain ($b=0$, $t=0$),

$$N_3(z,d,a,\nu) = d^{(\nu+1)} \sum_{k=0}^{\infty} \frac{(-ad)^k}{k!} \frac{1}{\pi^{1/2}} G_{1,3}^{3,0}\left(\left[\frac{z}{2}\right]^2 \frac{1}{d} \,\middle|\, {2+\nu+k \atop 1+\nu+k, 0, \frac{1}{2}}\right), \qquad (21)$$

$$N_3(z\to\infty,d,1,\nu) \approx d^{(\nu+1)} \left(\frac{z}{2d^{1/2}}\right)^{-1} e^{-2\left(\frac{z}{2d}\right)} e^{-d} \qquad (22)$$

(Vasil'ev et al. 1974, Haubold and Mathai 1986a). In a dense plasma each nucleus attracts neighboring electrons and repels neighboring nuclei, thus forming a weak screening cloud of electrons characterized by the Debye-Hueckel length κ which is a measure of the size of the ion cloud. We write ($d\to\infty$, $b=0$),

$$N_4(z,t,a,\nu) = t^{\nu+1} e^{at} \sum_{r=0}^{\nu} \binom{\nu}{r} \left\{ N_1(zt, at, \nu-r) - N_3(zt, 1, at, \nu-r) \right\}, \qquad (23)$$

$$N_4(z\to\infty,t,1,\nu) \approx 2\left(\frac{\pi}{3}\right)^{1/2}\left(\frac{z}{2}\right)^{1/3} e^{-3\left(\frac{z}{2}\right)^{2/3}} e^{t}\left[\left(\frac{z}{2}\right)^{2/3} - t\right]^{\nu}, \qquad (24)$$

where $t=Z_1Z_2e^2\kappa$ denotes the electron screening parameter (Harrison 1964, Haubold and Mathai 1986c).

9. A Numerical Result: Critchfield's Thermonuclear Functions

From the considerations in sections 6 and 7, respectively, it is evident that the nuclear reaction rate in eqs. (10) and (11) can be split up into two terms. One term contains the astrophysical cross section factor S(E) and its derivatives which can be determined through the laboratory approach to nuclear astrophysics (Rolfs and Trautvetter 1978). The other term is the integral part of the nuclear reaction rate so far independent from experimental results which we suggested to evaluate through a mathematical approach to nuclear astrophysics as represented by eq. (16), and all special cases following from $Ni_\nu(z,\delta,a,t,d)$ are thermonuclear functions in the sense of Critchfield's paper (1972). The following figure shows the numerical result for Critchfield's case of the thermonuclear function that is included in eq. (16) for $\delta=0$, $a=1$, $t=0$, $d\to\infty$, $\nu=0$ and thus represents the simplest case for eq. (17),

$$N_1(z,1,0) = \frac{1}{\pi^{1/2}} G^{3,0}_{0,3}\left(\left[\frac{z}{2}\right]^2 \Big|_{0,\frac{1}{2},1}\right), \tag{25}$$

the asymptotic representation of eq. (25) is given in (18) for $\nu=0$.

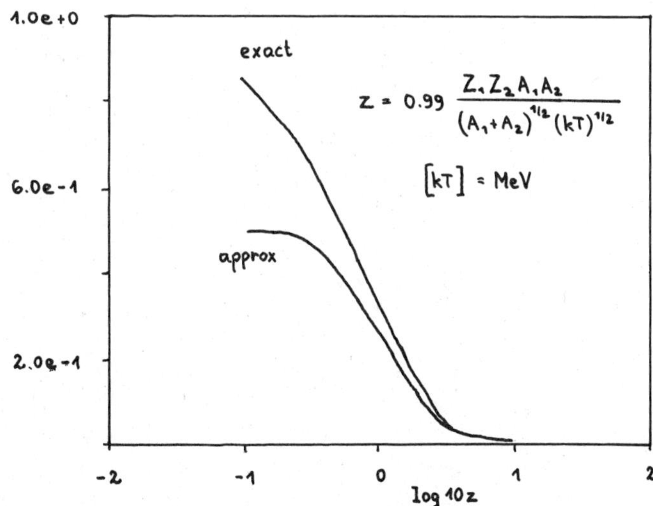

Figure: Numerical results for the thermonuclear function given in eq. (25) and the respective asymptotic representation given in eq. (18) for $\nu=0$.

For methods of deriving series representations of the G-function useful for numerical computation see Mathai and Saxena (1973) and Haubold and

Mathai (1985, 19 86a,b,d). As can be seen from the figure the closed-form evaluation of thermonuclear functions is of most importance for reactions between light nuclei (small electrical charges Ze or/and high temperatures T).

References

Atkinson, R.d'E., and Houtermans, F.G.: 1929, Z. Phys. 54, 656.
Bahcall, J.N.: 1966, Astrophys. J. 143, 259.
Bethe, H.A.: 1968, Naturwissenschaften 55, 405.
Critchfield, C.L.: 1972, in Cosmology, Fusion and Other Matters, (Ed. F.Reines), University of Colorado Press, Colorado, pp. 186.
Eder, G., and Motz, H.: 1958, Nature 182, 1140.
Fetisov, V.N., and Kopysov, Yu.S.: 1975, Nucl. Phys. A 239, 511.
Fowler, W.A.: 1974, Q. Jl R. astro. Soc. 15, 82..
Fowler, W.A.: 1984, Rev. Mod. Phys. 56, 149.
Harrison, E.R.: 1964, Proc. Phys. Soc. 84, 213.
Haubold, H.J., and John, R.W.: 1978, Astron. Nachr. 299, 225; 300(1979) 173.
Haubold, H.J., and Mathai, A.M.: 1985, Fortschr. Phys. 33, 623.
Haubold, H.J., and Mathai, A.M.: 1986a, Stud. Appl. Math. 75, 123.
Haubold, H.J., and Mathai, A.M.: 1986b, J. Math. Phys. 27, 2203.
Haubold, H.J., and Mathai, A.M.: 1986c, Astrophys. Sp. Sci. 127, 45.
Haubold, H.J., and Mathai, A.M.: 1986d, J. Appl. Math. Phys. (ZAMP) 37, 685.
Humblet, J.: 1985, J. Math. Phys. 26, 656.
Mathai, A.M., and Saxena, R.K.: 1973, Generalized Hypergeometric Functions with Applications in Statistics and Physical Sciences. Lecture Notes in Mathematics, Vol. 348, Springer-Verlag, Berlin-Heidelberg-New York.
Rolfs, C., and Trautvetter, H.P.: 1978, Ann. Rev. Nucl. Part. Sci. 28, 115.
Salpeter, E.E.: 1952, Phys. Rev. 88, 547.
Vasil'ev, S.V., Kocharov, G.E., and Levkovskij, A.A.: 1974, Izv. Akad. Nauk SSSR, Ser. Fiz. 38, 1827.

A microscopic approach to reactions of astrophysical interest

P. DESCOUVEMONT* and D. BAYE

Physique Théorique et Mathématique , CP 229, Université Libre de Bruxelles, Brussels, Belgium

Abstract

We present a microscopic model for investigating radiative capture and transfer reactions. We discuss the $^{12}C(\alpha,\gamma)^{16}O$ and $^{13}C(\alpha,n)^{16}O$ reactions, and the triple-α process.

1. Introduction

The study of stellar evolution requires the knowledge of cross sections at very low energies [1]. These astrophysical energies are so low with respect to the Coulomb barrier that direct measurements are in general impossible. Therefore a theoretical contribution is necessary. A first approach consists in parametrizing the available experimental data and extrapolating them to lower energies. This method has several drawbacks. In many cases, e.g. for reactions involving short-lived nuclei, experimental data do not exist. Moreover, the extrapolation is sensitive to the accuracy of the available data, and finally some effects, important at low energies, can be partly hidden in the energy range considered experimentally. A second kind of theoretical model consists in calculating the cross sections from the Hamiltonian of the problem. This approach is thus essentially independent of experiment. Microscopic models [2], which represent the topic of this paper, are classified in the second category. The advantages of microscopic theories are the following.

(i) All the information is obtained from a two-body nucleon-nucleon interaction; (ii) the Pauli principle, i.e. the indiscernibility of the nucleons of the target and the projectile is properly taken into account; (iii) bound, resonant and scattering states are defined in a unified way. This last property can be used to estimate the accuracy of the S-factor at low energies. The existing experimental transition probabilities in the spectrum of the unified nucleus are compared with their theoretical counterparts. If the agreement is satisfactory, the S-factor, calculated in the same manner, can be considered as reliable. The model employed in this work is the Generator-Coordinate Method (GCM), equivalent to the Resonating-Group Method (RGM) [3]. The first microscopic calculation of radiative capture reaction has been performed by Tang and coworkers [4] in 1981 for the $^{3}He(\alpha,\gamma)^{7}Be$ reaction. These last years, we have applied the GCM to a number of radiative capture reactions [5], among which $^{16}O(\alpha,\gamma)^{20}Ne$ [6], $^{12}C(\alpha,\gamma)^{16}O$ [7], $^{12}C(^{12}C,\gamma)^{24}Mg$ [8] or $^{15}O(\alpha,\gamma)^{19}Ne$ [9]. We have recently shown [10] that antisymmetrization effects can be important in the radiative capture process, even at low energies.

2. The microscopic model

All the information is obtained from the Hamiltonian

$$H = \sum_{i}^{A} T_i + \sum_{i<j}^{A} V_{ij} \qquad (1)$$

involving the A nucleons of the system. In (1), T_i is the kinetic energy of nucleon i, and V_{ij} comprises an effective two-body nuclear interaction and the exact Coulomb interaction. The wave functions of the system are antisymmetrized with respect to all nucleons, and are exact eigenfunctions of total angular momentum and parity. This latter requirement is fundamental in the study of the radiative capture process, in order to satisfy the correct selection rules. Let us consider a partition $A_{1c} + A_{2c}$ of the A nucleons of the system. The index c represents a channel corresponding to the partition (A_{1c}, A_{2c}) and to given internal states of the colliding nuclei. The relative coordinate between the clusters is defined as

$$\underset{\sim}{\rho}_c = \frac{1}{A_{1c}} \sum_i^{A_{1c}} \underset{\sim}{r}_i - \frac{1}{A_{2c}} \sum_i^{A} \underset{\sim}{r}_i \tag{2}$$

where $\underset{\sim}{r}_i$ is the coordinate of nucleon i. We also define $A_{1c} - 1$ and $A_{2c} - 1$ internal coordinates of clusters 1 and 2. The cluster j (j= 1,2) is described by a translation-invariant wave function $\phi^c_{I_j \nu_j \pi_j T_j}$ characterized by spin I_j, ν_j, parity π_j, isospin T_j and energy $E^c_{I_j \pi_j T_j}$. In the following the internal wave functions and the energies will be abbreviated as ϕ^c_j and E^c_j respectively.

The microscopic wave function of the system in the partial wave JMπ reads

$$\Psi^{JM\pi} = \mathcal{A} \sum_{c\ell I} [[\phi_1{}^c \otimes \phi_2{}^c]^I \otimes Y_\ell(\hat{\rho})]^{JM} \, g^{J\pi}_{c\ell I}(\rho) \tag{3}$$

In this expression, the angular momenta of the internal wave functions are coupled to give the channel spin I, which is coupled with the orbital momentum ℓ to provide the total angular momentum J of the system. In most calculations, only the entrance channel is taken into account. However, the inclusion of additional configurations improves the wave function. The occurrence of the antisymmetrization operator $\mathcal{A}$ in (3) represents an important complication. However, it allows one to use definition (3) for describing bound, resonant and scattering states of the system simultaneously. Since the internal wave functions ϕ^c_j are given, the relative functions $g^{J\pi}$ are the unknown quantities of the model. They are calculated in the GCM formalism associated to the microscopic R-matrix method. We refer the reader to refs [11,5] for information on these methods.

3. The $^{12}C(\alpha,\gamma)^{16}O$ reaction

The $^{12}C(\alpha,\gamma)^{16}O$ reaction plays a major role in stellar evolution since it determines the $^{12}C/^{16}O$ abundance ratio after helium burning [1]. In spite of important experimental and theoretical efforts, the cross section at the astrophysical energy 0.3 MeV remains uncertain (see ref [12] for a recent review). The cross section is expected to be essentially given by the contribution of the 1^-_1 and 2^+_1 weakly bound states. The S-factors strongly depends upon the reduced α widths of these states, which cannot be measured directly. The first estimate of Dyer and Barnes [13] provided an S-factor between 0.08 Mev.b and 0.17 MeV.b at 0.3 MeV for E1 transitions only. Kettner et al [14] give an higher S_{E1} (0.25 MeV.b) and suggest that the E2 multipolarity has to be taken into account. Recently, Redder et al [15] find 0.12 - 0.20 MeV.b for the E1 S-factor and 0.08 - 0.10 MeV.b for the E2 S-factor. The microscopic calculation of the $^{12}C(\alpha,\gamma)^{16}O$ cross section is rendered especially difficult since the E1 multipolarity, which dominates the cross section at low energies, exactly vanishes at the long wavelength approximation if one neglects isospin impurities. Moreover, the radiative width of the 2^+_1 bound state, which strongly influences the E2 component at astrophysical energies is largely underestimated in single-channel approaches [7]. We have reconsidered the $^{12}C(\alpha,\gamma)^{16}O$ reaction by including additional configurations [16]. We introduce the α + $^{12}C(2^+)$, p + $^{15}N_{gs}$ and n + $^{15}O_{gs}$ channels. The presence of the nucleon channels allows one to treat the E1 multipolarity in a microscopic way. With these conditions, the γ widths of the 1^-_1 and 2^+_1 states differ only slightly from the experimental values. Small effective charges are employed to provide the exact radiative widths [16]. The nucleon-nucleon forces for the ground state, for the positive-parity and negative-parity states are chosen in order to reproduce the 0_1^+, 1_1^- and 2_1^+ experimental energies respectively. They also provide the correct thresholds of the excited configurations. Beyond 1.5 MeV, the experimental capture cross section is governed by the properties of the 1_2^- resonance. We find an energy of 2.75 MeV, a dimensionless reduced α width of 50% (at 7.2 fm) and a $B(E1,1_2^-\rightarrow 0_1^+)$ equal to 1.33 10^{-4} W.U. Since the experimental counterparts (2.47 MeV, 34% and 6.0 10^{-5} W.U) respectively) are slightly different, one expects that the microscopic E1 S-factor should not be accurate enough. Moreover,the E2 S-factor is affected beyond 2 MeV by a broad barrier resonance. This resonance does not appear in the experimental d phase shift [17]; it is an inaccurately located approximation of a more excited state. In order to solve these problems, we have corrected the microscopic results in the following way. The capture cross section from the partial wave Jπ can be written as

$$\sigma^{J\pi} = \left| \sum_j \exp(i\varphi_j)\,(\sigma_j^{J\pi})^{\frac{1}{2}} \right|^2 \qquad (4)$$

where j represents a physical state or a background contribution. The phase term is equal to 0 or π, and σ_j is the capture cross section due to the jth state. For the lowest states it can be fairly well approximated by a Breit-Wigner expression [16]. In the present case, for the E1 multipolarity ,j= 1 represents the 1_1^- bound state, j= 2 the 1_2^- resonance, and the other terms simulate the background. For the E2 multipolarity ,j= 1 and j= 2 represent the 2_1^+ bound state and the broad 2^+ resonance respectively. Since the microscopic properties of the 1_2^- resonance do not exactly reproduce the experimental data, we have replaced σ_l^{1-} by a Breit-Wigner approximation with the experimental parameters. For the E2 multipolarity we have dropped the σ_l^{2+} contribution. The contribution of the bound states and of the background remain unchanged. Let us notice that this procedure is not a fit of the data. It consists in combining microscopic information with well known experimental data. The E1 and E2 S-factor obtained in this way are presented in fig.1.

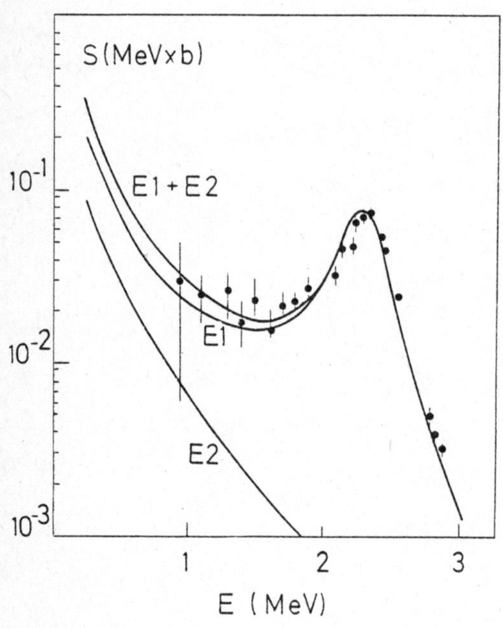

Fig.1 $^{12}C(\alpha,\gamma)^{16}O$ S-factor as a function of the cm energy. The experimental data are from ref. [15]

The agreement between the total microscopic S-factor and the experimental data of ref.[15] is excellent. Since our model describes correctly the properties of the 1_1^- and 2_1^+ bound states, and since it agrees with the available experimental data, we think that our results below 1 MeV should be reliable. At 0.3 MeV, we find

$$S_{E1} = 0.16 \text{ MeV.b}$$

$$S_{E2} = 0.07 \text{ MeV.b}$$

The E1 S-factor is intermediate between the two model extrapolations of ref [15], and is consistent with the new estimate of Barker [12]. The E2 S-factor, although non negligible is much smaller than the value proposed by Kettner et al [14].It is also smaller than the recent extrapolations performed by the Münster group [15] but larger than the value proposed by Barker [12]. Taking account of the cascade contributions (0.015 MeV.b [16]), we find a total S-factor of 0.24 MeV.b at 0.3 MeV.

4. The $^{13}C(\alpha,n)^{16}O$ reaction

The $^{13}C(\alpha,n)^{16}O$ reaction is of major importance in nuclear astrophysics since it is expected to be the main source of neutrons in stellar sites [18]. Until now the reaction rate used in astrophysical applications is derived from a linear extrapolation of existing experimental data [19]. However, as we have seen in the $^{12}C(\alpha,\gamma)^{16}O$ reaction, this procedure may be inaccurate when bound states or resonances of the unified nucleus exist in the vicinity of the threshold.

Let us first discuss the ^{17}O spectrum obtained in the microscopic approach [20]. The GCM wave functions (3) involve the n + ^{16}O and α + ^{13}C channels. The nucleon-nucleon interaction is adjusted in order to reproduce the experimental gap between these configurations and to provide good energies of ^{17}O low-lying states. The different states are classified according to whether they present a marked α clustering or not. The microscopic model gives rise to three ^{17}O states ($\frac{5}{2}^+_1$, $\frac{1}{2}^+_1$ and $\frac{3}{2}^+_1$) with small reduced α widths; the root mean square radius (2.57 fm), the magnetic moment ($-1.76\ \mu_N$) and the quadrupole moment (-2.38 e fm^2) of the ground state nicely agree with their experimental counterparts (2.67 fm, $-1.89\ \mu_N$ and -2.58 e fm^2 respectively). In addition to these states, the model predicts two bands $K^\pi = \frac{1}{2}^+$ and $\frac{1}{2}^-$ involving states with large θ_α^2-values. In fig.2, these states are displayed together with their experimental counterparts. From their energies, the $\frac{1}{2}^+_2$ (6.36 MeV), $\frac{3}{2}^+_3$ (7.20 MeV), $\frac{5}{2}^+_2$ (7.38 MeV) and $\frac{7}{2}^+_1$ (8.46 MeV) levels are valuable candidates for α-cluster states. A molecular band $K^\pi = \frac{1}{2}^-$ composed of the $\frac{1}{2}^-_1$ (3.06 MeV), $\frac{3}{2}^-_1$ (4.55 MeV), $\frac{5}{2}^-_1$ (3.84 MeV), $\frac{7}{2}^-_3$ (7.69 MeV) and $\frac{9}{2}^-_1$ (9.15 MeV) is also suggested.

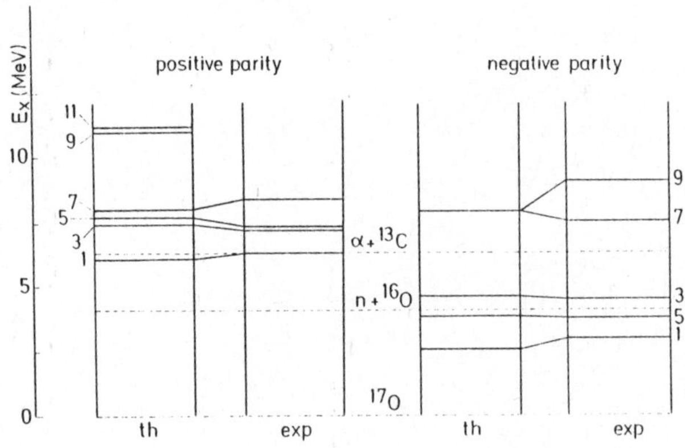

Fig.2 α-cluster states in ^{17}O. The spins have been multiplied by 2

The present model gives a good description of the $^{13}C(\alpha,\alpha)^{13}C$ elastic scattering at low energies [20]. Here, we focus on the $^{13}C(\alpha,n)^{16}O$ reaction. The transfer cross section is given by

$$\sigma_t = (\pi/2k^2) \sum_{J\pi} (2J+1) \mid U^{J\pi}_{\alpha n} \mid^2 \tag{5}$$

where $U^{J\pi}_{\alpha n}$ is a non-diagonal element of the collision matrix, and k is the wave number relative to the α + ^{13}C channel. Let us first discuss the data around the Coulomb barrier [21]. The microscopic cross section is plotted in fig.3.

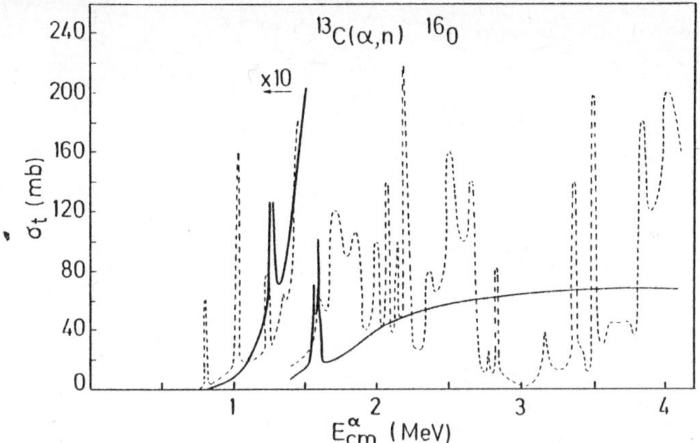

Fig.3 Microscopic (full line) and experimental (dashed line - ref [21]) $^{13}C(\alpha,n)^{16}O$ cross sections.

Below 1.5 MeV, the GCM non-resonant curve reproduces nicely the experimental data. The experimental peak at 1.02 MeV is due to the $\frac{5}{2}^+_2$ state; according to fig.2, this peak occurs in our calculation near 1.35 MeV. Beyond 1.5 MeV, the theoretical cross section raises slowly. Because of the lack of other reaction channels, the structures observed in the experimental data are missing, but our results are consistent with averaged values of the data.

The theoretical S-factor below 0.8 Mev is presented in fig.4 with the available experimental data [19]. In the considered energy range the S-factor is dominated by the contribution of the $\frac{1}{2}^+$ partial wave, and to a lesser extent by the contribution of the $\frac{3}{2}^+$ partial wave. With decreasing energies, the S-factor increases rapidly. This effect is well known in the $^{12}C(\alpha,\gamma)^{16}O$ reaction (see sect.3). It is due to the presence of a weakly-bound state with the considered spin and parity. In the present case the $\frac{1}{2}^+_2$ state (see fig.2) at E_x = 6.36 MeV, i.e. 4 keV below the α + ^{13}C threshold, is responsible for this enhancement.

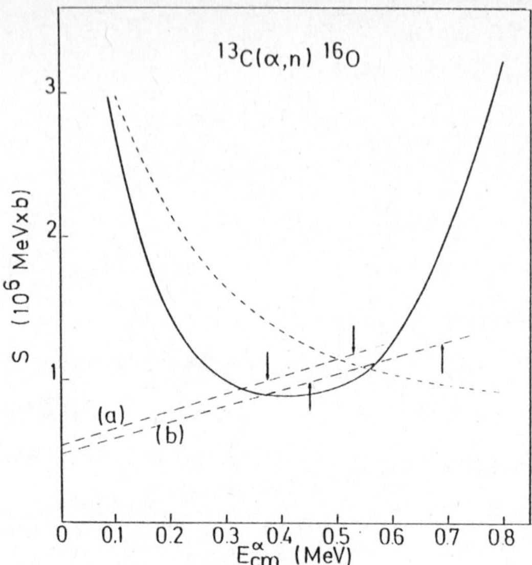

Fig.4 GCM (dotted line) and modified (full line) S-factors. The experimental data (dashed lines) are from ref.[19]. The arrows indicate the energy range where the experiments are carried out.

The microscopic S-factor presented in fig.4 does not accurately reproduce the experimental data, because the properties of the $\frac{1}{2}^+_2$ and $\frac{3}{2}^+_3$ states are not in quantitative agreement with experiment. In order to solve this problem, we have used a Breit-Wigner approximation of the $\frac{1}{2}^+$ and $\frac{3}{2}^+$ partial waves with the experimental data for the energies and widths. The only unknown quantity is the reduced α width of the $\frac{1}{2}^+_2$ bound state, for which we employ the microscopic result. The S-factor obtained in this way is presented in fig.4. It is seen that below 0.6 Mev, the agreement between theory and experiment is quite acceptable. Below 0.3 MeV, the S-factor is predicted to increase rapidly with respect to the linear extrapolations of the data. Experimental measurements below 0.3 MeV should give a confirmation of this proposal.

5. The triple-α process

The triple-α process, which leads to the production of ^{12}C nuclei is expected to occur in two steps [1]. The first one is the formation of ^{8}Be from two α particles. The second step involves α capture by ^{8}Be. Since the ^{8}Be nucleus is unstable with respect to α decay, usual capture models cannot be applied to the $\alpha(\alpha,\gamma)^{8}Be$ reaction : a bremsstrahlung calculation, involving two scattering states of the system, is required [22]. We have presented in ref.[22] a general framework for treating microscopically nucleus-nucleus bremsstrahlung. The long wavelength approximation of the electromagnetic multipole operator cannot be used. This leads to much more complicated expressions than in the calculation of capture cross sections.

Experimental data about the $\alpha(\alpha,\gamma)^{8}Be$ reaction do not exist. However,since the data on the $\alpha(\alpha,\alpha\gamma)\alpha$ bremsstrahlung reaction are well explained by our model [22], we think that the GCM capture cross section should be reliable. In fig.5,we present the microscopic $\alpha(\alpha,\gamma)^{8}Be$ cross section. The 2^{+} resonance of ^{8}Be appears as a peak centered around 2.5 MeV, where the cross section reaches 15 nb. The cross section presents a minimum near 4 MeV, and then increases monotonously.

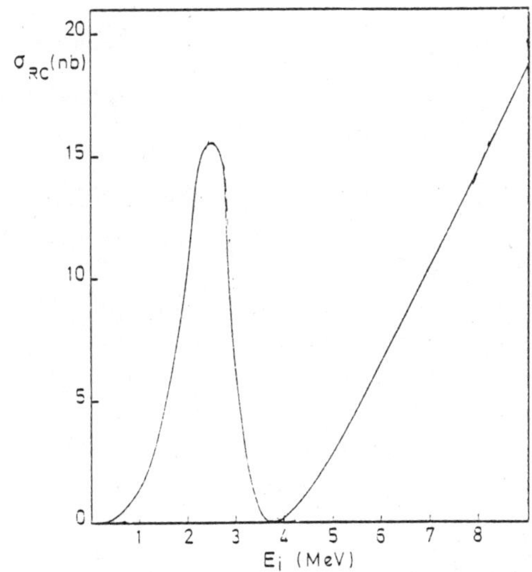

Fig.5 $\alpha(\alpha,\gamma)^{8}Be$ capture cross section as a function of cm energy

The second step of the ^{12}C formation crucially depends on the properties of the 0_2^+ excited state of ^{12}C [1]. The $^8Be(\alpha,\gamma)^{12}C$ reaction has been already considered in non microscopic models [24,25]. However,valuable wave functions of the ^{12}C nucleus must taken the α+α structure of ^8Be into account. We have therefore investigated the $^8Be(\alpha,\gamma)^{12}C$ reaction in a microscopic three-cluster model [23] (see fig.6).

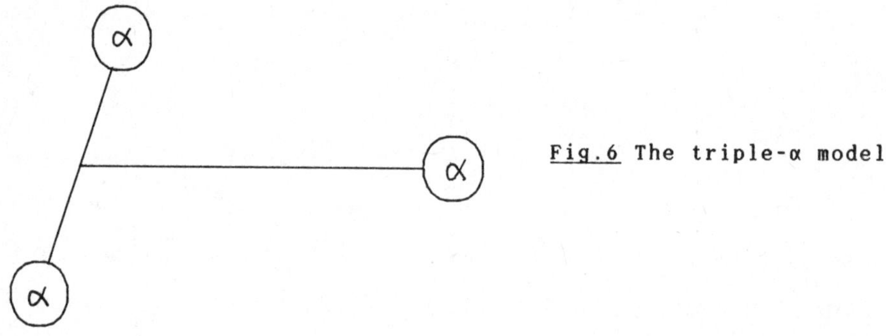

Fig.6 The triple-α model

With respect to usual two-cluster approaches, the three-cluster model represents an enormous increase of computation times. A test of the calculation is provided by the comparison between theoretical and experimental properties of bound states. The quadrupole moment of the 2_1^+ excited state of ^{12}C is found equal to 5.7 e fm^2, in nice agreement with experiment (6±3 e fm^2). For the $B(E2,0_2^+\rightarrow 2_1^+)$ value, which is important for the study of the $^8Be(\alpha,\gamma)^{12}C$ capture, we find 8.6 e^2 fm^4, while the experimental value is 13.4 e^2 fm^4. Let us mention that the quadrupole moment of the 2_1^+ state and the B(E2) value are completely different in a two-cluster calculation (-8.2 e fm^2 and 98.6 e^2 fm^4 respectively). This result stresses the importance of a three-cluster description of ^{12}C. We present in fig.7 the astrophysical S-factors for the $^8Be(\alpha,\gamma)^{12}C$ reaction towards the 0_2^+ and 2_1^+ states of ^{12}C. We use the effective charge δe = 0.12 e in order to reproduce the experimental $B(E2,0_2^+\rightarrow 2_1^+)$ value.

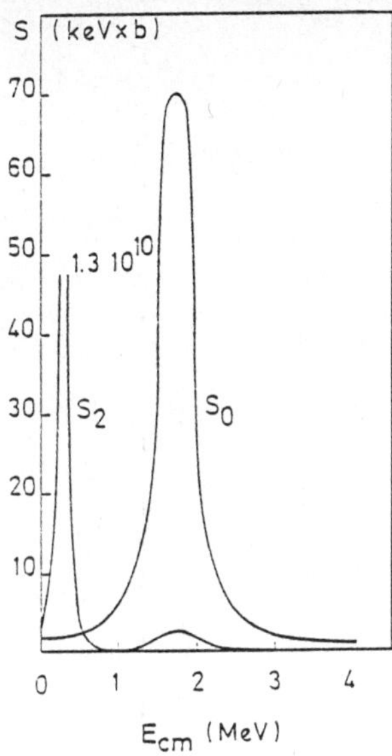

Fig.7 $^{8}Be(\alpha,\gamma)^{12}C$ S-factors towards the ^{12}C ground state (S_0) and first excited state (S_2).

The S-factor towards the ground state presents a peak near 1.75 MeV corresponding to the 2_2^+ resonance. The 0_2^+ resonance strongly influences the capture cross section towards the 2_1^+ state. The comparison of both S-factors indicates that the contribution of the ground state is not negligible when the energy tends towards zero. With these S-factors, we have estimated the reaction rate [23] at low temperatures. We find intermediate results between the reaction rates of Nomoto et al [24] and of Langanke et al [25].

Conclusion

In this contribution, we have presented different new results obtained in a microscopic cluster model. The method provides a unified description of bound, resonant and scattering states. This property offers an indirect test of the non-resonant cross sections. Of course, the lack of adjustable parameters is a problem when the properties of an important bound state or resonance are not in agreement with experiment. In the $^{12}C(\alpha,\gamma)^{16}O$ and $^{13}C(\alpha,n)^{16}O$ reactions we have solved this problem as consistently as possible with the microscopic theory by only modifying a few unsatisfactory physical quantities. We have extended the microscopic model to a three-cluster approach, and applied it to the $^{8}Be(\alpha,\gamma)^{12}C$ reaction. For reactions involving deformable nuclei, such as ^{8}Be, the three-cluster model has been shown to be necessary. The study of other reactions in this approach could be considered but the problem of enormous computation times must be overcome.

References

+ Chargé de Recherches FNRS.

1) C. Rolfs and H.P. Trautvetter, Annu. Rev. Nucl. Part. Sci. **26**(1978)115

B.W. Filippone, Annu. Rev. Nucl. Part. Sci. **36**(1986)717

2) K. Wildermuth and Y.C. Tang, "A unified theory of the nucleus", Vieweg, Braunschweig (1977)

3) Y.C. Tang, In topics in nuclear physics II, Lecture notes in Physics, Springer, Berlin, **145**(1981)572

4) Q.K.K. Liu, H. Kanada and Y.C. Tang, Phys. Rev.**C23**(1981)645

5) D. Baye and P. Descouvemont, Nucl. Phys. **A407**(1983)77

6) P. Descouvemont and D. Baye, Phys. Lett. **127B**(1983)286

7) P. Descouvemont, D. Baye and P.-H.Heenen, Nucl. Phys. **A430**(1984)426

8) D. Baye and P. Descouvemont, Nucl. Phys. **A419**(1984)397

9) P. Descouvemont and D. Baye, Nucl. Phys. **A463**(1987)629

10) D. Baye and P. Descouvemont, Ann. Phys. **165**(1985)115

11) D. Baye, P.-H. Heenen and M. Libert-Heinemann, Nucl. Phys. **A291**(1977)230

12) F.C. Barker, Austral. J. Phys. ,to be published

13) P. Dyer and C.A. Barnes, Nucl. Phys. **A233**(1974)495

14) K.U. Kettner et al. , Zeit. Phys. **A308**(1982)73

15) A. Redder et al. , Nucl. Phys.**A462**(1987)385

16) P. Descouvemont and D. Baye,to be published

17) R. Plaga et al., Nucl. Phys. **A465**(1987)291

18) A.G.W. Cameron, Phys. Rev. **93**(1954)932

19) C.N. Davids, Nucl Phys. **A110**(1968)619;

R. Ramström and T. Wiedling, Nucl. Phys. **A272**(1976)259

20) P. Descouvemont, to be published

21) J.K. Bair and F.X. Haas, Phys. Rev. **C7**(1973)1356

22) D. Baye and P. Descouvemont, Nucl. Phys. **A443**(1985)302

23) P. Descouvemont and D. Baye, Phys. Rev. **C35**(1987) in press

24) K. Nomoto, F.-K. Thielemann and S. Miyaji, Astron. Astrophys.**149**(1985)239

25) K. Langanke, M. Wiescher and F.-K. Thielemann, Zeit. Phys. **A324**(1986)147

THE ETFSI APPROACH TO THE NUCLEAR MASS FORMULA

J.M. Pearson* (Univ. München and Univ. de Montréal), F. Tondeur (Univ. Libre de Bruxelles), and A.K. Dutta (Univ. de Montréal)

We present an approach to the mass formula based on the extended Thomas-Fermi method with Strutinsky shell corrections. For extrapolating from known to unknown nuclei far from the stability line it is essentially as accurate as the Hartree-Fock method for a given form of force, but is so much faster that the construction of a complete mass table is feasible. Results of preliminary fits are presented.

1. Introduction

Whatever the site of the r-process of stellar nucleosynthesis, its elucidation will require the use of a reliable nuclear mass formula. This is because the evolution of this process depends critically on neutron-separation energies (S_n), beta-decay energies (Q_β), and fission barriers (among other properties) of many nuclei that lie so close to the neutron (n)-drip line that there is no possibility of being able to measure them in the laboratory. It thus becomes of the greatest importance to be able to make reliable extrapolations of binding energies away from the known region, close to the stability line, out towards the n-drip line.

If we are to have any confidence in the ability of our mass formula to extrapolate reliably it must not only give a good fit to the available data but also have a sound theoretical basis. Clearly, if two different mass formulas give comparable fits to the data but extrapolate differently, we would prefer the one with the better theoretical foundation.

We believe therefore that it is vitally important to develop mass formulas that are as rigorously based as possible. The ideal would be to be able to derive all nuclear properties from the "real" nucleon-nucleon interaction, but it will be quite impossible to do this in the forseeable future with anything like the required precision.

Mass formulas in use at the present time[1-5] are based rather on the so-called "macroscopic-microscopic" approach[6], in which the smoothly varying part of the binding energy is represented by one form or another of the drop (-let) model (DM), to which must be added microscopic corrections to take account of the fluctuations associated with shell-model and pairing effects. Now while this approach gives acceptable fits to the data (ref.3, for example, has rms errors of 0.835 MeV for 1323 masses, and 1.331 MeV for 28 fission barriers), it is open to theoretical criticism on two different counts:

a) It has been shown[7-9] that in the usual forms of the drop model the characteristic "leptodermous" expansion in powers of $A^{-1/3}$ is truncated prematurely. This difficulty appears to be rectified at least partially in the so-called "finite-range" droplet model[10,11], but there remains the problem of a premature truncation of the expansion in powers of $I = (N-Z)/A$. Our group investigated[11] this question by using the Skyrme-ETF method (see Section 2) to generate a large amount of "data" over the known region of the nuclear chart, and then fitting this "data" with various forms of the drop model. On extrapolating these fits out to the n-drip line discrepancies as large as 15 MeV with respect to Skyrme-ETF were found. Actually, ref.11 indicated that this problem might be resolved by the inclusion of higher-order "malacodermous" terms in I^2, but no systematic study of this has been made.

b) It is difficult to relate the macroscopic and microscopic parts coherently. To be specific, the calculation of the microscopic corrections involves the use of a single-particle (s.p.) potential, and since this must be generated by the distribution of nucleons in the nucleus it constitutes a link between the microscopic and macroscopic parts of the mass formula. Now the actual way in which the s.p. potential is generated from the nucleon distribution is by folding some two-body force over the latter, but in no form of the drop model is there an unambiguous prescription for choosing this force. Furthermore, the density distribution itself is determined only in a very crude way in the drop (-let) model.

Both these classes of difficulty are avoided when the binding energy is calculated by the Hartree-Fock (HF) method. In the first place there is no approximation based on a power-series expansion. Secondly, there is no separation of the total energy into macroscopic and microscopic parts, so complete consistency between the two is automatically guaranteed. A mass formula based on the HF method (or rather on the HF + BCS method, since pairing always has to be taken into account) will, therefore, be much more theoretically secure than one based on a drop (-let) model. This method represents, in fact, the most fundamental approach to a mass formula that has any chance of succeeding, even though it is much less rigorous than an approach based on the "real" nucleon-nucleon force.

The ideal procedure to be followed with this method would consist in taking some suitable form of effective interaction and fitting its parameters, along with those of the pairing force, to all the data on masses, fission barriers, radii, etc., as in the present DM fits. Unfortunately, the method suffers from the defect of requiring a very large amount of computer time, especially for deformed nuclei, with the result that its systematic application has been somewhat limited. In particular, the available HF effective forces have been fitted to relatively few of the available data, thereby detracting from the reliability of the method as a means of extrapolating.

In this paper we present an approach of intermediate complexity that we have developed: it is based on the extended Thomas-Fermi (ETF) method for the macroscopic part, with shell corrections calculated by the so-called Strutinsky-integral (SI) method.[12,13] When the underlying Skyrme-type force is fitted to the data the extrapolations to the n-drip line are very close to those given by the HF method, but being much more rapid computationally the method offers a practical approach to the ultimate task of constructing a mass table. We describe the method in Section 2 and compare its extrapolations to those of the HF method in Section 3 (these two sections are summaries of refs.12 and 13). In Section 4 we present the results of our first attempts to fit the data.

2. Skyrme-ETFSI Method

The basis of our method is a generalized Skyrme force:

$$\begin{aligned} V_{ij} &= t_o(1+x_oP_\sigma)\delta(\mathbf{r}_{ij}) \\ &+ t_1(1+x_1P_\sigma)\frac{1}{2\hbar^2}\{p_{ij}^2\delta(\mathbf{r}_{ij})+\text{h.a.}\}+t_2(1+x_2P_\sigma)\frac{1}{\hbar^2}\mathbf{p}_{ij}\cdot\delta(\mathbf{r}_{ij})\mathbf{p}_{ij} \\ &+ \frac{1}{6}t_3(1+x_3P_\sigma)\{\rho_{q_i}(\mathbf{r}_i)+\rho_{q_j}(\mathbf{r}_j)\}^{1/3}\delta(\mathbf{r}_{ij}) \\ &+ \frac{i}{\hbar^2}W_o(\boldsymbol{\sigma}_i+\boldsymbol{\sigma}_j)\cdot\mathbf{p}_{ij}\times\delta(\mathbf{r}_{ij})\mathbf{p}_{ij} \quad . \end{aligned} \tag{1}$$

To this we add the constraints $t_1 = -\frac{1}{3}t_2(5+4x_2)$ and $x_1 = -(4+5x_2)/(5+4x_2)$ in order that the effective and real nucleon masses be equal, $M_q^*=M_q$, a choice which allows a good fit to the s.p. energies near the Fermi surface without having to take particle-vibration coupling into account.

For the energy density $\mathcal{E}(\mathbf{r})$, which gives the total energy as

$$E = \int \mathcal{E}(r) d^3r \quad , \tag{2}$$

we now have

$$\begin{aligned} \mathcal{E}(r) = & \sum_q \frac{\hbar^2}{2M_q} \tau_q(\underset{\sim}{r}) + \mathcal{E}_{Coul}(\underset{\sim}{r}) + \frac{1}{2}t_o\{(1+\frac{1}{2}x_o)\rho^2 - (x_o + \frac{1}{2})\sum_q \rho_q^2\} \\ & - \frac{1}{4}t_2\{(1+\frac{1}{2}x_2)(\underset{\sim}{\nabla}\rho)^2 + (x_2 + \frac{1}{2})\sum_q(\underset{\sim}{\nabla}\rho_q)^2 + \frac{1}{3}(1+2x_2)J^2 + \frac{1}{3}(2+x_2)\sum_q J_q^2\} \\ & + \frac{1}{6}t_3\{(1+\frac{1}{2}x_3)\rho^{1/3}\rho_p\rho_n + \frac{1}{16}(1-x_3)\sum_q(2\rho_q)^{7/3}\} \\ & + \frac{1}{2}W_o\{\underset{\sim}{J}_n \cdot \underset{\sim}{\nabla}\rho_p + \underset{\sim}{J}_p \cdot \underset{\sim}{\nabla}\rho_n + 2\sum_q \underset{\sim}{J}_q \cdot \underset{\sim}{\nabla}\rho_q\} \end{aligned} \tag{3}$$

Here the index q denotes p or n, according as to whether the term in question relates to protons or neutron, respectively. Also we have introduced the number density ρ_q, the kinetic-energy density τ_q and the spin-current density $\underset{\sim}{J}_q$; we write $\underset{\sim}{J} = \underset{\sim}{J}_n + \underset{\sim}{J}_p$ and $\rho = \rho_p + \rho_n$. Finally, $\mathcal{E}_{Coul}(\underset{\sim}{r})$ is the Coulomb energy density. (Complete expressions for all quantities are given in ref.12.) It will be remarked that we retain the terms in J^2 and J_q appearing in $\mathcal{E}(\underset{\sim}{r})$. This complicates the formalism considerably but has a negligible effect on the computer time, and will certainly contribute to the reliability of extrapolation to exotic nuclei, insofar as it is more realistic to take a force rather than an energy density as the starting point.

Eq.(3) could serve as the basis of the HF method, imposing

$$\delta_\phi E = 0 \quad , \tag{4}$$

but instead our approach involves making a semiclassical approximation to the densities τ_q and $\underset{\sim}{J}_q$, expressing these quantities as functions of ρ_q and its gradients: we use the full fourth-order (in powers of $\hbar$) expansions of Grammaticos and Voros.[14,15] The resulting expressions are discussed fully in refs.12 and 13, where we consider in particular the problem of the spin-current terms; ref.13 also discusses our adoption of the (u,v) coordinate system of Brack et al (BGH)[9] for axially symmetric nuclei (we cannot handle triaxial nuclei, as such).

With the total energy E thus being expressed as a density functional it should in principle be minimized with respect to arbitrary variations in ρ_p and ρ_n:

$$\delta_\rho E = 0 \tag{5}$$

However, the resulting Euler-Lagrange equations are very difficult to solve, and we therefore parametrize the density distribution according to the Fermi form,

$$\rho_q(r) = \frac{\rho_{oq}}{1+\exp\left(\frac{r-C_q}{a_q}\right)} \quad , \tag{6}$$

and simply minimize E with respect to the parameters. (In ref.11 we took a generalized Fermi distribution but this is too time-consuming in the present mass-formula project.) Actually, Eq.(6) refers specifically to spherical nuclei; for deformed nuclei we have to make the additional Ansatz that ρ_q depends only on the u-variable, as in BGH.

But whether one solves the full Euler-Lagrange equations, or simply minimizes the energy with respect to the distribution parameters, it is inevitable that with the ETF method the energy will vary smoothly from one nucleus to another: the shell corrections are lost in truncating the semi-classical expansions. Thus we are forced back to a macroscopic-microscopic approach again, and have to add shell corrections.

However, compared to the DM mass formulas there is an important difference, since we can now determine quite unambiguously the s.p. fields $U_q(\underset{\sim}{r})$ and $\underset{\wedge}{W}_q(\underset{\wedge}{r})$ appearing in the s.p. Schrödinger equation,

$$\left\{ -\frac{\hbar^2}{2M_q}\nabla^2 + U_q(\underset{\sim}{r}) + \underset{\wedge}{W}_q(\underset{\wedge}{r})\cdot[-i\underset{\wedge}{\nabla}\times\underset{\wedge}{\sigma}]\right\}\phi(\underset{\sim}{r}) = \varepsilon\phi(\underset{\wedge}{r}) \quad , \tag{7}$$

the solutions to which are required for the determination of the shell (and pairing) corrections. One simply folds the same Skyrme-type force involved in the ETF functional over the density distribution obtained in the macroscopic part of the calculation. There is thus a high degree of coherence between the macroscopic and microscopic parts, the unifying factor being the Skyrme-type force that underlies both. With Eq.(7) solved the calculation of the shell and pairing corrections proceeds as follows.

Strutinsky-Integral Shell Corrections. Neglecting pairing, the total energy of a nucleus can be written as

$$E \cong E_{ETF} + \delta \quad , \tag{8}$$

where E_{ETF} is the macroscopic energy, derived as described above, and δ is the shell correction, given by

$$\delta = \sum_{\mu}{}' \varepsilon_{\mu} - \widetilde{\sum_{\mu}{}' \varepsilon_{\mu}} \tag{9}$$

The first term on the right-hand side of Eq.(9) represents the sum of the s.p. energies corresponding to Eq.(7) for all the occupied states, while the second term is an appropriately smoothed version of the first.

There are several ways for obtaining this smoothed quantity for a given s.p. potential, each of which gives a different approximation in Eq.(8). Perhaps the best known of these methods is that of Strutinsky smoothing (see ref.16 for a review), as used in refs.3-5, for example. Now it is well known that this method contains some ambiguities, mainly because of the continuum s.p. states, which means that they may be expected to become particularly troublesome towards the drip lines. However, in the present case, where we know the density distributions from which the s.p. potential is generated, these problems can be avoided, since another method, due to Chu et al[17], is available. This method, which we call the Strutinsky-integral method, is based on a direct application of the Strutinsky theorem (see, for example, Eq.(2.131) of ref.18), from which it is easy to show that

$$\widetilde{\sum_{\mu}{}' \varepsilon_{\mu}} = \sum_{q} \int d^3\hat{r}\,(\tilde{\tau}_q + \tilde{\rho}_q\, U_q + \tilde{\hat{J}}_q \cdot \hat{W}_q) + O\big((\delta\rho)^2, (\delta J)^2\big) \tag{10}$$

where $\hat{\tau}_q$, $\hat{\rho}_q$, and $\tilde{\hat{J}}_q$ are the smoothed densities that emerge from the minimization in the macroscopic part of the calculation, while U_q and $\tilde{W}_q$ are the corresponding fields, appearing in Eq.(7). The quantity $\delta\rho$ appearing in the error term represents the difference between the smooth density $\tilde{\rho}$ and the fluctuating density ρ corresponding to the s.p. states ϕ of the fields U_q and $\tilde{W}_q$ that are themselves generated by $\tilde{\rho}_q$ and $\tilde{\hat{J}}_q$; $\delta\hat{J}$ has a similar interpretation.

This SI prescription for shell corrections is very simple to apply and quite unambiguous, even at the drip lines. (It should be realized that this method can be used only when the distributions ρ_q and $\tilde{J}_q$ giving rise to the s.p. fields are known; this, of course, is the case here, but not with drop (-let) models.)

Executing this method requires that the s.p. field resulting from the macroscopic part of the calculation be diagonalized, in order to get the s.p. energies appearing in Eq.(9). Thus this method of calculating shell corrections may be regarded as doing one iteration of HF on top of ETF. Since this is by far the most time-consuming part of the whole calculation we see that the ETFSI method is an order of magnitude faster than HF.

Pairing. We handle pairing by doing BCS with a δ-function force,

$$v_{ij} = V_p \delta(\underset{\sim}{r}_{ij}) \quad , \tag{11}$$

this having been found to give a good fit to even-odd mass differences over the entire known region of the nuclear chart with a unique value of V_p, and without having to introduce any phenomenological I-dependence. The modification of the foregoing formalism necessitated by the introduction of BCS is completely standard: see ref.12 for details.

3. Comparison with HF+BCS

	t_o (MeV·fm^3)	t_2 (MeV·fm^5)	t_3 (MeV·fm^4)	x_o	x_2	x_3	W_o (MeV·fm^5)	V_p (MeV·fm^3)
HF	-1783·86	-86·0000	12696·3	0·400000	1·0	0·587228	107·000	-210·0
ETFSI	-1787·58	-93·4082	12726·9	0·400073	1·0	0·584416	113·716	-213·0

Table 1. Force parameters

To compare our method with the standard HF+BCS method we first perform HF+BCS calculations on 69 spherical nuclei and 6 deformed nuclei, all lying in the known region of the nuclear chart; we also calculate the fission barrier of ^{240}Pu. The force used is Skyrme-type, with $M_q^* = M_q$, and the parameters as given in the first line of Table 1. This force gives a good fit to the data, but this is not really important for the purposes of comparison of the HF and ETFSI methods.

When these same nuclei are calculated with the same force in the ETFSI approximation (with BCS) an overbinding of between 3 and 7 MeV results. This is characteristic of the ETF method, and in fact because of the Fermi form (6) adopted for the distributions the errors are a little smaller than might have been expected.

Now while errors this large would be unacceptable in modern mass formulas, it is not inevitable that this will pose a problem in practice, since with whatever method is being used the force will always be fitted to the data. Thus the crucial question is, when the ETFSI method is extrapolated out towards the n-drip line, will it agree with the HF extrapolation ?

We answer this question by first performing ETFSI+BCS calculations on these same nuclei, refitting the force parameters to the HF+BCS "data"-note that all these nuclei are in the known region of the nuclear chart. The rms error of this fit is 0.62 MeV, while the error in the ^{240}Pu fission barriers is about 0.5 MeV. The new parameters are shown in the second line of Table 1; we may regard the slight shift as a renormalization, absorbing the inherent errors of the ETFSI method, along with those associated with the use of Fermi distributions. (The fact that the renormalized force gives the same quality of fit to spherical and deformed nuclei means that there is a negligible deformation dependence of the error in the ETF functionals, and likewise negligible error arising from the u-Ansatz of BGH.)

We then use the two methods, each with its own interaction, to calculate 33 spherical nuclei and 1 deformed nucleus (^{262}U) lying close to the n-drip line. The rms discrepancy between the HF and ETFSI extrapolated binding energies is 0.93 MeV. As for S_n and Q_β, quantities of greater astrophysical interest, the rms discrepancy is only 0.16 MeV, which is of the order of the numerical accuracy of our calculations. The fission barriers of ^{262}U are calculated with the same accuracy.

We thus conclude that while the ETFSI method is an order of magnitude faster than the HF method it gives essentially the same extrapolations from known to unknown nuclei, for a given form of force.

4. Discussion

We are fitting our force parameters to the 1986 nuclear-mass compilation of Audi and Wapstra,[19] and require also a good overall fit to measured radii. Our best fit so far gives an r.m.s. error of 0.74 MeV to 578 spherical nuclei, including many recently measured highly unstable cases, with comparable results for 10 deformed nuclei. The second fission barrier of ^{240}Pu is about 1.2 MeV too low.

The corresponding r.m.s. errors in S_n and Q_β are 0.53 MeV and 0.77 MeV, respectively. These are considerably lower than the value of $\sqrt{2} \times 0.74 = 1.05$ MeV that we would have expected had the errors in the absolute binding energies been randomly distributed.

For the final fit of our force we intend to include several more fission barriers and deformed nuclei. Once our final force has been determined the construction of a complete mass table will be able to proceed via interpolation. This is possible with the ETFSI method, but not with HF, and in this way we expect to gain another two orders of magnitude in computer time, as compared to HF. The preliminary results reported here indicate that our final mass table will give at least as good a fit to the data as the DM mass formulas, and at the same time have the advantage of being much better based theoretically.

1. W.D. Myers, Atomic Data and Nuclear Data Tables 17 411 (1976)
2. H. von Groote, E.R. Hilf and K. Takahashi, Atomic Data and Nuclear Data Tables 17 418 (1976)
3. P. Möller and J.R. Nix, Nucl. Phys. A361 117 (1981)
4. P. Möller and J.R. Nix, Atomic Data and Nuclear Data Tables 26 165 (1981)
5. P. Möller and J.R. Nix, Atomic Data and Nuclear Data Tables (submitted 1986)
6. W.D. Myers and W.J. Swiatecki, Nucl. Phys. 81 1 (1966)
7. J.M. Pearson, Nucl. Phys. A376 501 (1982)
8. F. Tondeur, J.M. Pearson and M. Farine, Nucl. Phys. A394 462 (1983)
9. M. Brack, C. Guet and H.-B. Håkansson, Physics Reports 123 277 (1985)
10. P. Möller, W.D. Myers, W.J. Swiatecki and J. Treiner, Proc. 7th Int. Conf. on Atomic Masses and Fundamental Constants, AMCO-7, ed. O. Klepper (THD, Darmstadt 1985) p.457; Atomic Data and Nuclear Data Tables (to appear 1987)
11. A.K. Dutta, J.-P. Arcoragi, J.M. Pearson, R. Behrman and M. Farine, Nucl. Phys. A454 374 (1986)
12. A.K. Dutta, J.-P. Arcoragi, J.M. Pearson, R. Behrman and F. Tondeur, Nucl. Phys. A458 77 (1986)
13. F. Tondeur, A.K. Dutta, J.M. Pearson and R. Behrman, Nucl. Phys. (to appear 1987)
14. B. Grammaticos and A. Voros, Ann. Phys. 123 359 (1979)
15. B. Grammaticos and A. Voros, Ann. Phys. 129 153 (1980)
16. M. Brack and P. Quentin, Proc. Int. Conf. on Nuclear Self-Consistent Fields, Trieste (1975), ed. G. Ripka and M. Porneuf (North-Holland, Amsterdam, 1975) p.353
17. Y.H. Chu, B.K. Jennings, and M. Brack, Phys. Lett. 68B 407 (1977)
18. P. Ring and P. Schuck, The Nuclear Many-Body Problem (Springer-Verlag, New York, 1980)
19. G. Audi and A.H. Wapstra, Atomic Data and Nuclear Data Tables (to appear 1987)

---

* "Permanent" address: Laboratoire de Physique Nucléaire, Université de Montréal, Montréal H3C 3J7 Canada.

Work partially supported by NSERC of Canada

NUCLEAR-MATTER COMPRESSIBILITY FROM LOW-ENERGY NUCLEAR PHYSICS

M.M. Sharma*
Sektion Physik, Universität München, 8046 Garching, West Germany

INTRODUCTION

There has recently been much discussion on nuclear-matter compressibility. The knowledge of the nuclear-matter compressibility is important in the study of the behaviour of matter in neutron-stars and in supernova explosions as it is closely connected to the equation of state of nuclei in extreme excitations and of steller objects. In this talk, I will give a brief outline of the recent developments in the subject of nuclear compressibility and will discuss a method experimentalists adopt in Nuclear Physics to get insight into this fundamental property of matter. But first I summarize the definitions and the nomenclature involved.

NUCLEAR MATTER

In the liquid drop model, the binding energy of a nucleus is given by the mass formula which consists of volume, surface, Coulomb and asymmetry terms. If we neglect the Coulomb force between protons and assume the nucleus to be very large (neglecting the surface energy which goes as $A^{-1/3}$), a mixture of an equal number of protons and neutrons forms, in the limit of large A, a system known as nuclear matter. Such a system is hypothetical and does not exist in reality. The matter that is found in heavy nuclei and in steller objects is highly asymmetric with large excess of neutrons. We are thus interested in studying the properties and behaviour of the asymmetric nuclear-matter.

EQUATION OF STATE OF NUCLEAR MATTER

The properties and behaviour of nuclear matter is one of the fundamental problems in Nuclear Physics. This is described by the equation of state which expresses pressure in terms of density and temperature. It can be calculated from the total energy E of the interacting system. For a system of fermions with a short-range repulsion and a long-range attraction, it can be written as a sum of an interaction term and a thermal term:

$$E = E_{int} + E_{thermal} \tag{1}$$

At T=0 (zero temperature) the system saturates at an equilibrium density ρ_0 and has an equilibrium energy ϵ=E/N per particle. In nuclear matter at equilibrium these quantities are known to be:

$$E/N = -16 \text{ MeV} \quad \text{and} \quad \rho_0 = 0.15 \text{ fm}^{-3} \tag{2}$$

In the vicinity of the equilibrium state, the interaction energy can be expanded as a quadratic of the density:

$$E_{int} = E_0 + K/_{18} \cdot (\rho/\rho_0 - 1)^2 \tag{3}$$

where K is the compressibility defined as

$$K = 9\rho_0{}^2 . d^2E/_{d\rho}2\big|_{\rho=\rho_0} \tag{4}$$

The pressure P is related to the energy density by

$$P \;=\; -(\partial E/\partial V)_T = \rho^2(\partial E/\partial \rho)_T \tag{5}$$

The equation of state for nuclear matter at about zero temperature is obtained by evaluating the derivative (Eq. 5), which from Eq. 3 can be written as:

$$P \;=\; K/_9 . (\rho/\rho_0)^2(\rho-\rho_0) \;=\; P(\rho) \tag{6}$$

Here P is only the interaction pressure and the thermal pressure is neglegible

The equation of state P(ρ) is of great interest for possible occurence of phase-transition in nuclear matter at high densities and temperatures. The equation of state for steller matter at higher densities decides the mass and radius of a steller configuration in hydrodynamical equilibrium under various forces. In this discussion, the parameter K, the compressibility of nuclear matter plays a crucial role.

NUCLEAR COMPRESSIBILITY

The nuclear-matter compressibility defined by Eq. 4, when evaluated at the nuclear-matter saturation density, reduces to:

$$K_\infty = k_F{}^2 \; d^2(E/A)/dk_F{}^2 \tag{7}$$

where k_F is the Fermi momentum, and K_∞ refers to infinite nuclear-matter. The microscopic theory [1] of Landau-Fermi liquids leads to the expression for the compressibility as:

$$K_\infty = 6\epsilon_F(1 + F_0) \quad (8)$$

where $\epsilon_F = \hbar^2 k_F^2/2m*$ is the Fermi energy and F_0 is the correction parameter due to interactions and is referred to as Landau parameter. For an isotropic system, the effective mass m* is given in terms of nucleon mass

$$1/m = 1/m* \;(1 + F_1/3) \quad (9)$$

where F_1 is another Landau parameter. Equation 8 can be rewritten as:

$$K_\infty = 6.T_F(1 + F_0)/(1 + F_1/3) \quad (10)$$

Here T_F is the kinetic energy of a free nucleon having momentum k_F. It is obvious from Eq. (10) that the knowledge of the parameter K is linked closely to the understanding of the interactions at play in Landau-Fermi liquids and in nuclei.

Since we deal with finite nuclei in the laboratory, the compressibility of a finite nucleus is defined as curvature of binding energy per nucleon (analogous to Eq. 4) with respect to the radial coordinate at saturation:

$$K_A = r^2 d^2(E/A)/dr^2|_{r=r_o} \quad (11)$$

Giant monopole resonance (GMR), which corresponds to the compressional mode (nucleus undergoes density oscillations) of collective excitation of nucleus, is an important source of information on K_A. The frequency of vibration of this mode (also known as breathing mode) is related to K_A in a simple way in the liquid drop model as:

$$E_0 = \hbar\omega = \hbar\sqrt{\{K_A/(m<r_o^2>)\}} \quad (12)$$

Thus, information on K_A can be obtained from the experimental study of the GMR in nuclei. However, as we are interested more in K_∞, it is required to extrapolate K_A to obtain K_∞. This is achieved using the semi-empirical mass formula [2]. Taking the second derivative of E/A, we have:

$$K_A = K_\infty + K_S A^{-1/3} + K_\tau\{(N-Z)/A\}^2 + K_c Z^2 A^{-4/3} \quad (13)$$

where K_∞, K_S, K_τ and K_c are the second derivatives of coefficients of volume, surface, neutron-excess and Coulomb terms, respectively, with respect to the radial coordinate of the nucleus. The Coulomb term, K_c, can be obtained analytically [3]. Blaizot [4] has modified Eq. 13 taking into account the equilibrium condition of the nuclear ground-state as:

$$K_A = K_\infty + K_{S'}A^{-1/3} + K_{\Sigma'}\{(N-Z)/A\}^2 + K_C Z^2 A^{-4/3} \quad (14)$$

where K_C is now different from K_c of Eq. (13), and $K_{S'}$ and $K_{\Sigma'}$ are no more simply the second derivatives as in Eq. 13. The parameters K_∞, $K_{S'}$ and $K_{\Sigma'}$, which are unknowns in Eq. 14 are determined by a 3-parameter fit to the experimental GMR energies. This method has so far been the only source of experimental information on nuclear compressibility.

PRESENT STATUS ON COMPRESSIBILITY

Experimental data on the isoscalar GMR over a large number of nuclei have been used to fit Eq. 4 to obtain the value of various coefficients contributing to K_A. However, due to large variances in the available systematics, various values of K_∞ have been obtained. The values of $K_{S'}$ and $K_{\Sigma'}$ from various groups lie between -400 and -600 MeV and -250 and -600 MeV respectively. The value of $K_{\Sigma'}$, the asymmetry parameter, from a fit on various nuclei including ^{24}Mg, has been obtained [5] as (-285 ± 448) MeV. Thus, $K_{\Sigma'}$, which is important for the compressibility of asymmetric nuclear-matter, is poorly known.

Attempts have been made to obtain the compressibility theoretically. Blaizot and Grammaticos [6] have performed self-consistent RPA calculations on ^{16}O, ^{40}Ca, ^{90}Zr and ^{208}Pb nuclei employing various phenomenological Skyrme interactions. The RPA calculations reproduce the ground-state properties besides giving the compressibility of the nucleus and the nuclear matter. The results of the calculations for various interactions, taken from Ref. 4, are shown in Fig. 1. The resonance energies

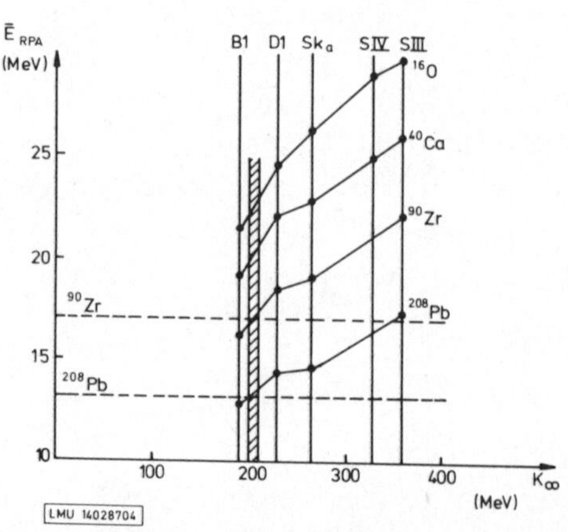

Fig. 1. The energy of the breathing mode vibration and the nuclear compressibility calculated using various interactions (Ref. 4)

calculated for the monopole operator are plotted against K_∞ obtained for different interactions for closed-shell nuclei. In a comparison of the ^{208}Pb calculations with the experimentally known results on the GMR, Blaizot [4] has extrapolated the value $K_\infty = 210 \pm 30$ MeV as shown by the hatched area in Fig. 1. This is presently known as the 'commonly accepted' value. However, this value is based upon comparison of only one nucleus with experiments. For a better comparison with experimental results, which are now available on many nuclei, the RPA calculations are required also for open-shell nuclei.

Brown and Osnes [7] using Landau-Fermi liquid theory have determined $K_\infty \leq 106$ MeV, employing an effective mass $m^* = 0.9$ m. Ainsworth et al [8] have recently modified these results [7] and obtain an equation of state with $K_\infty \simeq 190$ MeV (for symmetric matter). It may be noted that in a work on hydrodynamic simulation of Type II supernova, a value of $K_\infty \leq 180$ MeV has been required for prompt explosion [9]. This requirement is very close to the results of [8]. The experimental indications from the GMR from various groups (for a review see Ref. 12) for K_∞ are between 200 and 300 MeV. The divergence in the value of K_∞ is due to presence of large continuum background in the excitation spectrum, which results in large errors in the GMR parameters. With the present confusion on nuclear compressibility in mind, we have performed an experiment [10] which we discuss below.

THE PRESENT EXPERIMENT

With a view to pin down the compressibility parameters in Eq. 14, we have studied inelastic scattering of 120 MeV α-particles on isotopic

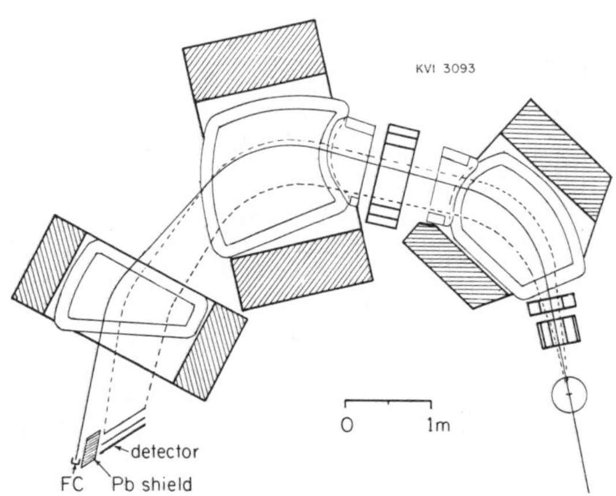

Fig. 2 The QMG/2 spectrograph

chains of Sn and Sm nuclei. The α-particle beam was obtained from the AVF cyclotron at KVI, Groningen. The targets used were self-supporting with thickness between 0.75 and 1.5 mg/cm^2. Since The GMR cross-section is maximum at 0°, it is necessary to perform the experiment at 0°. This requires the use of a spectrograph to separate the inelastically scattered α-particles from the outgoing beam. The QMG/2 spectrograph [11] at KVI, as shown in Fig. 2, was used in the present experiment. The spectrograph was set at 0°, with a full horizontal opening angle $\Delta\theta = 6°$ (-3° to +3°) and a vertical opening angle of $\Delta\theta = 4°$ (-2° to +2°). The α-particles were detected by a 52 cm long detection system [12]. The excitation energy covered by the spectrograph was 10-20 MeV.

The systematics of the GMR in Ref. 13 for Sn and Sm isotopes show that the results obtained at various laboratories differ from each other. These differences stem from different experimental background present and the different probes used. Also, systematics of the GMR in most of the works reviewed in [13] have been obtained by analysing the spectra containing large amount of underlying background and a significant GQR strength. In analysing our data we have used a method of difference spectrum, which helps throw out a substantial amount of the instrumental and the continuum backgrounds as well as the GQR component. This leads to a large reduction in the uncertainties associated with the systematics and thereby increases the reliability of the compressibility parameters obtained.

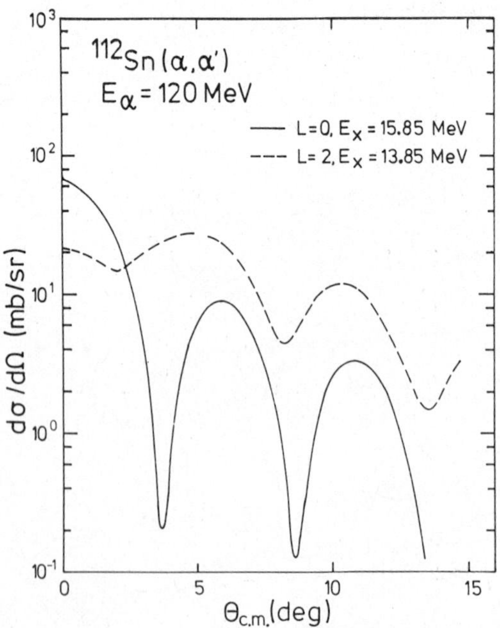

Fig. 3. The angular distribution of the GMR (L=0) and GQR (L=2).

DATA ANALYSIS AND RESULTS

The angular distribution of the GMR and GQR calculated using the distorted wave Born approximation (DWBA) for ^{112}Sn exhausting the full energy-weighted sum-rule (EWSR) are shown in Fig. 3. In the 0°-3° angular range covered in this experiment, the GMR has a characteristic angular distribution which peaks at 0° and drops by two orders of magnitude at 3°. On the other hand, the GQR which is the only other predominant mode of collective excitation in the excitation energy covered, has an almost flat angular distribution in this angular range. The angular range (0°-3°) has been software divided into two equal bins and the corresponding excitation-energy spectra have been obtained after eliminating most of the instrumental background. The spectra for ^{112}S are shown in Figure 4. The resonance bump in all spectra has been fitted to two Gaussians corresponding to the GMR and GQR by the method of minimization of chi squares. The underlying background is shown by dashed curve in all spectra. It can be seen that in Figs. 4 (b) and (c) the GMR and GQR peaks, respectively, are dominant, clearly in conformity with Fig. 3. The GQR parameters have been obtained from Fig. 4 c), which has a small GMR strength. The spectrum (d) has been obtained by subtracting the spectrum (c) from (b). It carries only a small GQR strength after subtraction as compared with a major contribution from the GMR. The parameters of the GMR have been obtained from this spectrum which has only a residual amount of the underlying continuum. Consequently, the method leads to an accurate determination of

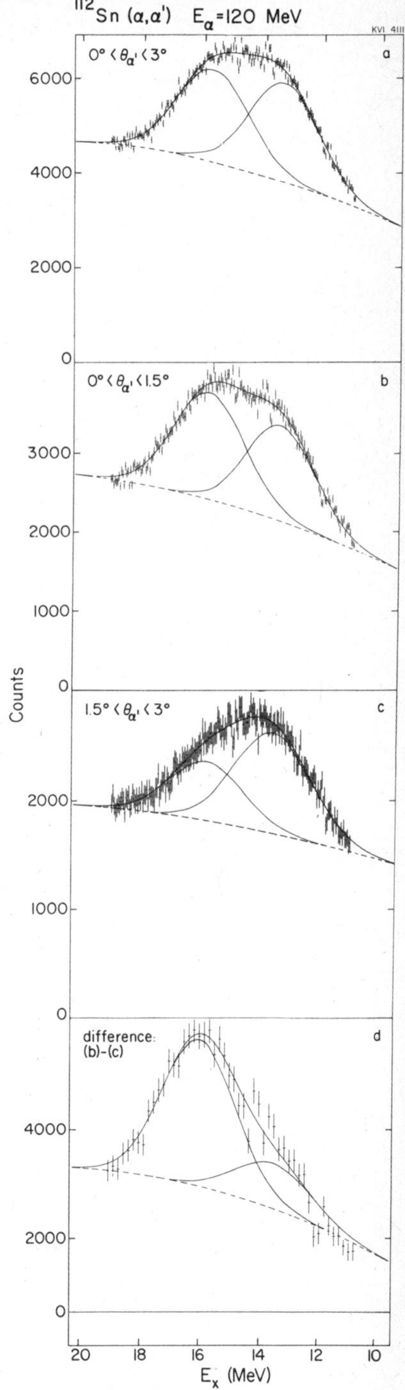

Fig. 4. The excitation energy spectra

the systematics of the resonance concerned. In this method, the consistency of the GMR and GQR parameters for all the spectra (a) to (d) has been observed. The ratios of the intensity of the GMR and the GQR for the two angular bins have been found to be in good agreement with the DWBA estimates.

The systematics of the GMR are shown in Fig.5. The centroid energy variation of the GMR is smooth, as shown by dots with error-bars which are roughly 80-100 keV, compared with the uncertainties of 200-400 keV in previous works on the GMR reviewed in [13]. The data from Grenoble [13]

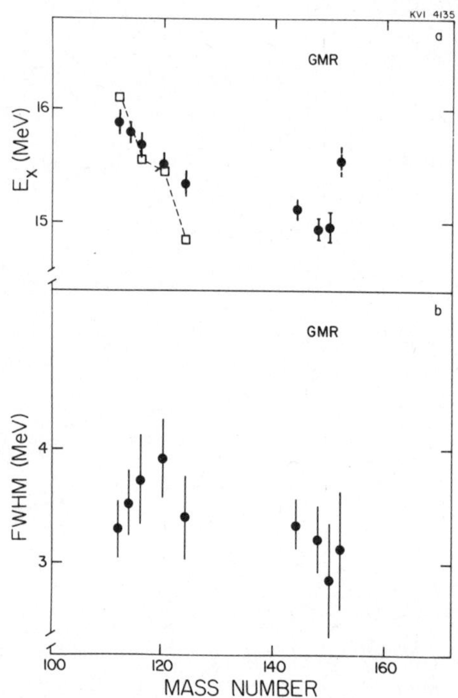

Fig. 5. The systematics of the GMR in Sn and Sm nuclei.

on the GMR of Sn nuclei is also shown by squares for sake of comparison. The dashed lines joining these points where the errors in energies are ±250 keV (not shown in figure) show a contrasting difference with our data. It may be noted that these data points have been obtained by analysing a spectrum similar to Fig. 4 (a) where full GQR strength and a huge continuum background are present. The systematics of the GQR have been found to vary smoothly with mass number [10], in agreement with the hydrodynamical model. The results on ^{150}Sm and ^{152}Sm are also shown in Fig. 5. Since these nuclei are deformed and have not been included in the fits for compressibility, they are not relevant for this discussion. While the GMR in all nuclei has been found to exhaust the full E0 EWSR, the GQR exhibits a strength a little more than the full E2 sum-rule. This may possibly be due to presence of some fraction of the L=4 ($2\hbar\omega$) resonance at the position of the GQR. The width of the GMR exhibits a behaviour somewhat like shell-effects. A similar effect has recently been found theoretically by Di Toro et al. [14], in the study of the effect of the ground state deformation on the width of the isovector giant dipole resonance in Sn nuclei.

THE FITTING OF DATA

The experimental centroid energies on Sn and Sm (A=144 and 148) nuclei have been employed to fit Eq. 14. However, the centroid energies used in calculating K_A in Eq. 12 are modified in the scaling model. According to Treiner et al. [14], by taking into account the width (Γ) of a Gaussian strength distribution E_x changes to $E_{x'}$ as:

$$E_{x'}{}^2 = E_x{}^2 + 3(\Gamma/2.35)^2 \qquad (15)$$

Using $E_{x'}$ in place of E_x in Eq. 12, a 3-parameter fit has been performed to obtain K_∞, $K_{S'}$ and $K_{\Sigma'}$. Since light nuclei show strong surface effects, we have included ^{24}Mg also in the fits. The results of fits on various sets of nuclei are shown in Table 1. Set (a) corresponds to the nuclei

TABLE 1

The nuclear compressibility parameters derived from fits on the GMR energies on various nuclei. The sets (a), (b) and (c) correspond to the present data and subsequent inclusion of ^{208}Pb and ^{24}Mg, respectively. The numbers in the parentheses denote the total number of nuclei included in the fit.

Sets of nuclei			Parameters obtained from fits		
			K_∞ (MeV)	$K_{S'}$ (MeV)	$K_{\Sigma'}$ (MeV)
(a)	Sn + Sm	= (7)	293 ± 12	-642 ± 55	-340 + 100
(b)	Sn + Sm + ^{208}Pb	= (8)	294 ± 15	-647 ± 69	-339 + 111
(c)	Sn + Sm + ^{208}Pb + ^{24}Mg	= (9)	273 ± 20	-551 ± 70	-302 + 118

studied in the present work. The K_∞ obtained is ~290 MeV and the $K_{S'}$ and $K_{\Sigma'}$ are -642 ± 55 MeV and -342 ± 100 MeV, respectively. Set (b) includes ^{208}Pb besides Sn and Sm nuclei. It can be noticed that the inclusion of a very heavy closed-shell nucleus like ^{208}Pb does not change the results of Set (a). In Set (c) ^{24}Mg has also been included. The results of Set (c) show a little change, though not far outside the error bars. Since for a light nucleus, the curvature term (which goes as $A^{-2/3}$) is important, we put the curvature term in Eq. 14 with a curvature coefficient about -300 MeV taken from Ref. 14. The results of the fit on Set (c) with curvature correction are much closer to the results of Sets (a) and (b) and the K_∞ obtained is ~290 MeV. This shows that with all important terms taken into account in Eq. 14, our Sn and Sm data fall in line with ^{208}Pb and ^{24}Mg and yield the nuclear compressibility parameters as:

$$\begin{aligned} K_\infty &= 290 \pm 20 \text{ MeV} \\ K_{S'} &= -640 \pm 70 \text{ MeV} \\ K_{\Sigma'} &= -335 \pm 100 \text{ MeV} \end{aligned} \tag{16}$$

CONCLUSION AND DISCUSSION

The value of the nuclear-matter compressibility according to our data is about 290 MeV, and is higher than the accepted value 210 ± 30 MeV [4]. The surface term comes out as -640 ± 60 MeV and the asymmetry parameter has been determined as -340 ± 100 MeV, as compared to the previous value -285 ± 448 MeV [5]. It is clear that we have been able to determine the asymmetry parameter with much higher accuracy than previously known [5].

The higher value of K_∞ implies that the equation of state of nuclear matter is stiffer than usually assumed. There have recently been some other indications in favour of this. Co' and Speth [16] have pointed out that the polarization of density distribution obtained from electron scattering data can also be used as a measure of the nuclear compressibility. By analysing charge-density difference between ^{208}Pb and ^{206}Pb and between ^{208}Pb and ^{207}Pb, Co' and Speth [16] have indicated that the experimental data show a compatibility to K_∞ higher than about 250 MeV. Clearly, our results [10,17] show an agreement with this analysis and both our results and the results of Ref. 16 seem to corroborate each other though the source of compressibility is different in each case.

Another indication for higher value of compressibility comes from the work of Glendenning [18] on relativistic mean-field theory calculations on neutron-stars. It has been observed that a value of $K_\infty \simeq 285$ MeV is consistent with the observed neutron-star masses. This work also shows a good agreement with our results, and supports a higher value of the nuclear-matter compressibility.

Acknowledgements

I thank my colleagues W.T.A. Borghols, S. Brandenburg, S. Crona, M.N. Harakeh, J. Meier and A. van der Woude of KVI Groningen who have collaborated on this work. I especially thank Prof. A. van der Woude and Prof. M.N. Harakeh for numerous discussions with them. I also thank Prof. J. de Boer for constant encouragement and support at the Sektion Physik, Universität München, where this work has been summarized. This work has been performed as part of research program of the Stichting voor Fundamenteel Onderzoek der Materie (FOM), the Netherlands.

* Fellow of the Alexander von Humboldt Foundation, West Germany.

REFERENCES

[1] A.B. Migdal, Theory of Finite Fermi Systems & Applications to Atomic Nuclei (Interscience, London, 1967).

[2] W.D. Myers and W.J. Swiatecki, Ann. Phys. (New York) 84 (1984) 211.

[3] V.R. Pandharipande, Phys. Lett. 31B (1970) 635.

[4] J.P. Blaizot, Phys. Rep. 64 (1980) 171.

[5] H.J. Lu, S. Brandenburg, R. de Leo, M.N. Harakeh, T.D. Peolhekken and A. van der Woude, Phys. Rev. C33 (1986) 1116.

[6] J.P. Blaizot and B. Grammaticos, Nucl. Phys. A355 (1981) 115.

[7] G.E. Brown and E. Osnes, Phys. Lett. 159B (1985) 223.

[8] T.L. Ainsworth, E. Baron, G.E. Brown, J. Cooperstein and M. Prakash, Nucl. Phys. A464 (1987) 740.

[9] E. Baron, J. Cooperstein and S. Kahana, Phys. Rev. Lett. 55 (1985) 126.

[10] M.M. Sharma, W.T.A. Borghols, S. Brandenburg, S. Crona, M.N. Harakeh J. Meier and A. van der Woude, to be published.

[11] A.G. Drentje, H.A. Enge and S.B. Kowalski, Nucl. Instr. Meth. 122 (1974) 485.

[12] J.C. Vermeulen, J. van der Plicht, A.G. Drentje, L.W. Put and J. van der Driel, Nucl. Instr. Meth. 180 (1981) 93.

[13] M. Buenerd, in Proc. of Int. Symp. om Highly Excited States and Nucl. Structure, Orsay, 1982, J. Phys. (Paris) C45 (1984) C4-115.

[14] M. Di Toro. M. Pisa and G. Russo, Phys. Rev. C34 (1986) 2334.

[15] J. Treiner, H. Krivine, O. Bohigas and J. Martroll, Nucl. Phys. A372 (1982) 253.

[16] G. Co' and J. Speth, Phys. Rev. Lett. 57 (1986) 547.

[17] A. van der Woude, W.T.A. Borghols, S. Brandenburg, M.M. Sharma and M.N. Harakeh, Phys. Rev. Lett. 58 (1987) 2383.

[18] N.K. Glendenning, Z. Phys. A326 (1987) 57.

II. Stellar Evolution, Nucleosynthesis and Isotopic Anomalies in Meteorites

EARLY NUCLEOSYNTHESIS, CHEMICAL EVOLUTION OF GALAXIES AND PARTICLE PHYSICS

Jean AUDOUZE
Institut d'Astrophysique du CNRS - Paris, France
and Laboratoire René Bernas - Orsay, France

ABSTRACT

After a brief review of the most recent works dealing with the primordial abundances of the very light elements (D, ^{3}He, ^{4}He and ^{7}Li) it is argued that the simple models of early nucleosynthesis predicting the existence of three neutrino flavors and a low baryonic density in the Universe imply specific galactic evolution schemes leading to a destruction of deuterium during the galactic history such that its present abundance is at most the tenth of its primordial one. Several models of chemical evolution of the Galaxy have been designed for that purpose. It has been found that models assuming a star formation rate (SFR) varying in time in two steps (the early epoch $t \leq 1Gyr$ where SFR is assumed to be constant and the rest of the galactic evolution when SFR is proportional to the gas density) are especially well suited to account for such a large deuterium destruction. By contrast, galactic evolution models assuming a bimodal SFR do not lead to such deuterium destruction effect.

Finally several very recent models attempting to account for the synthesis of the very light nuclear species in models such that $\Omega_B = 1$ (Ω_B being the baryonic cosmological parameter) are mentioned especially the recent scenario which proposes that the early nucleosynthesis is affected by the quark-hadron phase transition.

1. INTRODUCTION

The many implications on particle physics, cosmology, galaxy evolution, ... derived from the study of the early nucleosynthesis responsible for the formation of the lightest nuclear species (D, ^{3}He, ^{4}He and ^{7}Li) are described in many reviews (see e.g. references 1 and 2). In this short contribution we would like to focus our attention on four different issues (i) to provide the reader with the most recent developments regarding the determination of the primordial abundances of these elements (section 2), (ii) to discuss the simplest (canonical) models of early nucleosynthesis at the light of the primordial abundances (section 3), (iii) to review the most recent specific models of galactic evolution leading to a thorough destruction of D without an overproduction of ^{3}He. Special attention is devoted to the analysis of galactic evolution models with varying star formation rates such as those considered in references 3 and 4 (section 4), (iv) to mention the few cosmological scenarios taking into account recent developments of particle physics and assuming that the early nucleosynthesis, is able to account for the primordial abundances of the very light elements even if $\Omega_B \simeq 1$, i.e. with a baryonic density ρ_B corresponding to a de Sitter-Einstein Universe. In section 5 we point out specifically the models developed e.g. in references 5 and 6 assuming that the early nucleosynthesis could be favoured by the physical conditions derived from the quark-hadron phase transition. Our conclusion regarding the present value of the simplest models of early nucleosynthesis with respect to galactic evolution and/or particle physics considerations is developed in section 6.

2. RECENT DEVELOPMENTS REGARDING THE PRIMORDIAL ABUNDANCES OF D, ^{3}He, ^{4}He AND ^{7}Li

In this section we concentrate our attention mainly on the most recent contributions written since 1987 regarding these primordial abundances (see e.g. references 1 and 2 to complement this too brief discussion).

2.1 Deuterium

Reference 7 constitutes the most complete reference to-date regarding this element. One should mention the important observation of the D/H abundance performed by analyzing the absorption lines of a Z = 3 quasar (reference 8), the D/H ratio has found to be (4^{+4}_{-2}) 10^{-5} in the absorption system of Q 0420 -388 (Z = 3.08571). Although this D measurement is not free from possible confusion with a high velocity H cloud further D abundance determinations concerning such primitive (low metallicity Z) systems are badly needed and should be undertaken with the highest priority.

2.2 ^{3}He

R.T. Rood and his associates T.M. Bania and T.L. Wilson have persued their analysis of the $^3He^+/H$ abundance ratio in H II regions determined from the observations of the 3.46 cm radio line of this specific ion. Their most recent work (reference 9) indicates clearly that the maximum ^{3}He/H abundance ratio found in the present interstellar medium is 1.5 10^{-4} instead of 4 10^{-4} as stated in their previous analysis (reference 10).

2.3 Helium 4 (mass abundance ratio Y)

As discussed in reference 1, most authors (see e.g. references 11 and 12) quote $Y_p = 0.24 \pm 0.01$. Most recently a correlation between Y and the carbon abundance such that

$$Y_p = (0.235 \pm 0.012) + (0.02 \pm 0.01)(10^4 y_c)$$

has been proposed in reference 13 and agrees quite well with the Y, N/H correlation

$$Y_p = (0.238 \pm 0.005) + (2.9 \pm 1.5)10^3 (N/H)$$

established in reference 14. As noted in reference 13, ^{4}He, C and N are not produced in very massive stars : this is why these correlations could be more meaningful than the previous attempts to relate Y_p with the O abundances.

However, as emphasized by L. Vigroux (private communication), it is not appropriate to attempt to deduce a primordial He abundance by using together the determinations coming from blue compact galaxies and from galactic H II region which are not governed by similar physical conditions. Reference 13 is in default in that respect as many previous analyses.

Regarding the Solar abundance, it is worth mentioning the recent Y(Solar) value determined in the Uranus atmosphere by the IRAS Voyager experiment such that Y(Uranus) = 0.267± 0.048 (reference 15). This confirms (i) that ^{4}He is not fractionated in the Uranus atmosphere contrary to the Jupiter and Saturn cases, (ii) that Y(Solar) ~ 0.27-0.28.

2.4 Lithium 7

The early discovery by Spite and Spite (reference 16) that the primordial ^{7}Li/H abundance ratio could be as low as $\sim 10^{-10}$ is now confirmed by several different groups (references 17 and 18). The only difference between the different groups being in the interpretation of the data: Spite and Spite (reference 16) argue that the population II Li/H abundance is close to the primordial one while Hobbs and Duncan (reference 17) consider it only as a lower limit.

Table 1

Abundances of the lightest elements D, ^{3}He, ^{4}He and ^{7}Li

(Age(Gyr))	≈ 15	13–15	4,6	0
D	$3\,10^{-5}$ -10^{-4}	(2-8)10^{-5} [8]	(1-3)10^{-5} [7]	(0.7-1.3)10^{-5} [7]
^{3}He	$3\,10^{-5}$ -10^{-4}		(1-3)10^{-5}	<$2\,10^{-5}$-1.5 10^{-4} [9]
D+^{3}He	(0.6-2)10^{-4}		(2-4)10^{-5}	(0.1-1.5)10^{-4}
Y* (^{4}He)	0.24±0.01 [11]	(0.24±0.01)+AZ [13,14]	0.24-0.28 [15]	0.24-0.30
^{7}Li	(0.7-1.3)10^{-10} [16] (2-9)10^{-10} [17]	(0.8-2)10^{-10}	10^{-9}	10^{-9}

* by mass - all the other abundances are given by number.

[ ] designates the relevant reference.

3. SOME REMARKS ON THE SIMPLE (CANONICAL) MODEL OF EARLY NUCLEOSYNTHESIS

The canonical (simplest) models of early nucleosynthesis have been so often reviewed (see e.g. reference 1) that we will only here present a few comments about them.

The consequences of such models which provide limitations on the baryonic density (such that $\Omega_B < 0.2$) and on the maximum number of neutrinos flavors ($N_\nu \leq$ 3-4) are well known. The attention of the reader is called on a few recent contributions : (i) The early production of ^{7}Li has been calculated again in references 19 and 20 at the light of new nuclear reaction rates regarding especially T(α,γ) ^{7}Li (higher than previously) and ^{7}Li (p, α) ^{4}He (lower than before). This leads in particular the authors of reference 20 to conclude that

$$2\ 10^{-10} \leq \eta = (\frac{n_B}{n\gamma}) \leq 5\ 10^{-10}$$

i.e. $7\ 10^{-3} < \Omega_B < 7 10^{-2}$ which is a much lower ange value than what has been deduced in a previous references of this group (see e.g. ref 21);

(ii) The interpretation of most recent particle physics experiments provides also a stringent limit on the number of neutrinos flavours consistent with the implication of these early nucleosynthesis schemes (reference 22).

We should emphasize again and may be more strongly then before at the light e.g. of the most recent determinations of Y_p (see e.g. reference 13) the need for an appropriate destruction of D during the galactic history stronger than the one obtained in simplest models of galactic evolution like those of reference 23 in order to concile the predictions of the simple models of early nucleosynthesis and the presently adopted primordial abundances of light elements (see reference 1 for a thorough discussion of this point).

4. MODELS OF GALACTIC EVOLUTION LEADING TO A THOROUGH DESTRUCTION OF D

From the discussion of the previous section we (i.e. the Paris group and also colleagues like B.E.J. Pagel) argue that it is necessary to assume a more thorough D destruction during the galactic history than that coming from the classical models (reference 24). We have also shown (reference 23) that at least two types of specific models of galactic evolution could in principle achieve that goal : (i) those which assume a significant infall/inflow of already processed material; and (ii) those which take into account the stellar mass loss which could occur during the premain sequence phase. This second type of model is somewhat in difficulty after the report by reference 25 of strong limits on the rate of mass loss suffered by T Tauri stars.

I would like to devote most of that section to a recent analysis performed by E. Vangioni–Flam and myself (reference 4) where we have considered models where either the rate of star formation SFR is significantly higher in the past than now or those where the rate of star formation is bimodal. Models with high SFR in the past have already been mentioned in reference 3 while models with bimodal SFR have been proposed by several authors (references 26 and 27) in order to solve many crucial problems such as the classical F-G dwarf problem and the saturation of the metallicity content in the disk. One should recall that the proponents of this type of "bimodal" model have guessed that the D destruction rate should be large.

This is why we have compared the three types of models :

(i) Model I is the standard model with an initial mass function (IMF)

$$\phi(m) \propto m^{(-x+1)}, \quad x = 2 \ (0.4 \leq \frac{m}{M_\odot} \leq 100)$$

and a SFR proportional to the gas density σ.

(ii) Model II with varying SFR assumes that there are two regimes of SFR, such that

$$\psi_1(t) \ = \ \nu_1 \, \sigma \quad for \quad t \ \geq \ \tau Gyr$$

and

$$\psi_2(t) \ = \ \nu_2 \qquad for \quad t \ < \ \tau Gyr$$

and keeps the same IMF as in model I in the figures shown here : $\nu_1 = 0.3$, $\nu_2 = 0.5$ and $\tau = 2$ Gyr.

(iii) Model III follows the prescriptions outlined in reference 27 : the number of stars formed in the internal dm during a time dt is

$$d^2n(m,t) = \left[\phi_1(m)\psi_1(t) + \phi_2(m)\psi_2(t)\right] dmdt$$

with $\phi_1(m)$ and $\psi_2(m)$ having the same slope x but with different low mass limits ϕ_1 ranging from 0.1 to 100 $M_\odot$ and $\psi_1(t) = \nu_1 (\nu_1 = 0.06)$ while ϕ_2 ranges from 2 $M_\odot$ to 100 $M_\odot$ (much higher low limit) to 100 $M_\odot$ and $\psi_2(t) = \nu_2 \, e^{-t/\tau}$ (decreasing rapidly with time such that $\nu_2 = 0.3$ and τ = 2 Gyr.

Table 2 extracted from our paper shows the results obtained for the three models. One notes that model II with time varying SFR is able to fit the present gas density and metallicity, and is able to account for a large D destruction without overproducing ^{3}He. On the contrary Model III with bimodal star formation is unable to account well for the present gas density and metallicity and does not lead to any significant D destruction. The reason is that contrary to an a priori idea contained in most of the papers dealing with bimodal IMF models, a formation rate of massive

Table 2

	Primordial Phase A few minutes		Solar Systems at Birth $t = 8$ Gyr				Today $t = 12.5$ Gyr		
	Models I, II, III	Observations (a)	Model I	Model II	Model III	Observations	Model I	Model II	Model III
$\sigma = \frac{M_{gas}}{M_{total}}$	1		0.16	0.07	0.36	$0.05 \leq \sigma \leq 0.1$ (b)	0.07	0.05	0.13
H	0.76	0.70	0.71	0.70	0.70	~ 0.70	0.69	0.69	0.69
D	10^{-4}	$(3\pm1)10^{-5}$	$5.6\ 10^{-5}$	$2.1\ 10^{-5}$	$5.4\ 10^{-5}$	$5\ 10^{-6} \leq D \lesssim 1.5\ 10^{-5}$(c)	3.10^{-5}	$8.7\ 10^{-6}$	$4.9\ 10^{-5}$
$\frac{D_{prim}}{D_{pres}}$	1						3.3	11	2
^{3}He	$5\ 10^{-5}$	$(4\pm2)10^{-5}$	$4.5\ 10^{-5}$	$6\ 10^{-5}$	$2.8\ 10^{-5}$	$2\text{–}7\ 10^{-5}$ (d)	$5.5\ 10^{-5}$	$7.3\ 10^{-5}$	$2.7\ 10^{-5}$
^{4}He	0.24	0.28	0.27	0.29	0.29	~ 0.28 (e)	0.28	0.295	0.296
^{12}C		$3.8\ 10^{-3}$	$3.3\ 10^{-3}$	$3.9\ 10^{-3}$	$6.1\ 10^{-3}$	$\sim 4.5\ 10^{-3}$ (f)	$4.3\ 10^{-3}$	$4\ 10^{-3}$	$6.7\ 10^{-3}$
^{16}O		$8.5\ 10^{-3}$	$7\ 10^{-3}$	$8.7\ 10^{-3}$	$1.3\ 10^{-2}$	$\sim 10^{-2}$ (f)	$9.4\ 10^{-3}$	$9.4\ 10^{-3}$	$1.4\ 10^{-2}$
Z		0.02	0.018	0.023	0.034	~ 0.02 (f)	0.024	0.025	0.037
$\frac{\Delta Y}{\Delta Z}$							1.7	2.2	1.5

(a) Cameron, 1982

(b) Sanders *et al.*, 1984

(c) Vidal–Madjar, A., 1986

(d) Rood R.T., *et al.*, 1984

(e) Peimbert M., 1986

(f) Meyer J.P., 1985

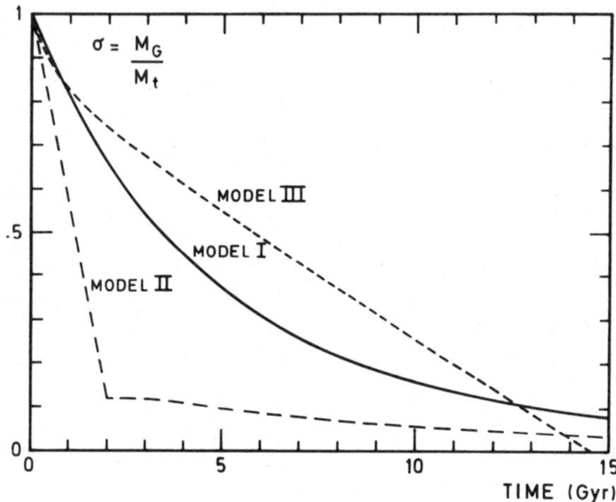

Figure 1. The evolution of the gas density calculated in galactic evolution models I (classical), II (time varying SFR) and III (bimodal SFR) (see text) (from reference 4).

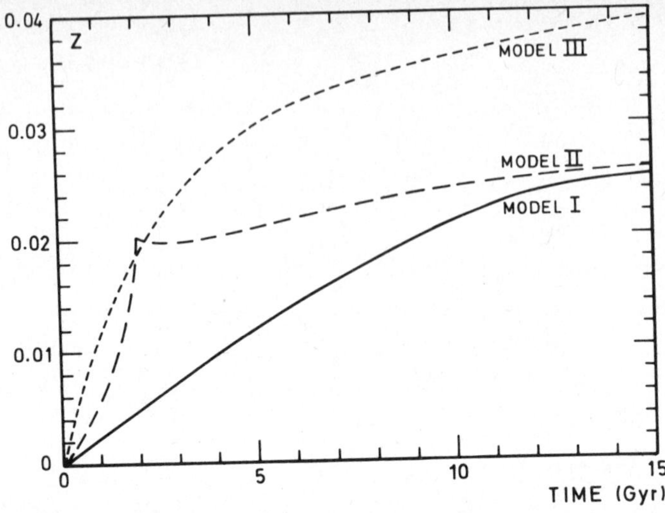

Figure 2. The evolution of the metallicity calculated in models I, II and III (see text).

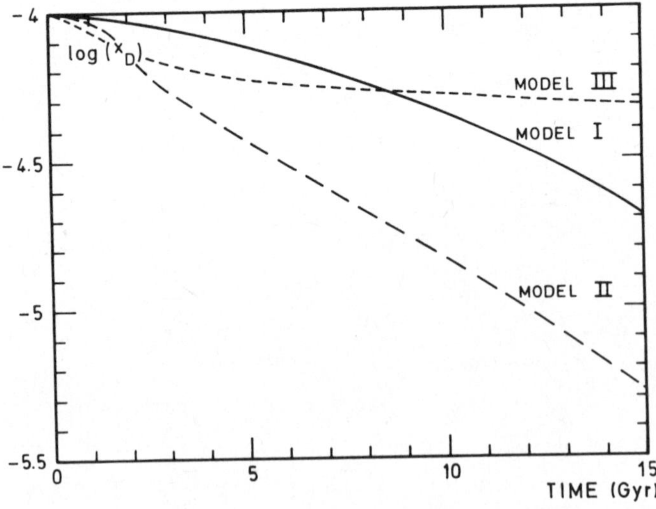

Figure 3. The evolution of the D abundance calculated in models I, II and III (see text).

stars only at the very beginning of the galactic evolution does not imply necessarily a large D destruction. Since D is destroyed at relatively low temperatures, its destruction does not depend on the stellar mass. It is then much better to let D destroyed by low mass stars which do not increase pathologically the metallicity. This is why models with varying SFR (type II) are much more satisfactory in that respect than bimodal models (type III) assuming the existence of a too large population of massive stars.

To end up this very brief presentation of such models we have outlined two conditions in order to account for the evolution of the gas density, the metallicity and the D abundance, (i) the rate of gas processing has to be high at the beginning of the galactic evolution and (ii) this processing should imply stars of <u>any</u> mass and not only massive ones which would overproduce the heavy elements.

In conclusion such models with varying star formation provide a fairly natural way to account for such effects. We have already shown that there are at least two observational tests regarding D destruction over the galactic evolution : either an accumulation of D abundance determinations concerning high redshift QSOs (reference 8) or the determination of such abundances in locations of the galaxy where the gas densities are different (reference 28) : if the D destruction rate is large, one expects to observe large D abundance variations in regions where the gas densities are different. So either we can apply only the most standard models of galactic evolution but then the canonical nucleosynthetic models could be in trouble or we have to call for such galactic models as those with varying SFR and then the early nucleosynthesis schemes become simpler and the consequences on cosmology and particles physics more certain.

5. EARLY NUCLEOSYNTHESIS AND RECENT DEVELOPMENTS IN PARTICLE PHYSICS

In this section I would like just to advertise a few contributions in which new particle physics processes and/or new families of particles have been invoked to alleviate some of the constraints coming from the early nucleosynthesis especially the stringent limit on the baryonic density deduced from the comparison between the primordial abundances of the very light elements and those computed in the frame of the canonical models considered previously.

Let me evoke two types of hypotheses in the first one which has been reviewed at length in references 29 and 30 partial photodesintegration of ^{4}He (and possibly also ^{7}Li) can be higgered by high energy photons like massive neutrinos, gravitinos (reference 29) or photinos (reference 30), in the second hypothesis developed especially in reference 5 and reanalyzed in reference 6, neutron-proton segregation occuring after the quark hadron phase transition could lead to interesting nucleosynthetic consequences.

In reference 5 it is shown that because of the larger neutron mean gree path (compared to the proton one), the existence of neutron rich bubbles just after the quark hadron phase transition would enhance the production of D, ^{3}He, ^{4}He, ^{7}Li and possibly heavier elements even in ($\Omega \geq 1$) Universe. The work reported in reference 6 confirms that the quark-hadron phase transition creates isothermal baryon number density fluctuations (but small). It confirms also the production of D, ^{3}He and ^{4}He in amounts consistent with the observations. Nevertheless ^{7}Li is overproduced (like in any inhomogeneous big Bang model) and there is an insignificant production of heavier elements.

From these analyses it is fair to write that these types of models are much more contrived than the standard models (even with specific requirements on chemical evolution of galaxies).

6. SUMMARY AND CONCLUSION

My present appraisal of the problems related to the early nucleosynthesis and therefore the formation of the lightest elements (D, ^{3}He, ^{4}He and ^{7}Li) is the following :

(i) There is still a lot of work to do concerning the proper determination of primordial abundances of such elements. Although a lot of progress has been achieved recently we are awaiting precise determinations of the primordial He abundances such as that of D and ^{3}He. Concerning D it will be exciting to obtain determination in various galactic locations where the gas densities varies (reference 28) such as for very metal poor and primitive quasars (reference 8). The primordial abundance of ^{3}He will remain a big puzzle for some time even after the very careful analysis reported in reference 9. For ^{7}Li all the different groups agree on its abundance in old population II stars, but there is still a large controversy about the possible destruction of this element during the life time of such stars.

(ii) As B.E.J. Pagel (reference 14), we claim in Paris contrary to the american school (Steigman, Schramm and collaborators) that the important consequences on (i) the present baryonic density (and therefore the influence of baryons on the overall dynamics of the Universe) and (ii) the maximum number of neutrinos (lepton) flavors can only be properly stressed if specific models of galactic models leading to noticeable D destruction can be invoked. In that respect the models such as those studied by E. Vangioni-Flam and myself (reference 4) assuming that the rate of star formation has been more intense during the early phases of the galactic history are especially encouraging.

(iii) Several attemps are currently proposed to build up models of early nucleosynthesis where the present baryonic density would correspond to cosmological parametersas high as $\Omega = 1$. The consequences of the existence of massive neutrinos, gravitinos and photinos have been considered in references 29, 30, 5 and 6. A promising scenario (references 5 and 6) consists to take advantage of the different diffusion mean free paths of photons and neutrinos following the quark-hadron phase transition which lead to the birth of neutron rich regions. In such regions, the early nucleosynthesis is favoured. But as noticed in reference 6 the resulting abundance of ^{7}Li is still too high.

To sum up the simplest models of early nucleosynthesis are still the most promising and also those which have the most interesting consequences. The price to pay is that they imply specific models of galactic evolution. This condition is not especially stringent since these evolution models have further virtues to solve several problems pertaining to the nucleosynthetic history of our Galaxy.

ACKNOWLEDGEMENTS

The research reported in this contribution has been supported in part by PICS n° 18 - Let me thank Wolfgang Hillebrandt for his generous hospitality, Elisabeth Vangioni-Flam for her careful reading of this manuscript and for having let me reporting our joint work and also Colette Douillet for her skilled presentation of this paper.

REFERENCES

[1] Audouze, J. 1987, in Observational Cosmology, eds. G.R. Burbidge, A. Hewitt and L.Z. Fang, Reidel Dordrecht (in press)

[2] Audouze, J., Spite, F. and Spite, M. 1988, Physics Report, to be published

[3] Audouze, J., Delbourgo-Salvador, P. and Vangioni-Flam, E. 1987, in Advances in Nuclear Astrophysics, ed. E. Vangioni-Flam et al, Editions Frontières, p. 47

[4] Vangioni-Flam, E. and Audouze, J. 1987, Astron. Astrophys. (submitted)

[5] Applegate, J.H., Hogan, C.J. and Scherrer, R.J. 1987, Physics Review D. (in press)

[6] Alcock, C.R., Fuller, G.M. and Mathews, G.J. 1987, Ap. J. (preprint)

[7] Vidal-Madjar, A. 1987, in Space Astronomy and Solar system exploration, ed. W.R. Burke, ESA-SP 268

[8] Carswell, R.F., Irwin, M.J., Webb, J.K., Baldwin, J.A., Atwood, B., Robertson J.G. and Shever P.A. 1987, preprint

[9] Bania, T.M., Rood, R.T. and Wilson, T.L. 1987, preprint

[10] Rood, R.T., Bania, T.M. and Wilson, T.L. 1984, Ap. J. 280, 629

[11] Kunth, D. 1986, PASP 98, 984

[12] Shields, G.A. 1986, PASP 98

[13] Steigman, G., Gallagher III, J.S. and Schramm, D.N. 1987, Ap. J. (submitted)

[14] Pagel, B.E.J., Terlevich, R.J. and Melnick, J. 1986, PASP 98, 1005

[15] Conrath, B., Gautier , D., Hanel, R., Lindal, G. and Marten, A. 1987, J. Geophys. R. to be published

[16] Spite, F. and Spite, M. 1982, Astron. Astrophys. 115, 357

[17] Hobbs, L.M. and Duncan, D.K. 1987, Ap. J., in press

[18] Beckman, J., Rebolo, R. and Molaro, P. 1987, in Advances in Nuclear Astrophys.

[19] Kajino, T., Toki, H. and Austin, S.M. 1986, preprint

[20] Kawano, L., Schramm, D.N. and Steigman, G. 1987, preprint

[21] Boesgaard, A.M. and Steigman, G. 1985, Annual Rev. Astron. Astrophys. 23, 319

[22] Cline, D.B., Schramm, D.N. and Steigmann, G. 1987, Comments in Nuclear and Particle Physics (in press)

[23] Delbourgo-Salvador, P., Gry, C., Malinie, M. and Audouze, J. 1985, Astron. Astrophys. 150, 53

[24] Audouze, J. and Tinsley, B.M. 1974, Ap. J. 192, 487

[25] Bouvier, J. 1987, Doctorat d'Astronomie thesis - Université de Paris 7, unpublished

[26] Larson, R.B. 1986, M.N.R.A.S. 218, 409

[27] Wyse, R.F.G. and Silk, J. 1987, Ap. J. (letters) 313, L11

[28] Delbourgo-Salvador, P., Audouze, J. and Vidal-Madjar, A. 1987, Astron. Astrophys. 174, 365

[29] Audouze, J., Lindley, D. and Silk, J. 1985, Ap. J. (letters) 293, L53

[30] Salati, P., Delbourgo-Salvador, P. and Audouze, J. 1987, Astron. Astrophys. 173, 1

CHEMODYNAMICAL MODELS OF GALACTIC EVOLUTION

A. Burkert and G. Hensler
Universitäts-Sternwarte München
Scheinerstr. 1, D-8000 München 80

Summary

Galactic evolution cannot be understood in its whole complexity if dynamical and chemical calculations are performed separately. Since the metal content determines the cooling of the interstellar medium (ISM), it influences also the dynamical evolution of galaxies. On the other hand, galactic dynamics affects the star formation rate and thus the corresponding nucleosynthesis. In order to allow for this interdependence, it is necessary to study processes in the ISM and the interaction between stars and the ISM in more detail.

As a first step towards a consistent galactic evolution model we consider in this study the temporal evolution of an isolated galactic region applying an advanced model of the ISM based on the three-component picture of McKee and Ostriker.

1. Introduction

Up to now galactic evolution has been studied on the basis of either dynamical or chemical models. Whereas dynamical models should, in principle, explain the formation of different components of galaxies, like metal-poor spheroidal halos and metal-rich rotating disks, calculations of the galactic chemical evolution trace specifically the history of element abundances and the interaction between stars and gas, e.g. by stellar mass loss and star formation.

Separate treatments of galactic dynamics and chemistry can certainly provide first insights into galactic evolution. However, in this paper, we want to emphasize the strong interdependence between galactic dynamics and chemistry. This interaction can be taken into account with the use of a detailed description of the interstellar medium (ISM). Therefore, we have constructed a model of the ISM which is based on the three-component picture of McKee and Ostriker (1977). This model taken together with a dynamical model (see Section 2.b) might then be a solid basis for detailed investigations of galactic evolution.

2. Dynamical models of galactic evolution

a) Dissipationless dynamical models

Dissipationless models seek to give an account of galactic evolution, relying on stellar dynamics alone (see e.g. Gott, 1973, 1975). However, the two-body relaxation time of the stellar component in galaxies by far exceeds the age of the universe, and violent relaxation (Lynden-Bell, 1967) only partially erases the memory of the initial state. Therefore, the results must strongly depend on the initial conditions. Each galaxy can in principle be modelled, if we combine the right number of stars with the right initial kinematics and metallicities.

Moreover, observations of high central densities in ellipticals (Fall, 1983) and the existence of galactic disks and metallicity gradients indicate, that dissipation plays an important role during the collapses of protogalaxies (Silk, 1985).

b) Dissipative dynamical models

The first extensive work on dissipative galactic evolution was done by Larson (1969, 1974, 1975, 1976). Since dissipation is strongly associated with the gaseous component, Larson treated stars and gas as two different fluids, interacting by means of star formation and reshoveling of metal-enriched gas from the stars into the ISM. Assuming a cloudy substructure, energy dissipation of the gaseous component was treated by inelastic cloud-cloud collisions.

Dissipative two-component models have the advantage that the initial conditions for the stars, like metallicity and kinematical properties, are consistently given by the properties of the gas from which the stars form. Moreover, the dynamics of the gaseous component might not depend sensitively on the initial conditions because dissipation due to cloud-cloud collisions erases the information of the initial state.

In his models, Larson assumed an initially gaseous, homogeneous, rigidly rotating sphere of zero metallicity and of temperature 10^4K. Gas and stars, which formed from the gas, were treated hydrodynamically by means of the isotropic Eulerian equation. In order to explain spheroidal components in his two-dimensional rotating systems, Larson introduced a very large viscosity (Larson, 1975). Scaling estimates show however, that viscosity cannot have played such an important role during galactic evolution (Gott and Thuan, 1976). Furthermore, because of the absence of encounters, the stellar component cannot be treated by means of the isotropic Eulerian equations.

In our first dynamical models (Burkert and Hensler, 1987a,b) we therefore, have adopted Larson's hydrodynamical approach, however, allowing for an anisotropic velocity distribution function in the stellar component and neglecting viscosity.

Figure 1 shows the time evolution of two models: a model where the stars are formed from the gas with an isotropic velocity dispersion (model 1) and a model, where we assume an anisotropic velocity dispersion for the new born stars (model 2). In both calculations, the initial conditions are a homogeneous, rigidly rotating, gaseous sphere of mass 10^{11} $M_\odot$, gas density 6 10^{-26}g/cm^3, temperature 10^4K, zero metallicity, and angular velocity $2(10^9\text{yrs})^{-1}$.

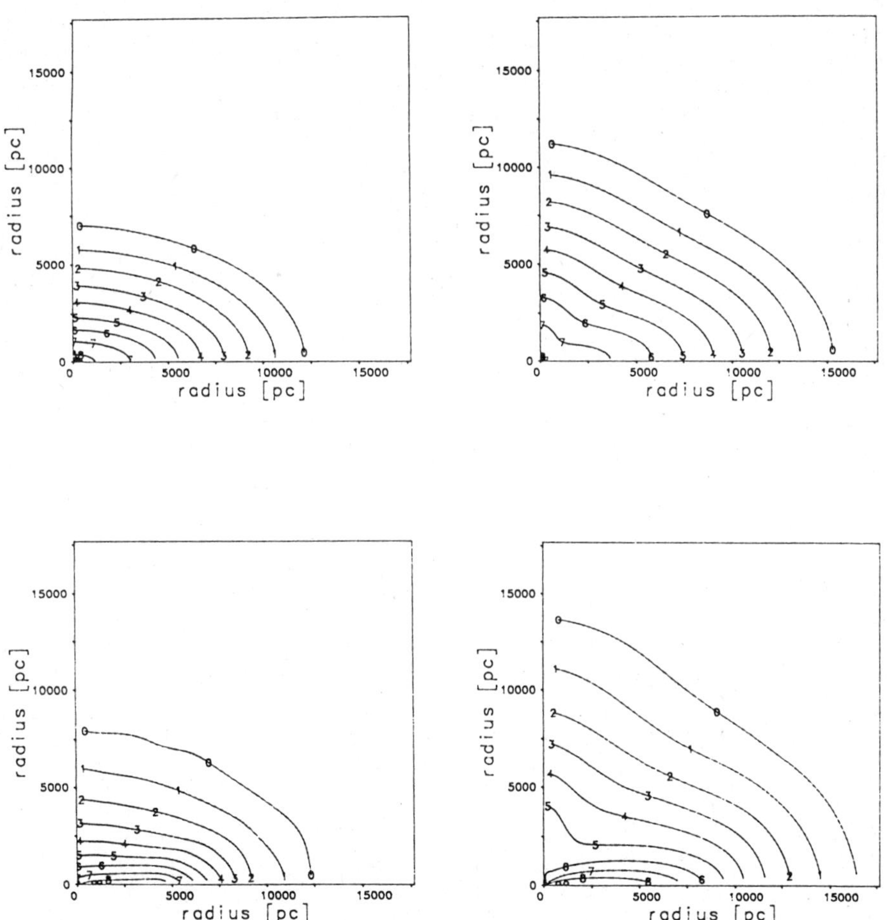

Fig. 1: Stellar isodensity contours in one quadrant of the meridional plane for a model with isotropic star formation (model 1 to the left) and a model with anisotropic star formation (model 2 to the right). The isodensity contours are spaced logarithmically from log ρ_{min} = -3.05 to log ρ_{max} = -1.35 for the upper pictures at an age $\tau \approx 2\tau_{ff}$ and to log ρ_{max} = -0.16 for the lower pictures at $\tau \approx 3\tau_{ff}$.

After two free fall times (τ_{ff} = 2.7 10^8yrs), in both models, a pressure-supported, metal-poor stellar halo is formed. In this spheroidal halo-potential, primordial gas with high angular momentum and metal-enriched gas from evolved metal-poor halo stars falls into the galactic plane, where it dissipates its kinetic energy by cloud-cloud collisions and forms a galactic disk. After ≈ 3 τ_{ff} this gaseous disk is condensed into metal-rich disk stars and the model-galaxy has reached its equilibrium state.

The metal-rich stellar disk, in the case of isotropic star formation, is visible in figure 2, which shows the iso-metallicity contours of the stellar component. Notice that the metallicity gradient perpendicular to the galactic plane is very steep, whereas in the equatorial plane metallicity decreases slowly outwards.

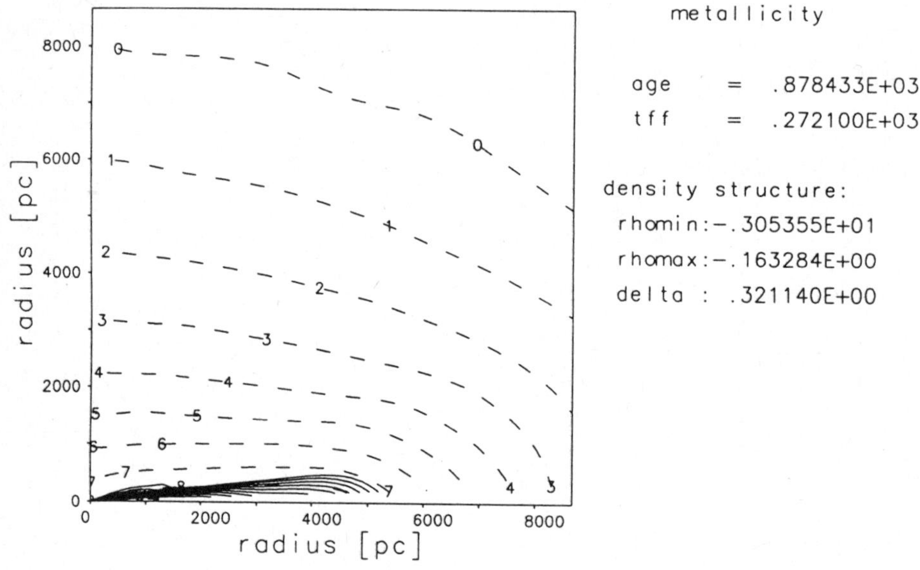

Fig. 2: Enlarged inner section of model 1. The labelled, dashed isodensity contours are the same as in figure 1. The contours of average stellar metal abundance (solid lines) show the thin galactic disk.

Although our first dynamical models might provide an explanation of the formation of some global structures of galaxies, like halos or disks, they are, nevertheless not sufficiently detailed to explain galactic chemistry. For example, heating and metallicity-dependent cooling processes in the ISM control the star formation time-scale and the collapse timescale of protogalaxies. Moreover, the collective interaction of many supernova remnants might cause galactic winds, which throw out metal enriched and primordial gas with high angular momentum, leaving behind a metal-poor elliptical galaxy.

3. Chemical models of galactic evolution

Chemical models study galactic metal enrichment by means of stellar nucleosynthesis. Therefore, the results strongly depend on the star formation rate, on the assumed initial mass function (IMF), its presumable dependence on metallicity and gas density, and on stellar evolutionary model-predictions of metal production and mass ejection as a function of the initial mass m and the metallicity Z of the stars.

In general, chemical models can be subdivided into models with or without galactic infall (Tinsley,1977; Lacey and Fall,1983; Clayton, 1984; Rana and Wilkinson, 1986).

a) Chemical models without galactic infall

Chemical closed box models might be applicable to nearly homogeneous, closed systems like the progenitors of globular clusters, small ellipticals, irregular or dwarf galaxies. They also might be useful for the study of certain regions in spiral galaxies, during epochs where infall and outflow are unimportant. In general, however, mass flow over the boundaries of the considered region or cell has to be taken into account. If, for example, a star burst occurs in the galactic center or in regions of the galactic disk, supernovae and stellar winds will increase the gas-pressure in the cell and will expel the residual gas, leaving behind a hot, thin bubble. The gas first has to cool and flow in again, before new stars can be born.

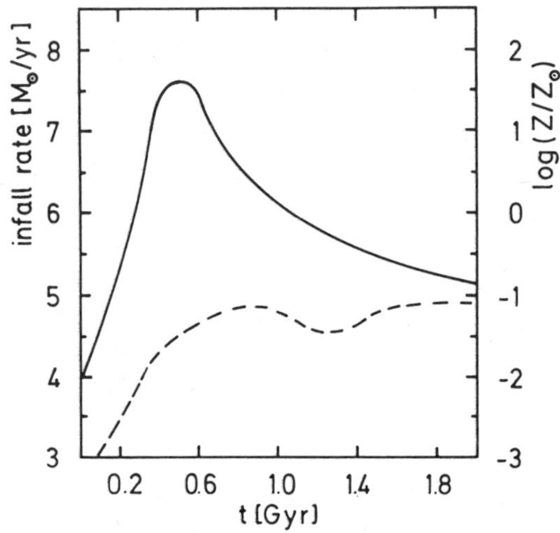

Fig. 3: Gas infall rate into the galactic plane as function of time of model 1 (solid curve). The dashed curve shows the normalized mean metallicity ($Z_\odot$ = 0.02) of the infalling gas.

b) Chemical models with galactic infall

If we allow for galactic infall, the results strongly depend on the time evolution of the infall rate and on the mean metallicity of the infalling gas. Figure 3 shows the gaseous infall-rate into the galactic disk and the mean metallicity of the infalling gas of our dynamical model 1 (see section 2.b). At least during the first 2 10^9yrs, infall should be taken into account. Notice further, that the mean metallicity of the infalling material is nearly constant.

Some information about the infall can in principle be given by dynamical models. However, galactic dynamics itself depends on the metallicity-dependent cooling and on the heating of the ISM. If we want to investigate the influence of these processes which link galactic chemical and dynamical evolution, we have to consider a several-component ISM and in detail its interaction with the stellar system by star formation and by stellar mass loss and energy loss.

4. An improved model of the ISM

a) The cloudy medium and the inter-cloud medium

Heating and cooling processes in the ISM lead to a strong connection between the different components of galaxies: the stellar system, the hot inter-cloud medium (ICM), and the cold cloudy medium (CM). For example, stars form out of the cold CM and give back metal-enriched gas to the ICM. This gas first has to cool and condense into clouds, before new, metal-enriched stars can form.

As a further step towards a reasonable model of galactic evolution, we have , therefore, improved the description of the ISM by applying the three-component picture of McKee and Ostriker (1977): a cold dense cloudy component is imbedded in a diffuse tenuous hot inter-cloud medium; stars can only form in the dense, cold cores of the clouds. CM and ICM are assumed to be in pressure equilibrium. As an example of the different interactions taken into account in Fig.4, the energy flux between stars, CM and ICM is shown.

We characterize the cold CM by a temperature $T \leq 10^4$K and the hot ICM by $T \geq 10^4$K. In each cell of the computational grid, the CM is described by its mean thermal energy e_{CM}^{th} or temperature T_{CM}, its mean density $\rho_{CM} = \langle m \rangle n_{CM}$, where $\langle m \rangle$ is the mean cloud mass and n_{CM} the number density of the clouds, its mean velocity $\underline{v}_{CM}$ and the mean kinetic energy $e_{CM}^{kin} = \rho_{CM} \langle (\underline{u}_{CM} - \underline{v}_{CM})^2 \rangle$, where $\underline{u}_{CM}$ is the peculiar velocity of the clouds.

The hot ICM with temperature T_{ICM}, density ρ_{ICM}, and mean velocity $\underline{v}_{ICM}$ is in pressure equilibrium with the CM. In general, the ICM has a large volume filling factor

$$f_{ICM} = \frac{e_{ICM}(\gamma_{ICM}-1)}{e_{ICM}(\gamma_{ICM}-1)+e^{th}_{CM}(\gamma_{CM}-1)} \qquad (1)$$

where γ_{CM} and γ_{ICM} are the adiabatic exponents of the CM and ICM respectively.

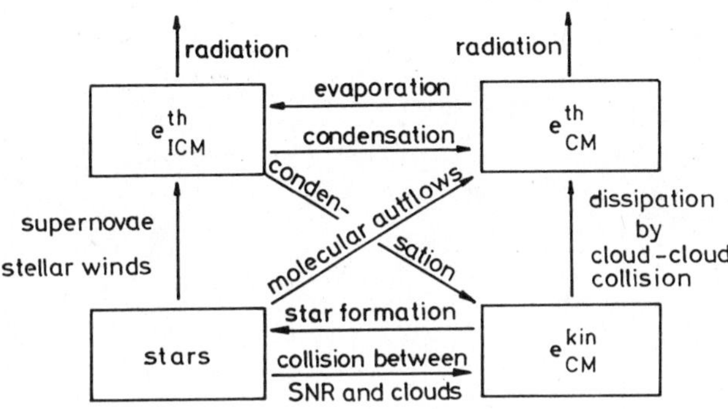

Fig. 4: Processes taken into account in the improved ISM model and energy-flux between the stellar and gaseous component. e^{th}_{CM} and e^{kin}_{CM}: mean thermal energy and mean kinetic energy of the clouds; e_{ICM}: mean thermal energy of the inter-cloud medium.

b) Processes in the ISM

For the cooling of the hot ICM, we have been provided with unpublished metal-dependent cooling functions by Dr. H. Böhringer, who has calculated the cooling of equilibrium plasmas by electron impact induced line emission according to Gaetz and Salpeter (1983), with additional lines compiled by Kato (1976). Since our models do not currently take into account detailed element abundances but only considered the total metallicity Z, cosmic element abundances (Anders and Ebihara,1982) and their scalings with fixed abundance ratios have been applied to the calculation of the cooling functions. For our purposes, it is sufficient to parametrize the cooling functions due to line emission by:

$$\Lambda(T,n_H,Z) = \Lambda_0(Z)\, T^{m(Z)} n_H^2 \qquad (2)$$

with: $\Lambda_0(Z) = 10^{-(22.0+4A(Z))}$

$m(Z) = A(Z) = 30Z$ for $10^4 \le T < 10^5$

and $\Lambda_0(Z) = 10^{-24.5}\, Z^{-11A(Z)}$

$m(Z) = A(Z) = (4 + 0.4\log(Z))/(5 - 11\log(Z))$ for $T \ge 10^5$.

If the ICM cools below 10^4K, it condenses into clouds with random velocities equal to the sound velocity of the ICM from which they form. The clouds dissipate their kinetic energy e^{kin}_{CM} by cloud-cloud collisions according to equation (6) of Larson (1969), thereby increasing their thermal energy as:

$$\frac{de^{kin}_{CM}}{dt} = - \frac{de^{th}_{CM}}{dt} \quad . \tag{3}$$

If clouds are heated up to temperatures greater than 10^4K, they are assumed to evaporate again.

c) Interactions between stars, CM and ICM

Stars form in the cold cores of the CM with a star formation rate (SFR) as given by Larson`s (1969) approach:

$$SFR = 0.55\ \rho^2_{SF}\ (M_\odot pc^{-3} yr^{-1}) \tag{4}$$

where the core mass ρ_{SF} of the CM is approximated by:

$$\rho_{SF} = \frac{10^4 - T_{CM}}{99\ T_{CM}}\ \rho_{CM} \quad , \text{ if } 10^2\text{K} \leq T_{CM} \leq 10^4\text{K}$$

$$\rho_{SF} = \rho_{CM} \quad , \text{ if } T_{CM} \leq 10^2\text{K} \quad .$$

The initial mass function IMF is set equal to Salpeter`s (1955) law

$$\frac{dN}{dm} \sim m^{-2.35} \tag{5}$$

with lower and upper mass limits of 0.01 $M_\odot$ and $100M_\odot$, respectively (Larson,1974). During their protostellar evolution, stars heat the CM by molecular outflows. If we assume that a fraction η_{MO} of the potential energy $E_{pot} = 0.5GM^2/R$ (G: gravitational constant; M: mass of the star; R: final stellar radius) lost during star formation heats the CM and take into account a mass-radius relation of $R \sim M^{0.6}$, the energy input into the CM amounts to

$$E_{MO} = \eta_{MO}\ 1.89\ 10^{48}\ M^{1.4} \quad (\text{erg}) \quad . \tag{6}$$

M in $M_\odot$ is the total mass converted into stars. ζ_{MO} typically is of the order of 0.01 (Margulis and Lada,1985).

We follow the stars during their HMS-lifetime τ_{HMS} (Chiosi et al.,1978; Franco and Shore,1984; Maeder,1987; Iben,1967)

$$\tau_{HMS} = 8\ 10^9/m^{2.8} \qquad \text{if } m \leq 10M_\odot$$

$$\tau_{HMS} = 5\ 10^7/m^{0.6} \qquad \text{if } m \geq 10M_\odot \tag{7}$$

where m in $M_\odot$ is the ZHMS-mass of the stars.

During τ_{HMS}, massive stars ($m \geq 10 M_\odot$) lose mass by radiative driven winds (Kudritzki et al., 1987)

$$dM/dt = -10^{-15}\ (L/L_\odot)^{1.6}\ (Z/Z_\odot)^{0.5} \qquad (M_\odot/yr)\ . \tag{8}$$

Taking $Z_\odot = 0.02$, a mass-luminosity relation approximated from Maeder (1981) and Chiosi (1978)

$$\log(L/L_\odot) = 1.4 + 2.42 \log(M/M_\odot) \tag{9}$$

and integrating equation (8) leads to a total mass loss by stellar winds during τ_{HMS} of:

$$\Delta M_{sw}(m) = \frac{m}{(1+1.72\ 10^{-4}\ Z^{1/2}\ m^{2.27})^{0.35}}\ M_\odot \tag{10}$$

where Z is the mean metallicity of the stars. From equation (10) we can conclude that, for massive stars with $m < 50\ M_\odot$, mass loss by stellar winds is negligible during their HMS life time.

The massive stars are attributed to become supernovae immediately after their HMS-lifetime, neglecting their post-main-sequence evolution with enhanced mass loss via stellar winds, but of very short duration ($\tau_{PMS} < 0.1\tau_{HMS}$). Assuming a remnant mass $M_{rem} = 3M_\odot$ and an internal core of metals M_Z of (Maeder,1987)

$$M_Z(m) = (0.357m - 2.2)\ M_\odot \tag{11}$$

the total mass of unprocessed material M_H^{ej} and metals M_Z^{ej} ejected by a supernova event amounts to ($m \geq 3$):

$$\begin{aligned} M_Z^{ej} &= (M_Z(m) - 3)\ M_\odot \\ M_H^{ej} &= (m - \Delta M_{sw} - M_Z^{ej} - 3)\ M_\odot\ . \end{aligned} \tag{12}$$

Each supernova delivers an energy input into the ICM and CM of

$$E_{SN} = 10^{51}\ \text{erg} \qquad .$$

Following the picture of McKee and Ostriker (1977), collisions between clouds and supernova shells increase the kinetic energy e_{CM}^{kin}. Here we allow for a fraction $0.2f_{CM}$ of E_{SN} (with f_{CM} the volume-filling factor of the CM) to be released as kinetic energy of the CM.

Low mass stars ($m < 10M_\odot$) are considered to return only unprocessed material immediately after their τ_{HMS}, by planetary nebulae (PN) leaving behind a mean remnant mass of $M_{rem} = 1M_\odot$. Energy input by PN into the ISM is neglected.

5. A closed model of chemical galactic evolution

As a test of our three-component model we have investigated a closed box, filled initially with an ICM of temperature 10^4K, zero metallicity, and density varying between 0.05 $M_\odot/pc^3$ (low density region) and 0.2 $M_\odot/pc^3$ (high density region). The validity of this model has already been discussed in Section 3.a.

In figure 5, the SFR as a function of time is shown for both, the high and the low density regions. We can distinguish three different epochs: a period of unstable star formation separates an epoch of steeply decreasing SFR from an epoch characterized by a slow decrease of SFR.

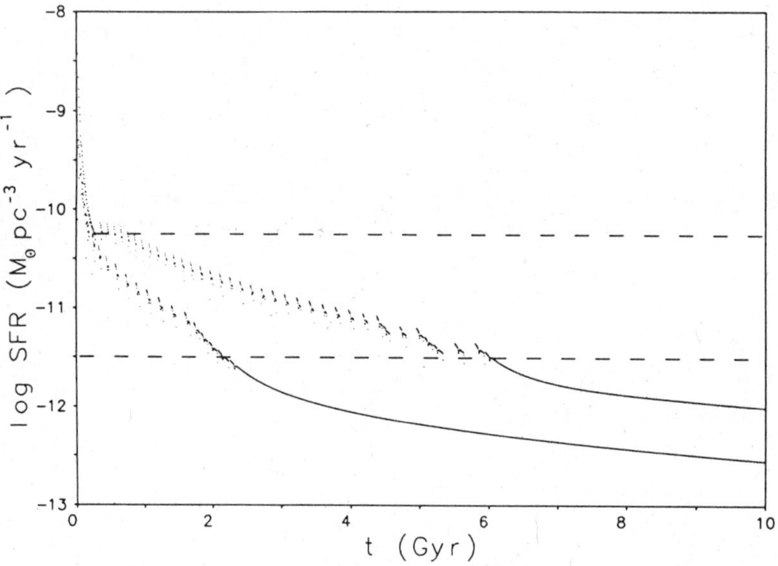

Fig. 5: Star formation rate as function of time for a high density region (upper curve) and low density region (lower curve). The two dashed lines enclose the region of unstable star formation.

Unstable star formation occurs if the SFR lies between 3 $10^{-12} M_\odot pc^{-2} yr^{-1}$ and 5 $10^{-11} M_\odot pc^{-3} yr^{-1}$, independent of the initial conditions or of whether stellar winds are taken into account. Star formation bursts of duration 3 10^7yrs are separated by periods of quiescence of duration 6 10^7yrs (figure 6).

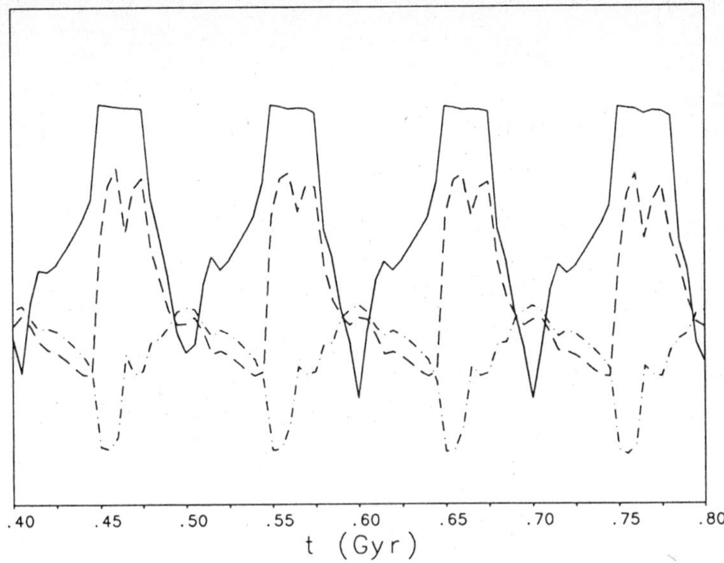

Fig. 6: Correlation between the star formation rate (solid curve), the mean pressure ⟨P⟩ (dashed curve), and the mean thermal energy e_{CM}^{th} of the cloudy medium (dot-dashed curve) during the period of unstable star formation in fiducial units.

During periods of quiescence, the CM cools (e_{CM}^{th} decreases) and, consequently, according to equation (4), the SFR increases. Soon the first supernovae increase e_{ICM} and by this the partial pressure ⟨P⟩ in the cell:

$$\langle P\rangle = e_{CM}^{th}(\gamma_{CM}-1) + e_{ICM}(\gamma_{ICM}-1) \quad . \tag{13}$$

The hot ICM compresses the clouds which then because of their high density, cool even better. Later on, when the CM is consumed more and more by star formation and evaporation, e_{CM}^{th} increases again, leading to a new period of quiescence.

The metallicity Z here is defined as the mass fraction in metals. Initially, due to a high supernova rate (10^{-11}-10^{-12} supernovae $yr^{-1}pc^{-3}$), the mean metallicity of the CM and of the stars increases rapidly, leading to a peak of Z_{CM} at solar metallicity after almost 2 10^8yrs (figure 7). Subsequently, mass loss from low-mass stars becomes more important and $\log(Z/Z_\odot)$ of the CM settles at about -0.1.

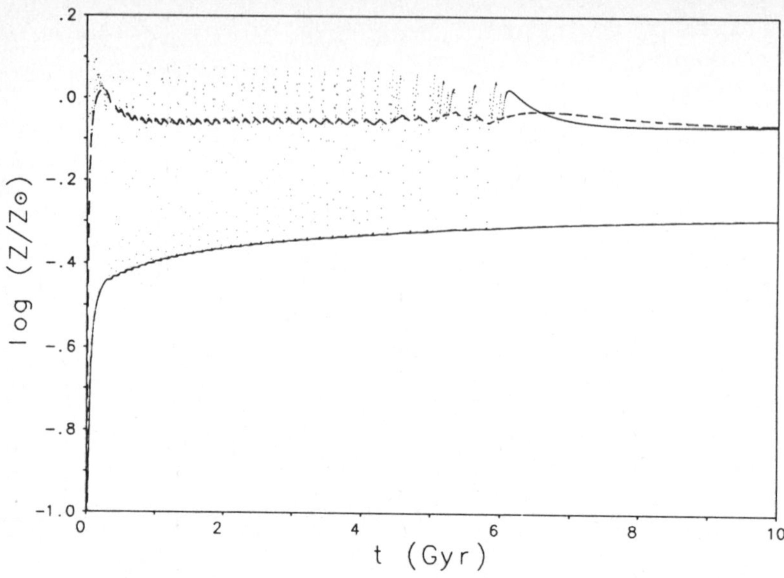

Fig. 7: The mean metallicity of the inter-cloud medium (dotted curve), the cloudy medium (dashed curve), and the low and high mass stars (solid curve) vs. time for the closed-box model.

During the period of unstable star formation, we observe a periodic change of the mean metallicity of the CM. Figure 8 shows the variation in time of the stellar density ρ_S as a function of $\log(Z/Z_\odot)$ of the CM from which these stars form. This picture accordingly reveals the time evolution of the mean metallicity of new born stars during the period of unstable star formation. Although there is a general trend of increasing metallicity with time, this increase proceeds in loops of 0.3 - 0.5 dex, because during a period of quiescence the CM is again rarified in its metallicity by metal-poor mass loss from low mass stars. As such a loop is run through in ≈ 3 10^7yrs, we can conclude, that stars which form during a period of 3 - 5 10^7yrs can show an intrinsic metallicity dispersion of about 0.5 dex. This model might explain observed metallicity dispersions of O-stars in OB associations within the galactic disk (Gehren et al., 1985).

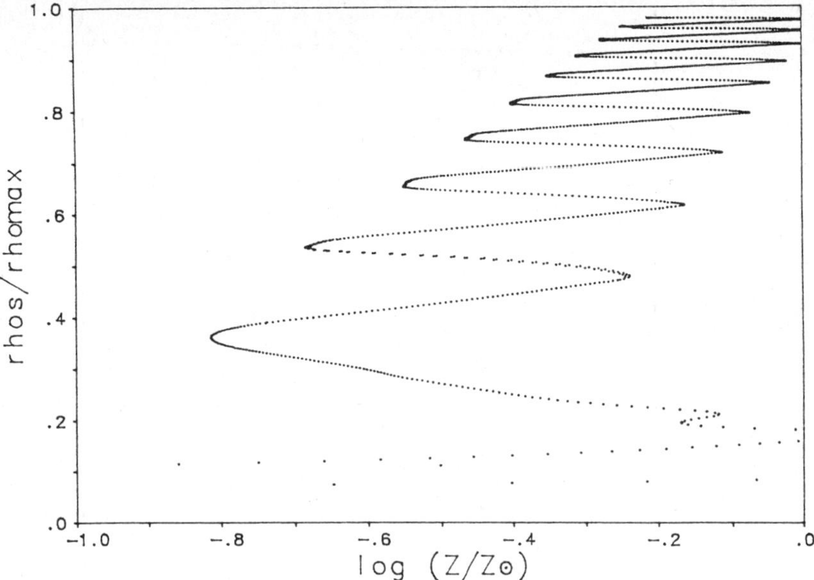

Fig. 8: Variation of the normalized stellar density vs. the mean metallicity of the cloudy medium, from which these stars form, during the period of unstable star formation.

6. Conclusions

Our calculations show that simple dissipative, dynamical models can explain the formation of spheroidal metal-poor halos and metal-rich disks of galaxies. However, in order to understand galactic dynamics and chemical evolution in more detail, an improved model of the ISM is needed which takes into account heating and cooling processes.

Our studies of the star-gas interaction based on a three-component model of the ISM shows, that a hot ICM with a large volume filling factor ($f_{ICM} > 0.9$) coexists with a cold cloudy component. Evaporation and condensation processes link the CM with the ICM. This interaction in our models allows the metallicity Z_{CM} of the CM to rise as fast as the metallicity Z_{ICM} of the ICM. On the other hand, a weaker interaction between CM and ICM would lead to a smaller Z_{CM} and by this to a metallicity content of the young stars smaller than that of the surrounding ICM.

Unstable star formation occurs, irrespectively of the chosen initial conditions and of the consideration of stellar winds. However, more parameter studies are necessary to investigate the influence of the different interactions taken into account. Furthermore, during a star burst the closed-box model seems rather questionable because the high pressure in the cell would throw out gas which, during a period of quiescence flows in again.

In future works, our model of the ISM will be combined with our dynamical model to give a chemodynamical model, which then might be a solid base for detailed investigations of galactic evolution and chemistry.

Acknowledgements

The authors are gratefully indebted to Dr. H. Böhringer from the MPI für extraterrestrische Physik in Garching/Munich for providing us with separately calculated cooling functions for various metallicities.
We acknowledge encouraging and helpful discussions with Professors T. Gehren and J. Truran. Furthermore, we thank W. Kley for the exchange of experience on numerics.
The calculations were performed on a Cyber 875 of the Leibniz-Rechenzentrum Munich.
The work was partly (A.B.) supported by the Deutsche Forschungsgemeinschaft under grant Ku 474/13-1.

References

Anders, A., Ebihara, M.: 1982, Geodim. Cosmodim. Acta 46, 2363
Burkert, A., Hensler,G.: 1987a, Monthly Notices Roy. Astron. Soc. 225, 21p
Burkert, A., Hensler, G.: 1987b, submitted to Astron. Astrophys.
Chiosi, C., Nasi, E., Sreenivasan, S.R.: 1978, Astron. Astrophys. 63, 103
Clayton, D.: 1984, Astrophys. J. 285, 411
Dalgarno, A., McCray, R.A.: 1972, Ann. Rev. Astron. Astrophys. 10, 375
Fall, S.M.: 1983, in IAU Symposium 100, "Internal Kinematics and Dynamics of Galaxies", ed. E. Athanassoula, Reidel, Dordrecht, p. 391
Franco, J., Shore, S.W.: 1984, Astrophys. J. 285, 813
Gaetz, T.J., Salpeter, E.E.: 1983, Astrophys. J. Suppl. 52, 155
Gehren, T., Nissen, P.E., Kudritzki, R.P., Butler, K.: 1985, Proceed. ESO Workshop "Production and Distribution of C,N,O Elements", eds. I.J. Danziger et al., p. 131
Gott, J.R.III.: 1973, Astrophys. J. 186, 481
Gott, J.R.III.: 1975, Astrophys. J. 201, 296
Gott, J.R.III., Thuan, T.X.: 1976, Astrophys. J. 204, 649
Iben, I.: 1967, Ann. Rev. Astron. and Astrophys. 5, 571
Kato, T.: 1976, Astrophys. J. Suppl. 30, 397
Kudritzki, R.P., Pauldrach, A., Puls, J.: 1987, Astron. Astrophys. 173, 293
Lacey, C.G., Fall, S.M.: 1983, Monthly Notices Roy. Astron. Soc. 204, 781
Larson, R.B.: 1969, Monthly Notices Roy. Astron. Soc. 145, 405

Larson, R.B.: 1974, Monthly Notices Roy. Astron. Soc. 166, 585

Larson, R.B.: 1975, Monthly Notices Roy. Astron. Soc. 173, 671

Larson, R.B.: 1976, Monthly Notices Roy. Astron. Soc. 176, 31

Lynden-Bell, D.: 1967, Monthly Notices Roy. Astron. Soc. 136, 101

Maeder, A.: 1981, Astron. Astrophys. 102, 401

Maeder, A.: 1987, Astron. Astrophys. 173, 247

Margulis, M., Lada, C.J.: 1985, Astrophys. J. 299, 925

McKee, C.F., Ostriker, J.P.: 1977, Astrophys. J. 218, 148

Rana, N.C., Wilkinson, D.A.: 1986, Monthly Notices Roy. Astron. Soc. 218, 497

Salpeter, E.E.: 1955, Astrophys. J. 121, 161

Silk, J.: 1985, Astrophys. J. 297, 9

Tinsley, B.M.: 1977, Astrophys. J. 216, 548

ABUNDANCE PATTERNS IN SOME OLD STARS

Bodo Baschek
Institut für Theoretische Astrophysik, Universität Heidelberg
Im Neuenheimer Feld 561, D-6900 Heidelberg

ABSTRACT
The abundance patterns of old extremely metal-poor stars are discussed with particular emphasis on the dwarf carbon star G 77-61 which has recently been analyzed by Gass (1985) and Gass, Liebert and Wehrse (1987). Its iron abundance, [Fe/H] = - 5.6, is the lowest found in a near main-sequence star up to now.

1. INTRODUCTION

Since the fundamental work by Burbidge et al. (1957) on the nucleosynthesis in stars, the chemical abundances of the oldest and most metal-poor stars have been of particular interest. For many years the most extreme metal-deficient population II stars known were the dwarf ('subdwarf') HD 140283, first analyzed in detail by Baschek (1959, 1962) and by Aller and Greenstein (1960), and the giant HD 122563, analyzed by Wallerstein et al. (1963), Pagel (1965), and Wolffram (1972). Their iron abundances (modern values) with respect to the sun are [Fe/H] = - 2.6 and - 2.7, respectively, where [M/H] = log (M/H) - log $(M/H)_\odot$. Stars with yet larger metal deficiencies, below [Fe/H] ≤ - 3, are very rare:

	[Fe/H]	T_{eff}[K]	log g[cm s^{-2}]	reference
BD +03°740	-3.1	6050	3.3	Magain,1987
G 64-12	-3.5	6350	4.0:	Carney and Peterson,1981
CD -38°245	-4.5	4800	2.0	Bessell and Norris,1984
G 77-61	-5.6	4250	3.9	Gass et al.,1987

G 77-61 is unique as it is a carbon star near the main sequence (Dahn et al., 1977) and a member of a binary system (Dearborn et al., 1987). Detailed analysis by model atmosphere techniques reveals an extremely low iron deficiency of [Fe/H] = - 5.6 (Gass 1985; Gass, Liebert, and Wehrse, 1987).
After a brief description of the general abundance patterns of extremely metal-poor stars, the recent results by Gass et al. are reported and discussed in this contribution.

2. ABUNDANCE PATTERNS OF THE EXTREME METAL-POOR STARS

The gross abundance pattern of the extreme population II stars is characterized (i) by approximately solar helium abundance, [He/H] $\simeq$ 0, in accord with the cosmological helium abundance, and (ii) by a deficiency of the bulk of the metals M (the elements heavier than He) relative to H by nearly the <u>same</u> factor, i.e. by [M/H] = η+[Fe/H] with $\eta \simeq 0$. There are, however, significant deviations from $\eta = 0$ for some elements or groups of elements: C is overdeficient for very low values of [Fe/H], N is sometimes overabundant with respect to Fe. Some (or all?) α nuclei such as Ca and Ti are in excess by about a factor of 3, and the s elements such as Sr and Ba are overdeficient with respect to Fe by a similar factor. Regarding Al, the situation is less clear. According to Magain (1987) Al is strongly overdeficient relative to Mg, the more the lower [Mg/H], whereas Bessell and Norris (1984) find no excess deficiency in CD $-38^{\circ}245$.

Bessell and Norris conclude that CD $- 38^{\circ}245$ with [Fe/H] = - 4.5 is an extreme member of Population II, in particular that "its relative abundances differ, if at all, in degree rather than in kind from those of Population II objects" and that there is no evidence for a new population with abundances produced by pregalactic objects.

We now discuss the peculiar carbon star G 77-61 in some detail. Its iron abundance is lower by one order of magnitude than that in CD $-38^{\circ}245$. According to Gass et al. (1987) this extreme abundance of Fe is primordial so that the abundance pattern of G 77-61 bears important information on the early element production.

3. THE CARBON STAR G 77-61 AND ITS ELEMENT ABUNDANCES

Astrometric, photometric and spectrometric observations (Dahn et al. 1977, Dearborn et al. 1986) show that G 77-61 is located near the main sequence and belongs kinematically to the Population II. The trigonometric parallax of 0.017" and the temperature around 4100 K result in an absolute magnitude $M_V = + 10.08$ and a luminosity $L/L_{\odot} = 0.018$. The membership in a binary system with a period of 245 days implies a mass of G 77-61 of about 0.35 $M_{\odot}$, whereas the mass of the invisible companion, probably a cool white dwarf, should be $\geq 0.55\ M_{\odot}$.

The gross features in the spectrum of the dwarf star G 77-61 are very similar to those of the giant carbon stars. As discussed by Gass et al. (1987) this is due to different combinations of the surface gravity g and the element abundances ε, mainly that of carbon. Reliable abundance determinations in a cool carbon star with molecular formation in its atmosphere require an elaborate analysis by model atmosphere tech-

niques. By this method Gass et al. obtain as parameters for G 77-61 an effective temperature T_{eff} = 4250 K and a surface gravity of log g(cm s^{-2}) = 3.9 in good agreement with Dearborn et al. (1986), and the element abundances given in Table 1.
The isotope ratio $^{12}C/^{13}C$ is fairly low in G 77-61. For the abundance of helium only an upper limit of [He/H] $\lesssim$ + 1.0 can be inferred from the presence of Hα; Gass et al. adopt [He/H] = 0, i.e. solar helium abundance. This is consistent with the position of G 77-61 in the HRD below the main sequence for normal composition stars, indicating somewhat enriched helium abundance, [He/H] $\simeq$ 0.4, and low metal content. The abundance pattern of G 77-61 is shown in Fig. 1. For the elements heavier than carbon it is fairly similar to that of CD $-38^{\circ}245$.

4. INTERPRETATION OF THE ABUNDANCE PATTERN OF G 77-61

First, the extremely low iron abundance can be regarded as characteristic of the stellar composition at birth. A depletion of Fe by diffusion in the largely convective cool dwarf as well as a reduction of Fe in the course of stellar evolution in combination with mass transfer from the companion seem impossible.
On the other hand, the most plausible mechanism to produce the CNO abundances with the carbon star characteristics, C/O > 1, is a contamination by mass transfer from the evolved primary. The mass of 0.35 $M_{\odot}$ is too low for helium burning to occur in G 77-61 itself. We may thus envisage the following scenario:
Primary as well as secondary (= G 77-61) both formed with about normal helium abundance and with extremely low metal abundances, in particular with [Fe/H] = - 5.6. The original CNO abundances were probably also very low, but their values cannot be reconstructed. The primary of $\lesssim$ 1 $M_{\odot}$ evolves through hydrogen and subsequent helium burning into a giant carbon star. The basic processes of element synthesis in giants in connection with thermal pulses, dredge-ups, envelope burning etc. are discussed by Iben and Renzini (1983) and Renzini and Voli (1981). In general, it is even easier for a low metal star to evolve into a carbon star than for a normal composition star. Qualitatively, one can obtain enrichment of N and C with C/O > 1 and a low $^{12}C/^{13}C$ ratio. It would be of interest to have calculations of the production of He, C, N, O, and the s elements available for such an extreme original metal abundance as observed in G 77-61. Furthermore the influence of the primary's evolution by a (close) companion may play an essential rôle.

In order to make the secondary appear as dwarf carbon star, the mass transfer of carbon rich material need not be very large. Assuming that

Table 1. Metal abundances in G 77-61 (Gass et al., 1987)

	log ε (log ε_H=12.0)	[M/H]
C	7.3	-1.2
^{12}C	7.2	-1.3
^{13}C	6.6	+0.1
N	5.2	-2.8
O	≤ 5.0	≤ -3.8
Na	3.1	-3.2
Mg	3.3	-4.1
K	≤ 3.2	≤ -1.8
Ca	≤ 1.5	≤ -4.7
Fe	2.0	-5.6

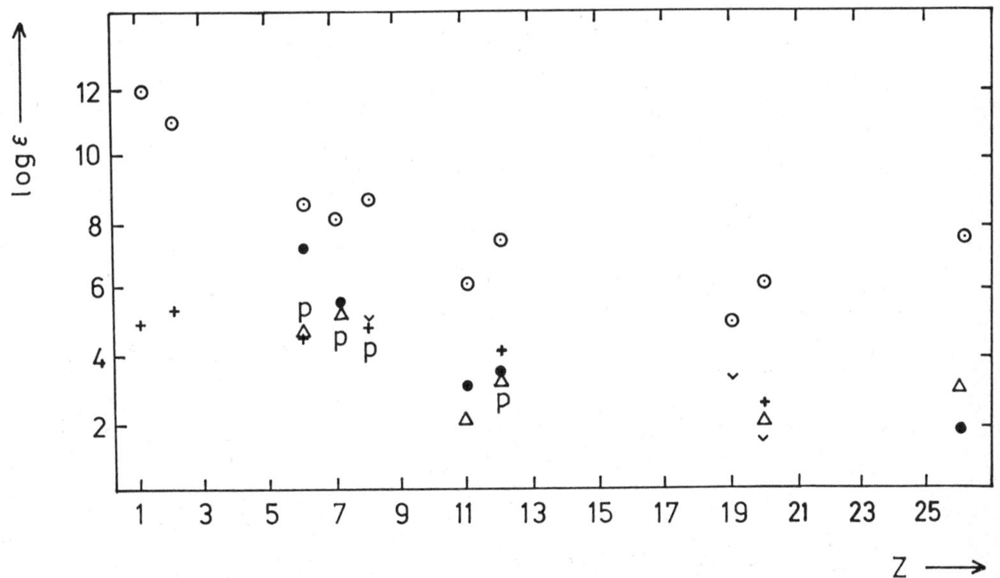

Fig. 1. Abundance patterns in extremely metal deficient stars (from Gass et al., 1987). Shown are the abundances of G 77-61 (filled circles, upper limits V; Gass et al., 1987) and of CD - 38° 245 (triangles; Bessell and Norris, 1984) together with the primordial composition (p; Wagoner et al., 1967) and the element production, normalized to Fe, by first generation massive stars or "hypernovae" (crosses; Woosley et al., 1984)

the secondary is completely convective, less than 10^{-5} $M_\odot$ in the form of carbon are required. So far no s elements have been found in the spectrum of G 77-61.
After some mass loss the primary finally evolves into a white dwarf. In order to be invisible at present, it must have cooled below about 5000 K so that the age of the system is at least 10^9 yrs.
The secondary is still on the main-sequence and probably fully convective. In principle, there is the possibility that the observed ratio $^{12}C/^{13}C \simeq 4$ could be achieved by CNO burning in G 77-61 itself over sufficiently long time scales. It remains, however, yet to be shown that the central temperature is high enough for the $^{12}C(p,\gamma)N^{13}(e^+\nu)^{13}C$ reaction to procede efficiently.
All elements heavier than C and N should be unaffected by nuclear burning during the evolution of the system and represent the composition at birth. Unfortunately, the few element abundances which could be obtained in G 77-61 (Na, Mg, Fe and upper limits for K, Ca) do not yet allow a clear-cut interpretation of the abundance pattern. Within the uncertainties the data seem compatible with the pattern of CD-38°245 (see Fig. 1) and would indicate that the characteristic pattern of extreme population II stars continues down to [Fe/H] = - 5.6. Taken at their face values, however, the abundances [M/H] decrease from Mg to Fe and hence would point towards some "extra" element production, perhaps by first-generation massive stars, hypernovae (cf. e.g. Woosley et al., 1984) or by primordial massive objects (Wagoner et al., 1967). As can be seen from Fig. 1, the calculated abundance patterns for these objects also are similar to that of G 77-61.
Clearly, it is highly desirable to derive abundances with good accuracy for more elements in the unique star G 77-61. Work based on spectra with better signal-to-noise ratios is in progress (R.C. Peterson, Liebert, Wehrse) to obtain further information on the early element production in our Galaxy.

ACKNOWLEDGMENT

I thank Rainer Wehrse for stimulating discussions and for generously putting the unpublished results on G 77-61 at my disposal.

REFERENCES

Aller, L.H., Greenstein, J.L.: 1960, Astrophys. J. Suppl. 5, 139
Baschek, B.: 1959, Z. Astrophys. 48, 95
Baschek, B.: 1962, Z. Astrophys. 56, 207
Bessell, M.S., Norris, J.: 1984, Astrophys. J. 285, 622

Burbidge, E.M., Burbidge, G.R., Fowler, W.A., Hoyle, F.: 1957, Rev. Mod. Phys. 29, 547
Carney, B.W., Peterson, R.C.: 1981, Astrophys. J. 245, 238
Dahn,C.C.,Liebert, J., Kron, R.G., Spinrad, H., Hintzen, P.M.: 1977, Astrophys. J. 216, 757
Dearborn, D.S.P., Liebert, J., Aaronson, M., Dahn, C.C., Harrington, R., Mould, J., Greenstein, J.L.: 1986, Astrophys. J. 300, 314
Gass, H.: 1985, Ph. D. Thesis, University of Heidelberg
Gass, H., Liebert, J., Wehrse, R.: 1987, Astron. Astrophys., in press
Iben, I., Jr., Renzini, A.: 1983, Ann. Rev. Astron. Astrophys. 21, 271
Magain, P.: 1987, ESO Messenger No. 47, 18
Pagel, B.E.J.: 1965, R. Obs. Bull. No. 104
Renzini, A., Voli, M.: 1981, Astron. Astrophys. 94, 175
Wagoner, R.V., Fowler, W.A., Hoyle, F.: 1967, Astrophys. J. 148, 3
Wallerstein, G., Greenstein, J.L., Parker, R., Helfer, H.L., Aller, L.H.: 1963, Astrophys. J. 137, 280
Wolffram, W.: 1972, Astron. Astrophys. 17, 17
Woosley, S.E., Axelrod, T.S., Weaver, T.A.: 1984, in Stellar Nucleosynthesis, C. Chiosi and A. Renzini (eds.), p. 263, D. Reidel

Evolution of Wolf-Rayet Stars

Norbert Langer
Universitäts-Sternwarte Göttingen
Geismarlandstraße 11, D–3400 Göttingen

Abstract. A scenario for the evolution of massive helium burning stars is presented, with special emphasis to the different WR stages. Several conclusions drawn from observations can be reconciled by this scenario, e.g. the high mean mass of WNL stars, the low mean mass of WNE and WC stars, the strange appearance of SN 1961v, and the suggestion that the Cas A progenitor was a WNL star. We conclude that the most massive stars terminate their life as WNL stars, and only less massive stars reach the WC stage during their evolution. Furthermore, we suggest that convective overshooting is an inefficient process in very massive H-burning stars.

1. Introduction

Wolf-Rayet (WR) stars are luminous massive stars which show strong and broad emission lines of nitrogen (WN stars) or carbon and oxygen (WC stars) as a result of heavy mass loss. Presently, the emission line strengths are interpreted to originate from significant chemical anomalies (Nugis, 1982; Willis, 1982; Garmany and Conti, 1982). This hypothesis is supported by stellar evolution calculations for massive stars which show that stellar mass loss and/or internal mixing processes may lead to the exposure of the ashes of hydrogen or helium burning at the stellar surface, giving rise to the WN or WC features, respectively (Maeder, 1982, 1983; Chiosi, 1982; Langer and El Eid, 1986; Langer, 1987).

Both WR categories, WN and WC, are divided into further subclasses, the WN series ranging from WN2 to WN9 and the WC series from WC4 to WC9, according to van der Hucht et al. (1981) and Conti et al. (1983). In this classification scheme higher numbers indicate lower degrees of ionisation of the ions responsible for the respective emission lines in the WR spectra, where for the WN stars lines of N III to N V and for the WC stars those of C II to C IV and O V can be observed. Thereby, WR stars showing higher degrees of ionisation are designated as 'early' types (WNE: WN2-WN5; WCE: WC4-WC6), in contrast to the 'late' types (WNL: WN6-WN9, WCL: WC7-WC9). We should mention that the so called WO stars correspond to subtypes of the WC sequence in the above scheme.

In this investigation we present a scenario for massive helium burning stars which leads to a model for the evolutionary status of the different WR subtypes WNL, WNE, and WC. This scenario is introduced in Sect. 2, while the required input physics for theoretical stellar models is discussed in Sect. 3, where its predictions are compared to observational constraints. Our conclusions are drawn in Sect. 4.

2. From chemical profiles to evolutionary sequences

In massive stars all evolutionary stages beyond central helium exhaustion proceed so rapidly that the probability of observing a star in one of these phases is negligible. On the other side, the presence of the so called Hubble-Sandage (HS) variables, which may be identified as progenitors of massive WR stars (Maeder, 1983; Humphreys, 1984) is a strong argument in favour of WR stars being in the post hydrogen burning stage of evolution (Langer and El Eid, 1986), since in case massive stars would evolve into WR stars already during central hydrogen burning, they would never reach as low surface temperatures as observed in the case of the HS variables, rather they would evolve towards the left of the zero age main sequence (ZAMS) in the HR diagram (cf. Prantzos et al., 1986). For these reasons we may well regard all observed WR stars to be in the central helium burning phase of evolution.

Evolutionary computations for massive stars show, that the spatial hydrogen profile, which originates from core hydrogen burning and is considerably altered in the phase between central hydrogen and helium burning, remains almost constant throughout the whole central helium burning phase (cf. e.g. Stothers and Chin, 1976; Langer et al., 1985). The small upwards shift of the lower edge of the H-profile due to shell hydrogen burning may be neglected here for simplicity. Then, by identifying the different possible post main sequence stages of massive stars by their surface chemical composition (Table 1), one may estimate the duration of these different stages for a given star by looking at its composition profile at He-ignition, and by attaching a certain rate of mass loss to each evolutionary phase (cf. Table 1). The surface mass fraction of the jth isotope $X^{(j)}$ at any time during central helium burning can be simply calculated according to

$$X^{(j)}_{surface}(t) = X^{(j)}(M_r); \quad M_r := M(t_0) - \int_{t_0}^{t} \mid \dot{M}(t') \mid \, dt', \tag{1}$$

where t_0 refers to the time of helium ignition, and $M(t)$ is the total stellar mass as a function of time.

On the basis of this consideration, Langer (1987) proposed a simple scenario for the post main sequence evolution of massive stars, which is schematically sketched in Fig. 1. In this picture, stars with ZAMS masses above a critical value ($M_{ZAMS} > M_{HS}$) evolve into a HS star after central H-exhaustion, where they get rid of their H-rich envelope in a very short time, thereby forming a WR star at the beginning of helium burning. Depending on the H-profile in intermediate layers, all the hydrogen containing layers may be lost in the HS stage leading to a WNE star ($X_{surface} \simeq 0$), or — in case a hydrogen poor zone exists within the envelope — just the H-rich part of the envelope is lost in the HS stage, which results in a WNL type star. In the case where no sufficiently H-poor zone exists in the envelope ($M_{HS} < M_{ZAMS} < M_2$), the star will reach certainly the WC stage in the course of its evolution, since after the HS stage only a few $M_\odot$ of matter (with a composition $X \simeq 0$, $Y \simeq 1 - Z$, and $^{14}N \simeq \frac{2}{3}Z$) separate the He/C/O-core from the stellar surface. On the other side, in the case that the mass ΔM of the H-poor zone of the envelope exceeds a critical value ($\Delta M_{crit} \simeq \langle \dot{M} \rangle_{WR} \cdot \tau_{He} \simeq 10\, M_\odot$) — which may be true for the most massive stars ($M_{ZAMS} > M_{WNL}$), cf. Langer (1987) — the star will even remain in the WNL phase of evolution up to central He-exhaustion, and it will therefore terminate its life as a WNL star.

type	surface abundances	mass loss rate
O-star	cosmic	e.g. Lamers (1981) rate
supergiant	~cosmic, evtl. $N\uparrow$, $\frac{^{12}C}{^{13}C}\downarrow$	Lamers/Reimers/???
HS-var.	$H\downarrow$ but still H-rich, $N\uparrow$, $C\downarrow$	very large ($\sim\infty$)
WNL	H-poor, but $H>0$, $He\uparrow$, $N\uparrow$, $C\downarrow$	$3\cdot10^{-5}\ M_\odot\ yr^{-1}$
WNE	$H=0$, $He\uparrow$, $N\simeq\frac{2}{3}Z$, $C\downarrow$	"
WC	$H=0$, $He\uparrow$, $N=0$, $C\uparrow$, $O\uparrow$, $Ne\uparrow$	"

Table 1: Different evolutionary stages of massive stars and their characteristic surface abundances and mass loss rates. The mass loss rate for Red Supergiants is extremely uncertain, while that for HS variables is sufficiently large in order to achieve a timescale of the HS stage which is small compared to the helium burning timescale.

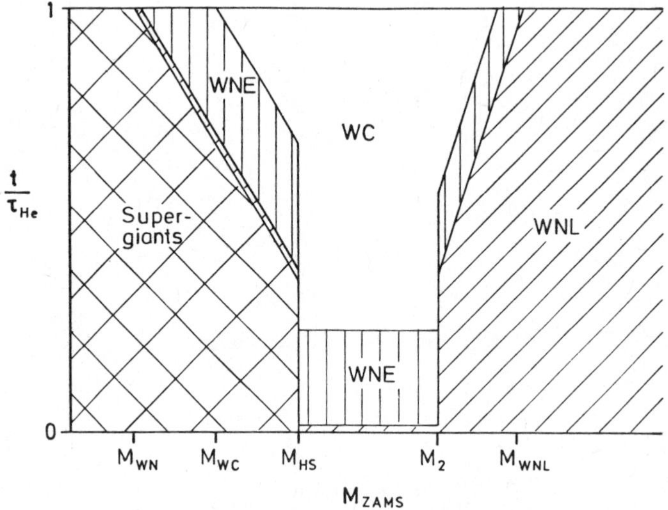

Fig. 1: Schematic 'phase'-diagram for the post main sequence evolution of massive stars according to our evolutionary picture. The vertical thickness of the differently shaded regions measures the fractional time of central He-burning spent in the respective evolutionary phase. The cross hatched region indicates the normal supergiant phase, diagonal hatched areas indicate the WNL phase, vertical hatching indicates the WNE phase, and the white area corresponds to the WC phase. A star of a given ZAMS mass (abscisse) evolves vertically upwards in this diagram. For example, a star with $M_{WC} < M_{ZAMS} < M_{HS}$ would spend approximately the first half of its He-burning time as a supergiant, evolving then through a very short WNL phase into a Wolf-Rayet star of type WNE and finally even into a WC star as its final evolutionary stage.

For stars with $M_{ZAMS} < M_{HS}$ the evolution has to proceed via the supergiant stage, with the consequence of the WR phase lasting considerably shorter than the helium burning lifetime τ_{He}. For sufficiently small ZAMS masses the WC-phase ($M_{ZAMS} < M_{WC}$) or even the WN-phase ($M_{ZAMS} < M_{WN}$) is never reached.

From observations (cf. references in Langer, 1987) one may deduce the following values for the critical ZAMS masses (which, however, are very uncertain): $M_{WN} \simeq 20 - 30\,M_{\odot}$, $M_{WC} \simeq 30 - 35\,M_{\odot}$, $M_{HS} \simeq 40 - 50\,M_{\odot}$, and $M_{WNL} \simeq 60 - 100\,M_{\odot}$.

3. Required input physics and observational evidence

Since the details in Fig. 1 depend sensitively on the input physics used in stellar model calculations — especially on the assumed amount of convective overshooting — we want to specify the input physics required in order to construct Fig 1. Moreover, we line up observational evidence in favour of the evolutionary sequences outlined in Fig. 1, which then in turn justifies the choice of input physics parameters.

A basic condition for the validity of Fig. 1 is, that for H-burning massive stars it is

$$-\dot{M} < -\dot{M}_{core}, \tag{2}$$

where M is the total stellar mass and M_{core} is the mass of the H-burning convective core, since otherwise the thickness ΔM of the H-poor part of the stellar envelope at He-ignition as well as its hydrogen content are not monotonic functions of the ZAMS mass (cf. Langer, 1987). We want to emphasize that Eq. (2) is equivalent to the condition that the mass of the hydrogen containing stellar envelope is growing during central hydrogen burning. Eq. (2) is determined firstly by the stellar mass loss rate, and secondly by the efficiency of convective overshooting, which is known not only to increase the absolute value of M_{core} but also to significantly reduce the rate of core mass decrease $\dot{M}_{core}$ (cf. Langer and El Eid, 1986; Prantzos et al., 1986). Recent stellar evolution computations for massive stars in the range of $15 - 100\,M_{\odot}$ (Langer, 1986; Langer and El Eid, 1986) show that both, stellar mass loss and convective overshooting during central H-burning, can be adjusted in a way to be consistent with observations of massive main sequence stars while reconciling the evolutionary picture of post main sequence evolution outlined in Fig. 1. In order to recover Eq. (2) a reduced effectivity of convective overshooting is required (i.e. the overshooting distance must be much smaller than one pressure scale height, which is in agreement with Mermilliod and Maeder, 1986) when the mass loss rate of Lamers (1981) is used.

Several predictions can be extracted from the evolutionary scenario for massive helium burning stars as drawn up by Fig. 1:

1) The separation of the types WNL and WNE is rather distinct.
2) The most massive WR stars are of type WNL.
3) Low mass WR stars are of type WNE or WC.
4) Most WNE stars do evolve into WC stars.
5) The most massive WNL stars do not evolve into WNE or WC stars.

Points 1) to 5) are in good agreement with several observations concerning WR stars (cf. Langer, 1987, for details), as there are **(a)** a very high mean mass of WNL stars derived from binary systems ($60 - 70\,M_\odot$), compared to that of WNE ($7\,M_\odot$) and WC ($15\,M_\odot$) stars (Moffat, 1982; Niemela, 1983); **(b)** very high bolometric (Smith and Willis, 1983) and visual (Breysacher, 1986) luminosities for WNL stars compared to WNE and WC stars; **(c)** a monotonic mass-luminosity relation for WR stars, derived from theory (Maeder, 1983) and observations (Lundström and Stenholm, 1984); and **(d)** ZAMS-masses for different WR subtypes derived by a comparison of cluster turnoffs and stellar evolutionary tracks for clusters containing WR stars by Schild and Maeder (1984). Furthermore, the existence of the HS variables and their position in the HR diagram fits nicely in the above scheme. Two additional arguments supporting especially point 5) are the following:

The supernova remnant Cas A originates definitely from the explosion of a massive star, since its mass has been estimated to be larger than $15\,M_\odot$ (Fabian et al., 1980). The lack of observations of the supernova explosion in ~1600 indicates a very low peak luminosity of the supernova outburst, which is in good agreement with what one expects from an exploding WR star (Schaeffer et al., 1987). The WR progenitor hypothesis for Cas A is confirmed by an analysis of the chemical composition of the remnant, which indicates further that the Cas A progenitor was a WNL star (Fesen et al., 1987). Finally, a comparison of composition and amount of the oxygen burning products of massive stars explosions with the chemical composition of the so called 'fast moving knots' of Cas A (Johnston and Yahil, 1984; El Eid and Langer, 1986) indicate, that the ZAMS mass of the progenitor star was probably of the order of $\sim 100\,M_\odot$.

The second example is the peculiar supernova 1961v in NGC 1058. Observations indicate, that the progenitor star might have been as massive as $2000\,M_\odot$, that its envelope was nitrogen enriched, and furthermore that hydrogen was present but underabundant (Utrobin, 1984; Cowan and Branch, 1985). One may therefore identify the progenitor of this supernova also with a very massive WNL type star. The evolution of a $1000\,M_\odot$ star has been explored by Klapp (1983): he found the rate of core mass decrease during central hydrogen burning to be $-\dot{M}_{core} \simeq 2 \cdot 10^{-4}\ M_\odot\ yr^{-1}$. Adopting the mass loss rate of Lamers (1981) for such an object yields a mass loss rate due to stellar winds of $-\dot{M} \simeq 1 \cdot 10^{-4}\ M_\odot\ yr^{-1}$. Therefore, Eq. (2) may be valid for such massive stars: on the ZAMS the envelope mass may be neglected (the star is nearly fully convective), but at central H-exhaustion it would amount to $1 \cdot 10^{-4}\ M_\odot\ yr^{-1} \cdot \tau_H \simeq 200\,M_\odot$ ($\tau_H \simeq 2 \cdot 10^6\,yr$). Probably, an amount of $200\,M_\odot$ will not be lost during the later burning stages, and so there is a good chance that at the end of its life the star will be in the WNL phase. We have to mention here, that stars as massive as $2000\,M_\odot$ may explode due to the $e^{\pm}$-pair instability in case of finite angular momentum (Glatzel et al., 1985). Furthermore, the vibrational instability which is known to occur in homogenous very massive stars has possibly only the effect of slightly increasing the stellar mass loss rate (Appenzeller, 1986). For these reasons, in case a $2000\,M_\odot$ star is formed, it may well evolve as described above.

Both, Cas A and SN 1961v, are examples of exploding very massive WNL stars, supporting therefore the argument that those stars do usually not reach the WNE- or WC-phase in their evolution.

4. Conclusion

From Fig. 1 we can derive the following evolutionary sequences for massive stars as a function of the ZAMS mass:

$M < M_{WN}$: OB-star → supergiant → SN II
$M_{WN} < M < M_{HS}$: O-star → supergiant → WNE (→ WC) → SN 'I'
$M_{HS} < M < M_2$: O-star → HS-var → WNE → WC → SN 'I'
$M_2 < M < M_{WNL}$: O-star → HS-var → WNL → WNE (→ WC) → SN Typ 'I'
$M_{WNL} < M$: O-star → HS-var → WNL → SN Typ 'II'

This evolutionary picture is supported by several observations (cf. Sect. 3), therefore restricting the overshooting distance in core H-burning massive stars to values much smaller than one pressure scale height. We emphasize that the phases designated as SN 'I' and SN 'II' in the above scheme corresponds to possible supernova explosions of WR stars, which still contain ('II') or do not contain ('I') hydrogen in their surface layers. These kinds of supernovae have not yet been identified in observed explosion events (see, however, Begelman and Sarazin, 1986; Branch, 1986; Branch and Nomoto, 1986; Branch and Venkatakrishna, 1986; for a discussion of the peculiar type I supernovae called type Ib).

A open problem concerning the evolution of WR stars is the question which stars evolve into early WC's (WCE) and which into late ones (WCL), and what is the physical effect dividing the WC sequence into early and late types. Schild and Maeder (1984) conclude that late WN's evolve into early WC's and vice versa from studies of WR stars in open clusters. This is confirmed by Langer et al. (1987), who suggest that effects of partial recombination of carbon and oxygen on the opacity and equation of state in WC envelopes may be responsible for the existence of the subclasses WCE and WCL. Further work on this topic is in progress.

Acknowledgement: This work has been supported in part by the Deutsche Forschungsgemeinschaft through grant no. Fr 325/22-2.

References

Appenzeller, I.: 1986, in: Proc. workshop on *Instabilities in Luminous Early Type Stars*, Lunteren, C. de Loore, H. Lamers, eds., Reidel, in press

Begelman, M.C., Sarazin, C.L.: 1986, Astrophys. J. (Letters) **302**, L59

Branch, D.: 1986, Astrophys. J. (Letters) **300**, L51

Branch, D., Nomoto, K.: 1986, Astron. Astrophys. (Letters) **164**, L13

Branch, D., Venkatakrishna, K.L.: 1986, Astrophys. J. (Letters) **306**, L21

Breysacher, J.: 1986, Astron. Astrophys. **160**, 185

Chiosi, C.: 1982, in: IAU-Symposium No. 99 on *Wolf-Rayet stars: Observations, Physics, Evolution*; C. de Loore, A.J. Willis, eds.; p. 323

Conti, P.S., Leep, E.M., Perry, D.N.: 1983, Astrophys. J. **268**, 228

Cowan, J.J., Branch, D.: 1985, in: Proc. 3rd workshop on *Nuclear Astrophysics*, Ringberg, W. Hillebrandt et al., eds., p. 66

El Eid, M.F., Langer, N.: 1986, Astron. Astrophys. **167**, 274

Fabian, A.C., Willingdale, R., Pye, J.P., Murray, S.S., Fabbiano, G.: 1980, M.N.R.A.S. **193**, 175

Fesen, R.A., Becker, R.H., Blair, W.P.: 1987, Astrophys. J. **313**, 378

Garmany, C.D., Conti, P.S.: 1982, in: IAU-Symposium No. 99 on *Wolf-Rayet stars: Observations, Physics, Evolution*; C. de Loore, A.J. Willis, eds.; p. 105

Glatzel, W., El Eid, M.F., Fricke, K.J.: 1985, Astron. Astrophys. **149**, 413

van der Hucht, K.A., Conti, P.S., Lundström, I., Stenholm, B.: 1981, Space Sci. Rev. **28**, 227

Humphreys, R.M.: 1984, in: IAU-Symposium No. 99 on *Wolf-Rayet stars: Observations, Physics, Evolution*; C. de Loore, A.J. Willis, eds.; p. 279

Johnston, M.D., Yahil, A.: 1984, Astrophys. J. **285**, 587

Klapp, J.: 1983, Astrophys. Space Sci. **93**, 313

Lamers, H.J.G.L.M.: 1981, Astrophys. J. **245**, 593

Langer, N.: 1986, Ph. D. thesis, Göttingen University

Langer, N.: 1987, Astron. Astrophys. (Letters) **171**, L1

Langer, N., El Eid, M.F., Fricke, K.J.: 1985, Astron. Astrophys. **145**, 179

Langer, N., El Eid, M.F.: 1986, Astron. Astrophys. **167**, 265

Langer, N., Kiriakidis, M., El Eid, M.F., Fricke, K.J., Weiss, A.: 1987, submitted to Astron. Astrophys. (Letters)

Lundström, I., Stenholm, B.: 1984, Astron. Astrophys. Suppl. **58**, 163

Maeder, A.: 1982, Astron. Astrophys. **105**, 149

Maeder, A.: 1983, Astron. Astrophys. **120**, 113

Mermilliod, J.C., Maeder, A.: 1986, Astron. Astrophys. **158**, 45

Moffat, A.F.: 1982, in: IAU-Symposium No. 99 on *Wolf-Rayet stars: Observations, Physics, Evolution*; C. de Loore, A.J. Willis, eds.; p. 263

Niemela, V.S.: 1983, in: Proc. *Workshop on Wolf-Rayet Stars*, Paris-Meudon, M.C. Lortet, A. Pitault, eds.; p. III.3

Nugis, T.: 1982, in: IAU-Symposium No. 99 on *Wolf-Rayet stars: Observations, Physics, Evolution*; C. de Loore, A.J. Willis, eds.; p. 127

Prantzos, N., Doom, C., Arnould, M., de Loore, C.: 1986, Astrophys. J. **304**, 695

Schaeffer, R., Cassé, M., Cahen, S.: 1987, Astrophys. J. Letters, in press

Schild, H., Maeder, A.: 1984, Astron. Astrophys. **136**, 237

Smith, L.J., Willis, A.J.: 1983, Astron. Astrophys. Suppl. **54**, 229

Stothers, R., Chin, C.-W.: 1976, Astrophys. J. **204**, 472

Utrobin, V.P.: 19894, Astrophys. Space Sci. **98**, 115

Willis, A.J.: 1982, in: IAU-Symposium No. 99 on *Wolf-Rayet stars: Observations, Physics, Evolution*; C. de Loore, A.J. Willis, eds.; p. 87

Advanced Phases and Nucleosynthesis in Very Massive Stars

M.F.El Eid[1], N.Prantzos[2], and N. Langer[1]

1. Universitäts-Sternwarte Göttingen, FRG

2. Institut d'Astrophysique, Paris, France

Abstract. A brief discussion of current problems in very massive star ($M \geq 60\,M_\odot$) evolution is presented. In particular we focus on those stars which encounter the electron- positron pair creation instability at oxygen ignition. A star of initial mass of $100\,M_\odot$ is an important candidate for this kind of instability, and may be used to explore the role the pair instability in determining the fate of the known most luminous stars. Using up-dated input physics and a detailed nuclear reaction network, we present in this contribution results concerning the carbon and neon burning phases for this star. The ensuing dynamical evolution through explosive oxygen burning and the supernova stage are under way, and will be given elsewhere.

1. Introduction

Stars more massive than $60\,M_\odot$ are called very massive stars (VMS), and they represent the most luminous stars known in the Milky Way and in several nearby galaxies. Famous examples are η Car, P Cyg in The Milky Way, S Doradous in the large Magellanic cloud, and the Hubble–Sandage variables in M31 and M33.

A composite Hertzsprung–Russel (H-R) diagram for the most luminous stars (Humphreys and Davidson, 1984) show the interesting feature that stars with luminosities in excess of $10^6 L_\odot$ do not exhibit effective temperatures less than $\sim 1.5\ 10^4\ K$. This observational feature indicates that some kind of instability occurs when a VMS evolves rightwards across the upper part of the H-R diagram causing a drastic increase in mass loss which may reduce the star to a Wolf–Rayet (W-R) star. A W-R star which then exhibits strong stellar wind has a hot surface which is hydrogen- deficient but helium–rich and either nitrogen–rich (spectral type WN) or carbon/oxygen–rich (spectral type WC). However, mass loss is not the only important process leading to the formation of W-R stars (cf. Langer; Meynet and Maeder, in this volume). Larger convective cores due to overshooting from these cores may also influence this formation process.

Many uncertainties are still involved in the theoretical description of the combined effect of mass loss and overshooting during the quasistatic evolution of massive stars. In particular, the mechanism of mass loss is not well known. Surface radiation pressure certainly plays a key role, but not in all evolutionary phases. Turbulent pressure near the surface may also cause heavy mass loss during the supergiant stages of VMS as suggested by de Jager (1984). Evolutionary calculations with detailed description of this effect are not simple, and still not available.

Concerning overshooting the situation is controversial, since its extent is uncertain, although there are several indications (cf. Langer and El Eid, 1986 (LE86); Meynet and Maeder; Kuhfuss, this volume) supporting a moderate amount of overshooting of less than one pressure scale hight.

Current observations indicate that W-R stars originate from stars initially more massive than 40 $M_\odot$. Therefore, it seems to be important to follow the final evolution of these W-R stars through their possible supernovae stages. In recent works LE86 and El Eid and Langer (1986) (hereafter EL86) have followed the whole evolution of a 100 $M_\odot$ pop I star, and have found that the mass of the resulting W-R star at the end of He burning is rather sensitive to the assumed degree of overshooting. Evolving through the carbon and neon phases the W-R star collapsed at the onset of oxygen burning, since it became dynamically unstable due to conspicuous creation of electron–positron ($e^{\pm}$) pairs by the radiation field. The pair instability occurred even in the case without any overshooting. The effect of such instability is that oxygen burning proceeds explosively and reverses the collapse into explosion (pair creation supernova PCSN).

Depending sensitively upon the mass of the W-R star (more precisely upon the mass of the oxygen core at He exhaustion) explosive oxygen burning may lead to complete disruption of the star or it may initiate violent pulsations accompanied by mass ejection (pulsational pair instability as proposed by Woosley and Weaver (WW86), 1986).Thus there are many interesting problems related to the explosion of W-R stars. Extended hydrodynamical computations of these phases with realistic input physics are required, in order to figure out whether such explosions could have been the progenitors of some oxygen- rich supernovae like Cas A, SN 1985f, SN 1961v, and similar objects. These calculations should involve detailed nucleosynthesis as well as radiation transport which is needed to obtain the light curves of such explosions.

We also mention the impact of such computations on the nucleosynthetic yield from W-R stars. Recently Prantzos et al. (1987) have calculated the s–process products during core He burning of W-R stars originating from stars in the mass range $50 \leq M/M_\odot \leq 100$, and have emphasized that the contribution of these W-R stars to the s–process is not only due to stellar winds, but also to mass ejection during their possible explosions.

Keeping this background situation in mind, we have enough motivations to continue elaborating on the evolutionary models of VMS, especially on their advanced burning phases. This is the primary aim of this contribution, which is organized as follows. In Sect. 2 some improvements of the input physics incorporated in the present calculations are presented. Sect. 3 contains new results for the carbon and neon burning phases of a 45 $M_\odot$ W-R star originating from a 100 $M_\odot$ pop I star. Concluding remarks are given in Sect. 4.

2. Advanced burning Phases of a 100 $M_\odot$ star

In connection with the $e^{\pm}$ pair instability a 100 $M_\odot$ star is an interesting case, since it seems to lie on the border line separating complete disruption by explosive oxygen burning and pulsation with mass ejection. During the quasistatic evolution of this star up to the end of the He burning phase the role of mass loss and overshooting is uncertain. Consequently, the masses of the resulting W-R star spans a relatively wide range as seen in Table 1, where a comparison among several calculations is made indicating that the calibration of the W-R mass to the initial mass is not at all straight-forward.

Table 1. Comparison of the remaining mass M_α at the end of He burning for a star of initially 100 $M_\odot$ obtained by several authors. The remaining star may be identified with a W-R star. d_{over}/H_p is the overshooting distance in terms of the pressure scale hight H_p . M_O is the mass of the oxygen core.

Authors	1	2		3	4
$M_\alpha/M_\odot$	50	45	61	65	42
$M_O/M_\odot$	42	38	52	54	36
d_{over}/H_p	0.25	0.	1.90	Roxbourgh Crit.	0. (probably)
W-R Type	WC	WN	WC	WC	WN

1: Maeder and Meynet (1987), 2: Langer and El Eid (1986)
3: Prantzos et al. (1986), 4: Woosley and Weaver (1986)

Table 1 shows that the Roxbourgh criterion for convection leads to the largest core mass, and is comparable to an overshooting distance of at least 1.9 H_p (cf. LE86). Whether the Roxbourgh criterion is the wright approach for determining the degree of overshooting or not is questionable (cf. Kuhfuss this volume; Baker and Kuhfuss, 1986). Presently, the tendency points towards a moderate amount of overshooting. According to the recent work of Maeder and Meynet (1987) an overshooting distance of about 0.25 H_p should be more realistic. If so, then a 100 $M_\odot$ star would evolve to a W-R star of 50 $M_\odot$, while a 120 $M_\odot$ star leads to 64 $M_\odot$ W-R star in agreement with the estimate of LE86.

The rest of this contribution will deal with the advanced stages of the 45 $M_\odot$ W-R star. Although, this star has been evolved through the supernova stage by EL86, its final fate is a bit controversial in light of the results obtained by WW86 indicating pulsational pair instability rather than complete disruption. In order to resolve this discrepancy, we are redoing the calculations for this star with several modifications in the input physics.

The first important modification concerns the opacity. We have relaxed the previous assumption (EL86) that the opacity is only due to Thomson scattering. Indeed for the thermodynamic conditions encountered in the final phases of pair unstable stars the Thomson scattering opacity should be corrected by including inelastic Compton scattering in the presence of $e^\pm$ pairs and degeneracy of the electrons. Compton scattering without degeneracy in the electron distribution has been calculated by Sampson (1959). An extension of this work to include degeneracy have been done by Buchler and Yueh (1976). We have refitted Sampson's results up to $T_9 = 9.0$ by the following expression:

$$\kappa(T_9) = \kappa_{Th}\ (1.001 + 2.394\ T_9 - 0.347\ T_9^2)^{-1}, \tag{1}$$

where κ_{Th} is the Thomson opacity. This expression, which is a slightly modified version of that used by Fuller et al. (1986), clearly indicate that the opacity at high temperatures decreases well below the Thomson scattering(elastic) value (see Sect. 3). At temperatures larger than

$T_9 = 9.0$ the correction factor has been calculated according to Buchler and Yueh (1976) as follows:

$$\kappa(T,\eta) = \sigma_0 \; n \; G(T,\eta)^{-1}/\rho, \tag{2}$$

where $\sigma_0 = 6.652 \; 10^{-25} \; cm^2$ is the Thomson cross section, $n = n_+ + n_-$ denotes the combined number density of electrons and positrons, and ρ is the mass density. The function G has been taken from Buchler and Yueh (1975), and will not be given here.

Another modification in the present calculations concerns the network of nuclear reactions. EL86 have restricted the network to α- nuclei between He and Ni. While with that network, the nuclear energy generation rate is fairly reproduced, it is not sufficient to account for detailed nucleosynthesis. However, the modified version shown in Fig. 1 should be more adequate, since it includes the (p,γ), (α,p), (α,γ), (n,γ), and (α,n) reactions.

Concerning the equation of state for the $e^{\pm}$ pairs no improvement is required, since the relativistic Fermi-Integrals are evaluated numerically if necessary (cf. EL86).

Finally, we have modified the calculations of the neutrino energy loss rates according to the recent work of Munakata et al. (1985) who have calculated these rates in the framework of the Weinberg–Salam theory allowing for a variable number of neutrinos other than electron neutrinos.

Using the above modifications, we have followed the evolution of the 45 $M_{\odot}$ star starting at the end of He burning up to the end of Ne burning, which coincides with the onset of the collapse phase initiated by the pair instability. The hydrodynamical code described by EL86 has been used to perform the present computations. The further evolution of the star is under investigation, and will be published elsewhere

3. Carbon and Neon Phases

The network shown in Fig.1 has been utilized up tp ^{38}Ar for these phases. The compositions at the end of He burning are those given by EL86 for the elements C, O, and Ne. For the other nuclear species we have adopted the average values calculated by Prantzos (1986) to account for the nuclear transmutation during He burning.

The mass fraction of carbon started decreasing at the center when $T_9 > 8.0$ and $\rho > 2.5 \; 10^4 g \; cm^{-3}$. The effective energy generation rates (nuclear minus neutrinos) were negative during this phase. Hence, no convective core develops under these conditions. Fig. 2a shows the nuclear transmutations during carbon burning, which has occurred within 20% of the mass. The nuclear species comprising about 98% in mass fraction are ^{16}O, ^{20}Ne, ^{24,25,25}Mg. ^{27}Al and ^{28}Si have been enhanced by a factor of ~20 and ~14 respectively compared with the initial values. At the end of the carbon burning phase the central temperature, density, and dimensionless entropy were respectively:$T_c = 1.41 \; 10^9$ K, $\rho_c = 2.08 \; 10^5$ g cm^{-3}, $S_c = 3.84$. We note that the $e^{\pm}$ pairs were not important during this phase.

Neon burning commenced near to $T_c = 1.46 \; 10^9 K$ and $\rho_c = 2.35 \; 10^5 g \; cm^{-3}$. The global effective energy generation rate during this phase was also negative due to the enhanced neutrino energy losses. Fig. 2b displays the composition profiles at the end of Ne burning, where $T_c = 2.03 \; 10^9 K$ and $\rho_c = 6.81 \; 10^5 g \; cm^{-3}$. The major nuclear constituents are ^{16}O, ^{24,25,26}Mg

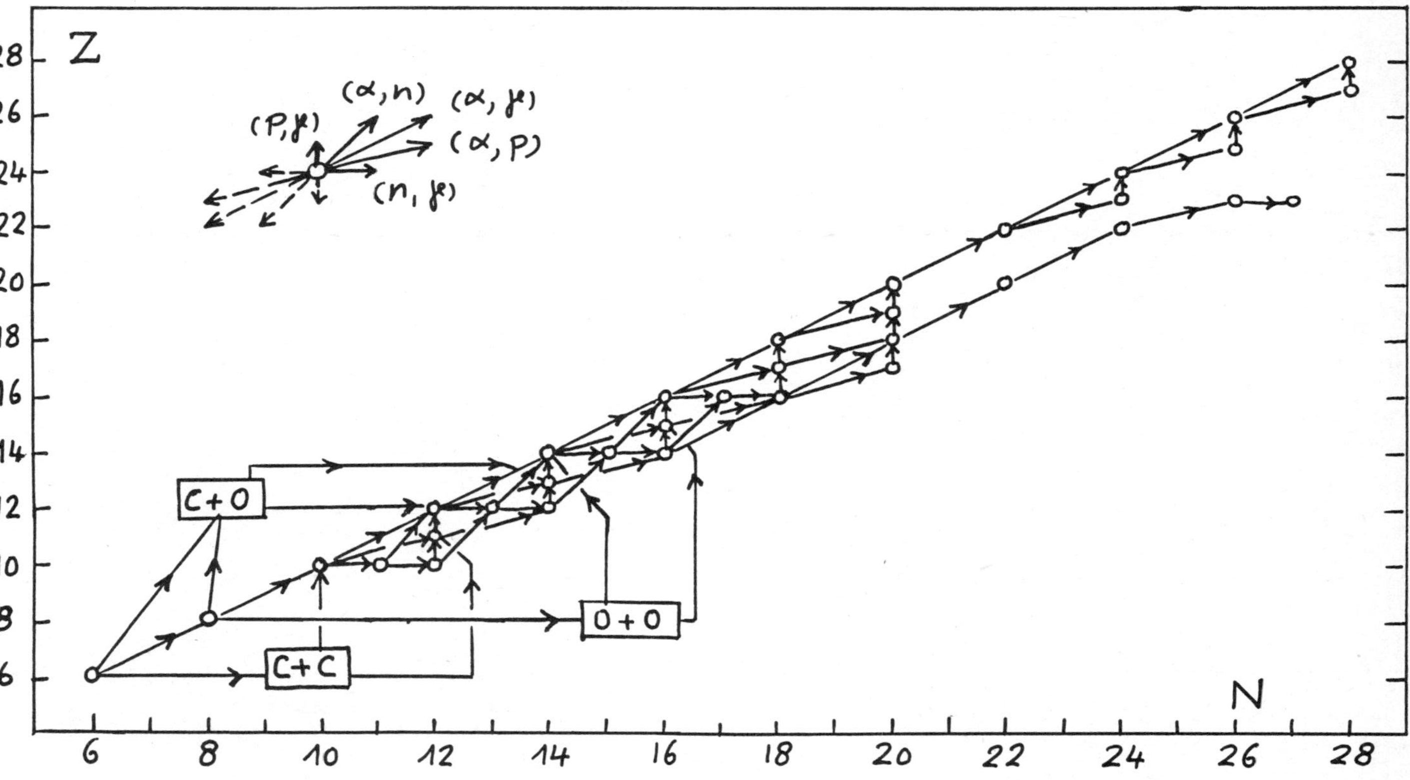

Fig. 1: Nuclear reaction network carbon, neon, and oxygen burning. The tripel alpha reaction and the neutron channels of the C+C, O+O, and C+O reactions, as well as all inverse reactions have been included in the calculations, but are not shown in the Figure. For carbon and neon burning only nuclei up to ^{38}Ar are needed.

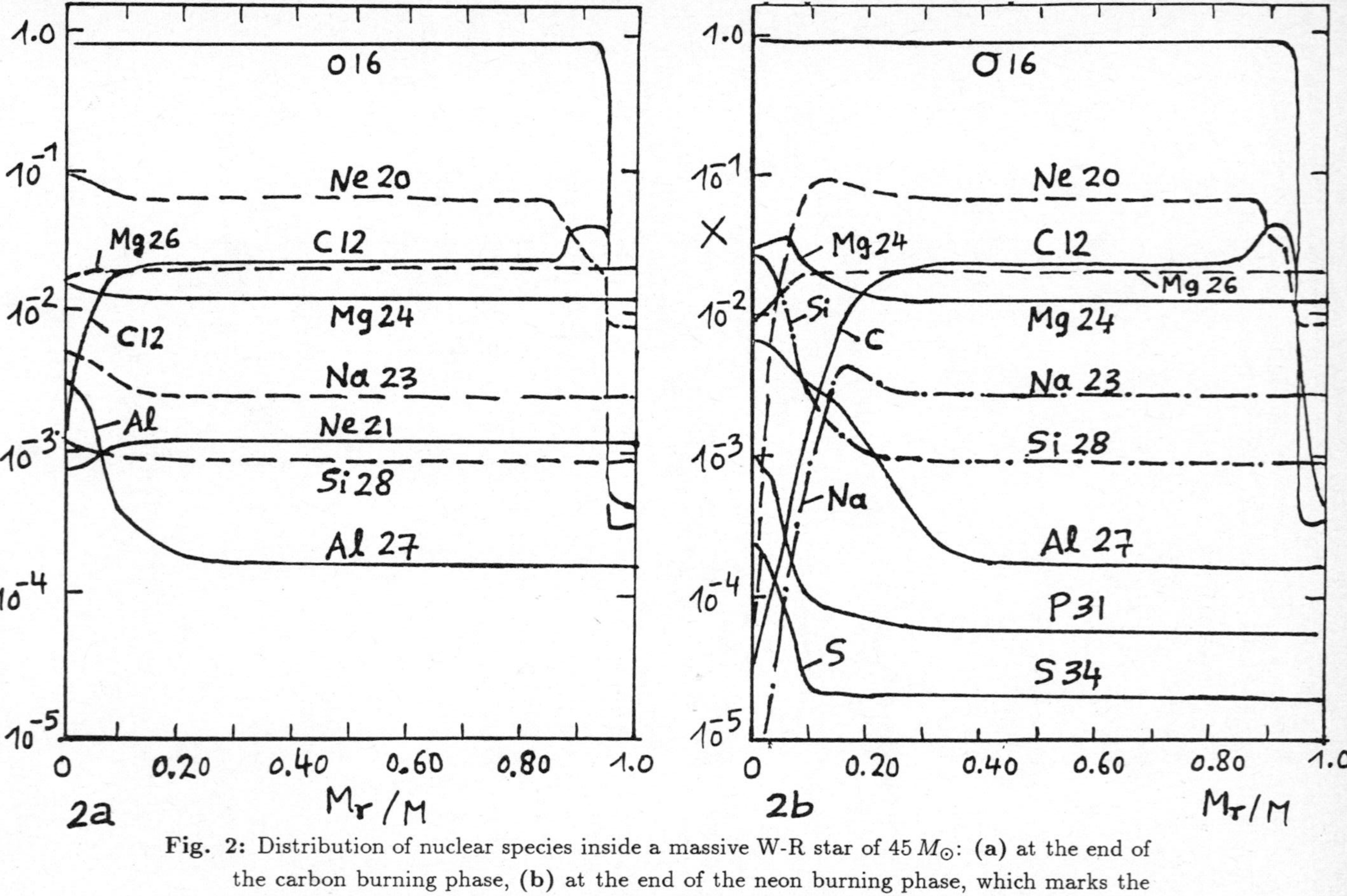

Fig. 2: Distribution of nuclear species inside a massive W-R star of $45\,M_\odot$: **(a)** at the end of the carbon burning phase, **(b)** at the end of the neon burning phase, which marks the onset of collapse due to $e^{\pm}$ pair creation, which comprises about 40% of the mass at this instant.

^{27}Al, ^{28,29,30}Si. ^{31}P and ^{34}S have been enhanced. In Fig. 3, the profiles for the entropy and the degeneracy parameter η_e of the electrons are shown. The decrease of the entropy towards the center indicate stability against convection (cf. Cox and Giuli, 1968 Sect. 13.4). In Fig. 3 η_e is close to -1.0 at the center, and will increase towards zero during the ensuing oxygen burning . This will affect the equation of state as well as the opacity, which should than be calculated by Eq. (2) given in Sect. 2. We note that at the end of Ne burning the opacity had a value of 0.044 $cm^2\ g^{-1}$ at the center and 0.17 $cm^2\ g^{-1}$ at the surface of the star. The low central value is the remarkable effect of the correction by Eq. (2). The distribution of the opacity inside the star will play a key role in determining the final outcome of the supernova calculations (see recent calculations by El Eid , 1986).

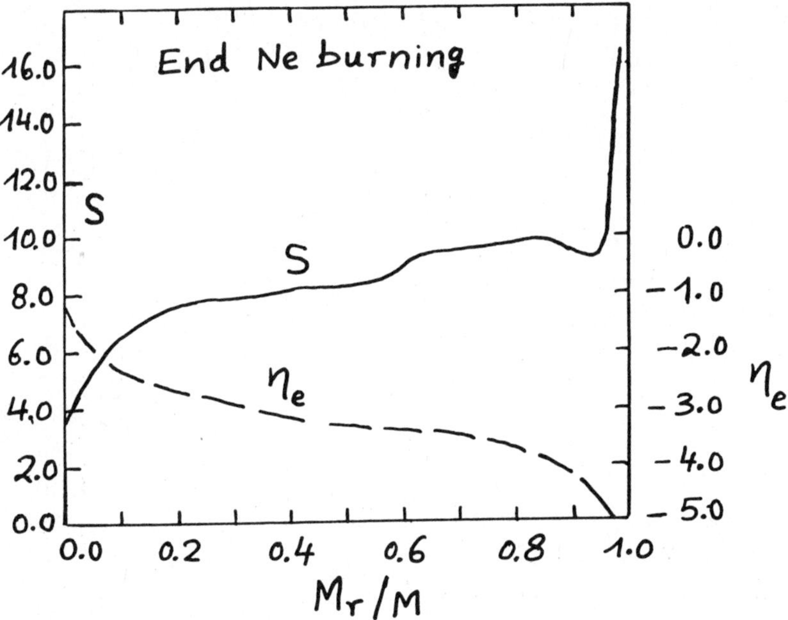

Fig. 3: Distribution of the total dimensionless entropy S (in units of the Boltzmann constant), and the degeneracy parameter η_e of the electrons for the star as in Fig. 2 at the end of neon burning.

4. Concluding remarks

We have described (Sect. 2) some improvements in the input physics required to follow the evolution of very massive stars, especially their final phases. These stars are important not only as possible progenitors of some peculiar oxygen–rich supernovae, but also as an interesting site of nuclear processing (cf. N. Prantzos; M. Rayet, this volume). We have presented some preliminary results (Sect. 3) for the carbon and neon burning phases of a massive W-R star of $45\,M_\odot$, which may originate from a $100\,M_\odot$ ZAMS star. The final evolution of this star and other massive W-R star through the supernova stage is in progress.

Acknowledgement: The calculation have been partly done with the Cray-XMP at the CEN, Saclay. M.F.E. is very grateful for the hospitality and support he has received from the service d'astrophysique at CEN. He thanks W.Hillebrandt for supporting his stay at the nice Ringberg castle. The authors would like to thank M.Arnould, and M. Rayet for many critical discussions.

References

Baker, N.H., Kuhfuss, R., 1986, Preprint, MPI für Physik und Astrophysik, no **256**

Buchler, J.R., Yueh, W.R., 1976, Astrophys. J. **210**, 440

Cox, J.P., Giuli, R.T., 1968, Principal of Stellar Evolution, vol II, Cordon and Breach

De Jager, C., 1984, Astron. Astrophys. **138**, 252

El Eid, M.F., Langer, N., 1986, Astron. Astrophys. **167**, 274

El Eid, M.F., 1986, in "Advanced in Nuclear Astropysics", eds. E. Vangioni–Flam et al., edition Frontieres, Paris, France

Fuller, G.M., Woosley, S.E., Weaver, T.A., 1986, Astrophys. J. **307**, 675

Humphreys, R.M., Davidson, K., 1984, Science **23**, 243

Langer, N., El Eid, M.F., 1986, Astron. Astrophys. **167**, 265

Langer, N.: 1987, Astron. Astrophys. (Letters) **171**, L1

Maeder, A., Meynet, G., 1987, Astron. Astrophys. in press

Munakata, H., Kohyama, Y., Itoh, N., 1985, Astrophys. J. **296**, 197

Prantzos, N., 1986, Ph.D. Thesis, Universite De Paris VII, France

Prantzos, N., Arnould, M., Arcoragi, J–P., 1987 Astrophys. J. **315**, in press

Prantzos, N., Doom, C., Arnould, M., DE Loore, C., 1986 Astrophys. J. **304**, 695

Sampson, D.H., 1959, Astrophys. J. **129**, 734

Woosley, S.E., Weaver, T.A., 1986 IAU Coll. **89**, eds. D. Mihalas, and K.H. Winkler

OVERSHOOTING AND ELECTRON-POSITRON PAIR INSTABILITY

G. Meynet and A. Maeder
Geneva Observatory
CH-1290 Sauverny, Switzerland

I. INTRODUCTION

In recent years many calculations have been performed in order to understand the effects of mass loss and overshooting from convective cores on the evolution of the stars (Chiosi and Maeder, 1986 and see references therein). Drawing the lessons of these previous works we tried to establish a reference grid of massive star models (Maeder and Meynet, 1987). Particularly great attention was paid to the choice of the overshooting parameter. Its value is very important since it governs the size of the convective cores and therefore the quantity of available fuel for the nuclear reactions active in the center of the star.

In the following we present at first the method used to determine the value of the overshooting parameter and new observable properties sensitive to overshooting. In the second part we investigate the effect of electron-positron pair creation instability on the models evolved with mass loss and overshooting.

II. CHOICE OF THE OVERSHOOTING PARAMETER AND OBSERVATIONAL CONSEQUENCES

As for the moment there is no well accepted theoretical way to estimate the amount of overshooting from convective cores, we chose the value of the overshooting parameter which reproduces at best the observable properties sensitive to it. In this observational approach of the problem one property appears to be very useful : the location of the top of the main sequence is very sensitive to the amount of overshooting from convective cores (cf. Maeder, 1976; Bressan et al., 1981; Stothers and Chin, 1981; Doom, 1982ab). In the HR diagram in figure 1 we show various theoretical envelopes of the main sequence (MS) computed with different values of the overshooting parameter. This parameter is equal to the overshooting distance, d_{over}, measured in units of pressure scale heights, H_p. The hatched areas correspond to the observed sequences, as found by Humphreys and Davidson (1979), Mermilliod and Maeder (1986). We see that the MS given by the models without overshooting is too narrow for the early B-type stars and too wide for the O-type stars. The models computed with the Roxburgh criterium, that is to say with a very large overshooting distance, do not fit at all the observed sequences. The present models with d_{over}/H_p = 0.25 provide a good agreement, both for the upper envelope of O-stars as well as for the upper envelope of B-stars in the HR diagram. Thus we decided to perform the calculation of our grid of massive star models with a value of the overshooting distance equal to 0.25 H_p. In this context it is interesting to note that calculations now in progress show that the same value of the overshooting distance reproduces well the observed width of the main sequence for the lower part of the HR diagram, at least for masses greater than 3 $M_\odot$.

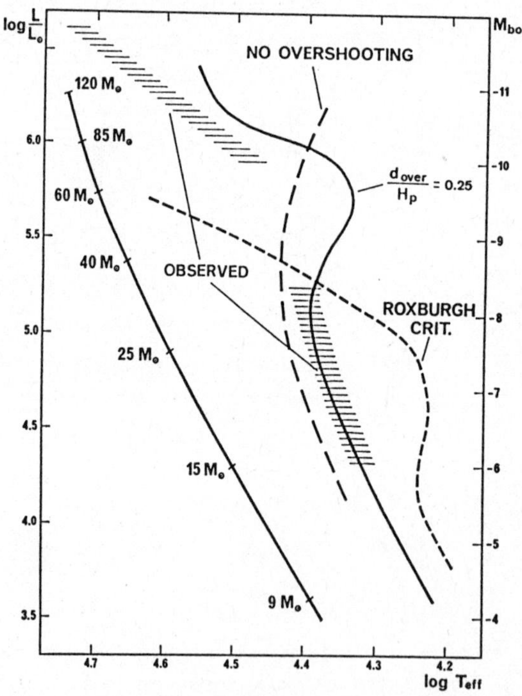

Figure 1 : The comparison of the main sequence width for various star models with the observations (hatched bands, cf. Humphreys and Davidson, 1979; Mermilliod and Maeder, 1986). Models without overshooting (case B, Maeder, 1981) or models with a very large overshooting (Doom, 1985) do not reproduce the observations. The best fit is realized for models with (d_{over}/H_p) = 0.25 when Los Alamos opacities are used.

The HR diagram obtained with this value of the overshooting distance is shown in Fig. 2. One major result of the models with mass loss (cf. Chiosi and Maeder, 1986) was the discovery of the existence of three evolutionary sequences according to the range of initial stellar masses considered (cf. also Fig. 1). This finding is also confirmed by the present models, even though the mass limits between the three ranges considered were slightly changed:

- the upper mass limit for the formation of red supergiants has been decreased from 60 $M_{\odot}$ to about 40 $M_{\odot}$, a figure which is in better agreement with the observations (cf. Humphreys and Davidson, 1979; Humphreys, 1983);

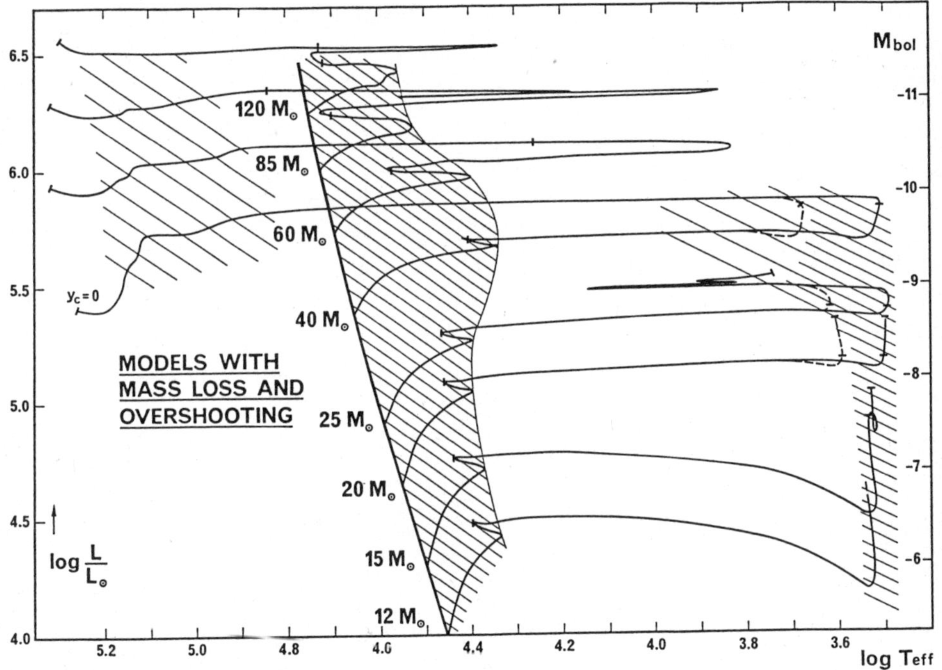

Figure 2 : Evolutionary tracks of massive stars with initial composition X = 0.70 and Z = 0.02 evolved with mass loss by stellar winds and overshooting. Hatched areas indicate the main sequence band and the He-burning phase. For red supergiants, the broken lines indicate the tracks computed with $\alpha_{\rho} = 1/H\ \rho = 0.3$ while the continuous lines refer to the usual mixing length theory with $\alpha_{p} = 1/H_{p} = 1.5$. The first slash along the evolutionary tracks indicates the central exhaustion of H, the second, the beginning of the He-burning phase and the last the central exhaustion of He.

- the WR stars form from initial masses larger than about 30-40 $M_\odot$; thus the mass range for WR stars as post-red supergiants is relatively small, as suggested by Humphreys et al. (1985);

- for the 15 $M_\odot$ model we noticed (cf. fig. 2) that no blue loop is produced which crosses the cepheid instability strip. This is consistent with the results by Matraka et al. (1982) who showed that overshooting may reduce or even suppress the loops. This is consistent also with observations of galactic cepheids since the brightest cepheids (e.g. SV Vul) observed in the Galaxy are less luminous than $M_{bol} = -6.5$. Calculations now in progress show that extended blue loops are predicted for models with mass equal to 12 $M_\odot$ or lower and computed with the same physics as here.

Among the new observable properties sensitive to overshooting the ratio of the lifetimes of the WN and WC stars t_{WN}/t_{WC} provides a good confirmation of our choice of the overshooting distance. We consider that a star enters the WR stage when the surface abundance X_s in hydrogen is below 0.30 and the logarithm of the effective temperature is superior to 4.5 on the final bluewards tracks. This last condition is required in order to avoid considering stars in the LBV region as WR stars. This restriction is not critical since the LBV stage is short. The modification of the limiting value of X_s for defining the entry in the WR stage generally does not affect the WR lifetime very much. As an example, for the 60 $M_\odot$ model, the limits $X_s = 0.40$ and 0.20 give respectively WR lifetimes of 0.604 and 0.537 instead of 0.542 for the adopted limit $X_s = 0.30$. The ratio of the lifetimes of the WN et WC stages are easily estimated by considering the manner in which the chemical abundances change at the surface. The transition between both stages is clearly defined, while the beginning of the WN stage is subject to the same remarks as above.

The ratio t_{WN}/t_{WC} is in fact sensitive to both mass loss and overshooting. High mass loss rate and large overshooting reduce the WN phase and cause a more rapid entrance into the WC stage; thus both effects favour small t_{WN}/t_{WC}. In our calculation, for the 40 and 60 $M_\odot$ models, i.e. in the range where most WR stars are likely to originate (cf. Humphreys et al., 1985), the ratios t_{WN}/t_{WC} amount respectively to 1.22 and 1.34. The comparison with the models of Prantzos et al. (1986) is very interesting. The mass loss rates these authors used are nearly the same as ours, but their convective cores are very large. For models of initial mass 50 and 60 $M_\odot$ they obtain $t_{WN}/t_{WC} = 0.34$ and 0.21. This clearly illustrates the sensitivity of this ratio to overshooting. The observed ratio of the number of WN stars to the number of WC stars in the Galaxy is 1.3 (cf. Vanbeveren and Conti, 1980). This is in remarkable agreement with our results, and for the adopted mass loss rates well supports our choice of the overshooting parameter $d_{over}/H_p = 0.25$ as found from the HR diagram of young associations. This is particularly interesting since the t_{WN}/t_{WC} ratio seems rather sensitive.

III. THE ELECTRON-POSITRON CREATION PAIR INSTABILITY

It is well known that, at sufficiently high temperatures, an appreciable fraction of the photons have energies higher than twice the rest mass energy of the electrons and that some of these very energetic photons give birth to pairs of electrons and positrons. This process may considerably change the thermodynamic quantities describing the state of the stellar matter and thus induce instabilities. In this context an important quantity is the adiabatic index $\Gamma_1 = \partial \ln P/\partial \ln \rho|_S$ where P and ρ are respectively the pressure and the density, the derivation being performed at constant entropy S. The value of Γ_1 is critical for the stability of the stars against adiabatic perturbations. When a sufficiently large part of the star (about 30% by mass) has a value of Γ_1 below 4/3, the star becomes dynamically unstable. The effect of pair creation favours such instabilities at the end of massive star evolution (Fowler and Hoyle, 1964). The range of initial stellar masses for such an instability to occur is always subject to discussion. It depends on the mass loss rates and the amount of overshooting. Let us call M_{min} the value of the minimum initial mass for the occurrence of the pair creation instability. Mass loss, by reducing the mass of the core at the end of the He-burning phase, enhances M_{min}. On the contrary, overshooting enlarges the convective core and therefore lowers this limit. Thus both effects act in an opposite way. In what follows we examine the value of this mass limit in the context of the grid of massive star models presented in the second paragraph.

When thermodynamic equilibrium is achieved and the electronic gaz is non degenerate, non relativistic, the number density n_+, n_- of positrons and electrons may be written as follows :

$$n_+ = n_1 e^{-\Phi} \quad (1)$$

$$n_- = n_1 e^{+\Phi} \quad (2)$$

where Φ is the chemical potential in units of kT and n_1, a function of temperature. For n_1 we used the following expression :

$$n_1 = \frac{2}{\pi^2}\left(\frac{kT}{\hbar c}\right)^3 \overline{K_2}(Z) \qquad (3)$$

with $Z = m_e c^2/kT$ and $\overline{K_2}(Z) = \frac{1}{2}Z^2 K_2(Z)$, where $K_2(Z)$ is the modified Bessel function of second order. The other symbols have their usual meaning (see also Fowler and Hoyle, 1964). $\overline{K_2}$ was approximated by the following expression

$$\overline{K_2} \simeq \frac{9.08}{T_9^{3/2}} \exp\left(\frac{-5.93}{T_9}\right) (1+0.316\ T_9+0.023\ T_9^2) \qquad (4)$$

which reproduces very well $\overline{K_2}$ for $T_9 \leq 4$. (T_9 is the temperature in units of 10^9K). The condition of charge neutrality implies that

$$n_o = n_- - n_+ \qquad (5)$$

where n_o is the number density of ionisation electrons. This equation can be used to determine the chemical potential Φ which then may be written

$$\Phi = \ln\left(\frac{n_o}{2n_1} + \sqrt{\left(\frac{n_o}{2n_1}\right)^2 + 1}\right) \qquad (6)$$

Using (1)-(2)-(6) and the expressions for the total pressure P and the internal energy u presented below, all the relevant thermodynamic quantities may be computed.

$$P = (n_+ + n_- + n_{ions})kT + 1/3\ aT^4 \qquad (7)$$

$$u = 3/2(n_+ + n_- + n_{ions})kT + aT^4 + (n_+ + (n_- - n_o))m_e c^2 + u_o \qquad (8)$$

where n_{ions} is the number density of the ions and u_o, a constant.

Let us examine the overall effect of pair creation on the model structure. The presence of pairs favours the contraction of the star since pair creation subtracts energy that otherwise provides pressure support (Barkat et al., 1967). Therefore, in the region where the pair creation effect is important, that is to say in the central parts of the star, the temperatures and densities are enhanced. In the rest of the star no changes occur. As a numerical example, in the 120 $M_\odot$ model, at the beginning of the C-burning phase (the mass fraction of ^{12}C at the center, $X(^{12}C)_c$ = 0.0105), the central temperature and density are respectively enhanced by 0.7% and 9% relatively to their values computed without the presence of pairs. Enhancement of the density is superior to 5% in the central three percent of the total mass. At the end of the C-burning phase ($X(^{12}C)_c$ = 0.003) the situation is still the same.

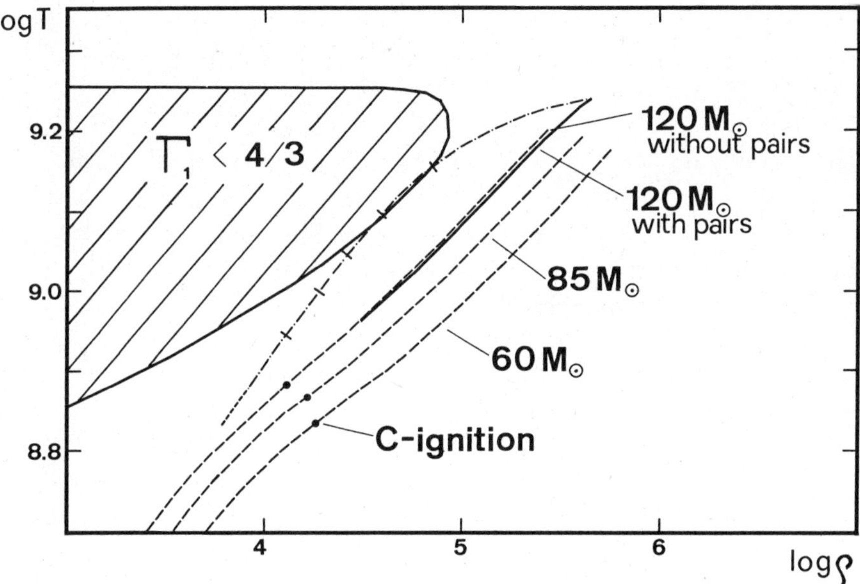

Figure 3 : The dashed lines represent the evolution of models of initial mass 60, 85 and 120 $M_\odot$ in the log T_c vs. log ρ_c plane (central temperatures vs. densities) when the effect of pair creation is neglected. The big dots on these lines indicate the beginning of the C-burning phase. The continuous line shows the evolution of the 120 $M_\odot$ model in the log T_c vs. log ρ_c plane when the pair creation effect is included. The dot-dashed line is the run of temperature and density inside the 120 $M_\odot$ model at the end of the Neon-photodesintegration phase when the pair creation effect is present. At this stage the actual mass of the star is 64 $M_\odot$ and the mass fraction of ^{20}Ne at the center is equal to 0.0162. The small slashes indicate the way the mass of the star varies along this line, two consecutive slashes encompass ten percent of the total mass of the star. The hatched region corresponds to the zone where Γ_1 is below 4/3. The limit of this zone was computed assuming a value of 2 for the mean molecular weight of the ionisation electrons and a value of 17 for the mean molecular weight of the ions.

The expressions used here to include the pair creation effect are very simple and may be easily incorporated into an evolutionary code. For temperatures and densities involved in our calculation (see Fig. 3) our approximate expressions for n_+ and n_- are correct with a precision of 2 to 3% during most of the evolution and for the largest part of the star. From this statement and the considerations made above on the overall effect of pair creation we can deduce that the precision of the thermodynamic quantities describing the internal structure amounts to a few thousandths.

When temperature is superior to about 1.3 10^9K the photodesintegration of the Neon becomes active : at the center of the star the energy produced by the reaction $^{20}Ne(\gamma, \alpha)^{16}O$ represents more than 5% of the total nuclear energy. During this phase Γ_1 becomes lower than 4/3 in the layers surrounding the core, which is in agreement with the results obtained by El Eid and Langer (1986)(see Fig. 3). It is interesting to note here that the number of positrons is not very important at this stage. As a numerical example, when the mass fraction of ^{20}Ne at the center = 0.0162, that is to say for the stage represented by the dot-dashed line on Fig. 3, the maximum of the ratio of the number of positrons to the number of ionisation electrons in the star is only 0.08. As evolution proceeds the fraction of the stellar mass inside the instability region becomes more and more extended and finally the star becomes dynamically unstable. Thus the 120 $M_\odot$ star undergoes the pair creation instability at the end of the Neon-photodesintegration phase. By considering the position of the evolutionary track for the central conditions in the 85 $M_\odot$ model we see that this star will probably not undergo such an instability; this statement is supported by the fact that the run of temperatures and densities inside the 120 $M_\odot$ model at the end of the Neon-photodesintegration phase just grazes the instability region. Thus, for our choice of the overshooting parameter and with the adopted mass loss rates we find a value between 100 and 120 $M_\odot$ for the minimum initial mass which will undergo the pair creation instability. This mass limit is well in the range of values proposed by many authors (Arnett, 1972; Ober et al., 1983; Woosley and Weaver, 1986). This is not surprising since both effects, mass loss and overshooting, act in an opposite way. Mass loss, by decreasing the mass of the core, enhances the mass limit for pair instability. On the contrary, overshooting by increasing the core mass lowers this mass limit. Thus, with their presently adopted sizes, both effects more or less compensate each other. Only in case of extreme mass loss rates (due for example to binary mass transfer) would this limit occur at considerably larger initial masses.

IV. CONCLUSION

The main results presented here are the following :

1) Our choice of the overshooting parameter $d_{over}/H_p = 0.25$, as found from the HR diagrams of young associations, reproduces very well some observable properties sensitive to the overshooting as, for instance, the ratio t_{WN}/t_{WC} of the durations of the WN and WC stages.

2) For this value of the overshooting parameter and for mass loss rates given by the observations, the minimum initial mass which undergoes the pair creation instability is between 100 and 120 $M_\odot$. This value does not differ from that generally given in the literature. The moderate amount of overshooting, which tends to lower this limit, compensates the effect of mass loss, which tends to enhance it.

REFERENCES

Arnett, W.D., 1972, *Ap. J.* **176**, 699
Barkat, Z., Rakavy, G., Sack, N., 1967, *Phys. Rev. Lett.* **18**, 379
Bressan, A.G., Bertelli, G., Chiosi, C., 1981, *Astron. Astrophys.* **102**, 25
Chiosi, C., Maeder, A., 1986, *Ann. Rev. Astron. Astrophys.* **24**, 329
Doom, C., 1982a, *Astron. Astrophys.* **116**, 303
Doom, C., 1982b, *Astron. Astrophys.* **116**, 308
Doom, C., 1985, *Astron. Astrophys.* **142**, 143
El Eid, M.F., Langer, N., 1986, *Astron. Astrophys.* **167**, 274
Fowler, W.A., Hoyle, F., 1964, *Ap. J. Suppl.* **9**, 201
Humphreys, R.M., 1983, *Ap. J.* **269**, 335
Humphreys, R.M., Davidson, M.K., 1979, *Ap. J.* **232**, 409
Humphreys, R.M., Nichols, M., Massey, P., 1985, *Astron. J.* **90**, 101
Maeder, A., 1976, *Astron. Astrophys.* **47**, 389
Maeder, A., 1981, *Astron. Astrophys.* **102**, 401
Maeder, A., Meynet, G., 1987, *Astron. Astrophys.* in press
Matraka, B., Wassermann, C., Weigert, A., 1982, *Astron. Astrophys.* **107**, 283
Mermilliod, J.C., Maeder, A., 1986, *Astron. Astrophys.* **158**, 45
Ober, W.W., El Eid, M.F., Fricke, K.J., 1983, *Astron. Astrophys.* **119**, 61
Prantzos, N., Doom, C., Arnould, M., de Loore, C., 1986, *Ap. J.* **304**, 695
Stothers, R., Chin, C.W., 1981, *Ap. J.* **247**, 1063
Vanbeveren, D., Conti, P.S., 1980, *Astron. Astrophys.* **88**, 230
Woosley, S.E., Weaver, T.A., 1986, *Ann. Rev. Astron. Astrophys.* **24**, 205

S - PROCESS PRODUCTION IN THE CENTRAL HELIUM BURNING OF LARGE MASSES ($M \geq 15\ M_\odot$)

D. Bencivenni (1), V. Castellani (2), A. Chieffi (1)
(1) Istituto Astrofisica Spaziale, CNR, 00044 Frascati, Italy
(2) Istituto Astronomico, Universita' " La Sapienza",
Via Lancisi 29, 00161 Roma, Italy.

I. INTRODUCTION

It is well known that one of the "classical" sites where s-elements may be produced, is the central He burning phase of large mass stars ($M \geq 15\ M_\odot$) through the chain:

$$N^{14}(\alpha,\gamma)F^{18}(\beta^+)\ O^{18}(\alpha,\gamma)\ Ne^{22}(\alpha,n)\ Mg^{25}$$

It was firstly recognized that in such a way it is possible the production of s-elements till the first neutron closure shell (N=50). The classical work of Lamb et al.(1977) showed that stars in the mass range 15 - 50 $M_\odot$ may produce elements till the Yttrium - Zirconium - Stronthium group in "noticeable" amount.

A scenario which is worth to be revisited, since we have now:

a) - new cross sections for n-capture (Almeida and Käppeler, 1983)

b) - new evaluation of the efficiency of convection during central He-burning (Castellani et al., 1987)

c) - new cross section for $C^{12}(\alpha,\gamma)O^{16}$ (Caughlan, Fowler, Harris and Zimmerman, 1985)

This paper will report the results of numerical experiments concerning the influence of point 1) and 2). The effects of increasing $C^{12}(\alpha,\gamma)O^{16}$ will be finally discussed.

As a matter of the fact, the recent reevaluations of the cross sections of several reactions (like $Ne^{22}(n,\gamma)Ne^{23}$ ...) already led several authors to reexamine the possible importance of the central He burning phase of large masses, as a possible site for the production of some of the s-elements observed in nature. Busso and Gallino (1985) (but also Arnett and Thielemann,

1985) found that the net effect of the new rates, is to significantly reduce (a factor of 4) the production of s-elements during the central He-burning phase of the core masses in the range ($4 \leqslant M\alpha \leqslant 32$). But these authors used central temperatures and densities of stellar models (Arnett 1972), which were computed with old reaction rates. In this scenario our goal is to examine again the production of s-elements in the He-burning core of large masses using updated evolutionary tracks computed in Frascati.

Moreover Castellani et al. (1987) have recently reanalized the last part of the central He-burning addressing in particular the existence of the convective instabilities that they called Breathing Pulses. The net effect of a B.P. is that of rejuvenating the star, bringing fresh He and N^{14} in the center of the star. The real existence of these Pulses is still very uncertain but, in our opinion, it may be of interest to see what would be the production of s-elements if the lifetime of the He-burning phase is extended due to convective phenomena (B.P., overshooting, semiconvection).

The evolutionary code used, is the latest version of the FRANEC (Frascati Raphson Newton Evolutionary Code), which includes :

1) - Latest Reaction Rates (Caughlan,Fowler,Harris and Zimmerman 1985, Harris,Fowler,Caughlan and Zimmerman 1983)

2) - Latest Opacities from the " Los Alamos Opacity Libraries".

The s-process Network includes 200 nuclei (from Ne^{20} to Eu^{153}), the main branching points are taken into account and the (n,α) and (n,p) reactions are included too.
The set of differential equations which describe the temporal evolution of the various nuclei is solved by means of a Raphson Newton Tecnique.

II. EFFECTS OF A BREATHING PULSE ON THE EVOLUTION OF A 15 $M_\odot$ Y=0.28 Z = 0.02

We have computed two evolutionary tracks. The first one takes fully into account the formation of the B.P., the second one inhibites the growth of the convective core when Yc drops below 0.1. The main effects of the B.P. appear:

a) the lifetime of central He - burning phase is increased by 25 % ;

b) the helium abundance at which the temperature of 250 millions °K of degree is reached, raises from 0.008 to 0.02.

As a consequence the "active" phase of the production of the s-elements is increased by a factor of 2. So, the s-production for the stellar model in which the breathing pulses are taken in account is more efficient by a factor 4-5 (Fig.1) for the elements in the range of mass number 60÷70 .

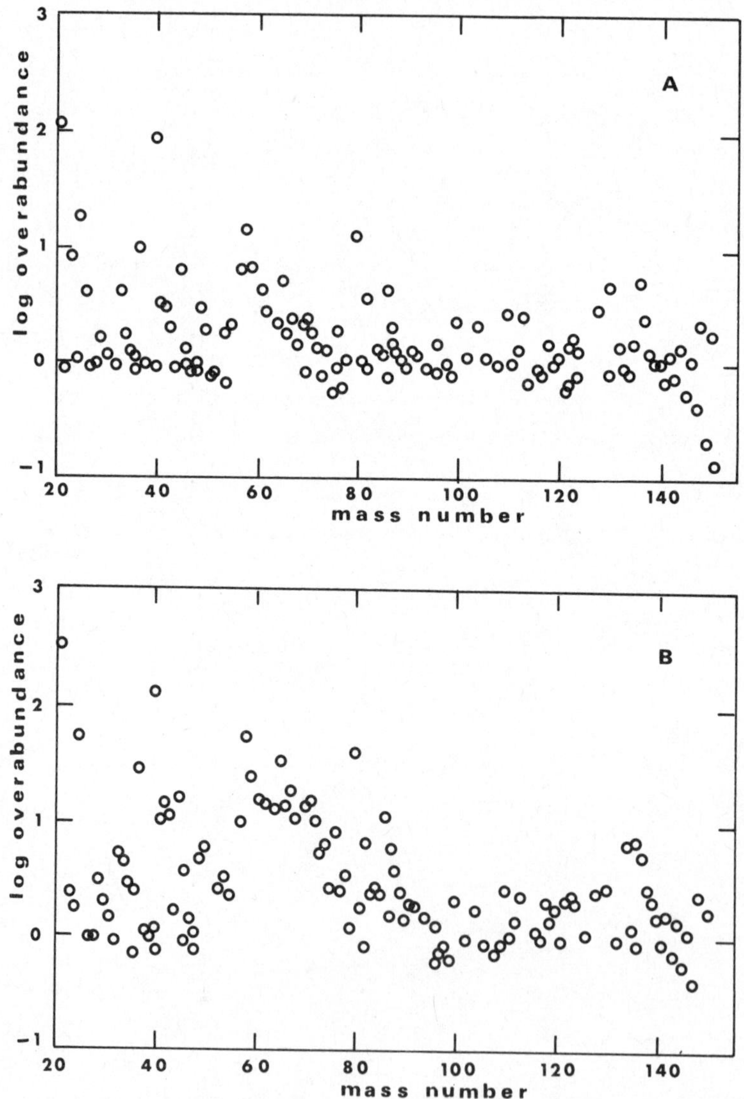

Fig.1. Overabundances in the convective core of a M = 15 M⊙ Y = 0.28 Z = 0.02 star at the end of core He - burning for two cases: a) without breathing pulses, b) with breathing pulses. It is evident that the s-production is more efficient for the b) case in the range of mass number 60 ÷ 70.

III. EFFECTS OF THE NEW C^{12} $(\alpha, \gamma)O^{16}$ REACTION ON THE EVOLUTION AND ON THE S - PROCESS PRODUCTION OF A STELLAR MODEL OF 25 M⊙ Y=0.25 Z=0.003

The main effect is that the lifetime of the last part of the central He-burning is increased when the larger cross section is used. This is due to the fact that the Q value of the 3α reactions is very similar to the Q value of the $C^{12}(\alpha,\gamma)$ reaction. This implies that C^{12} takes earlier the control and, in turn, it bursts longer since to produce the same amount of energy 1 α particle instead of 3 α has to be burnt.

We have computed two evolutionary tracks:the first one with the old $C^{12}(\alpha,\gamma)O^{16}$ (Fowler,Caughlan,Zimmerman,1975) the second one with the new carbon-α rate (Caughlan,Fowler,Harris and Zimmerman 1985). When the C^{12} (α, γ) rate is increased (by a factor 3), the results show that the final part of the He-burning is prolonged (therefore favouring the s-process production), but in the same time the average temperature is slightly lowered (Fig.2). So, we expect a different production for s-elements. In fact the s-production for the stellar model with new carbon-α rate, is slightly increased, by a factor 2, in the mass number range 60÷70 (Fig.3).

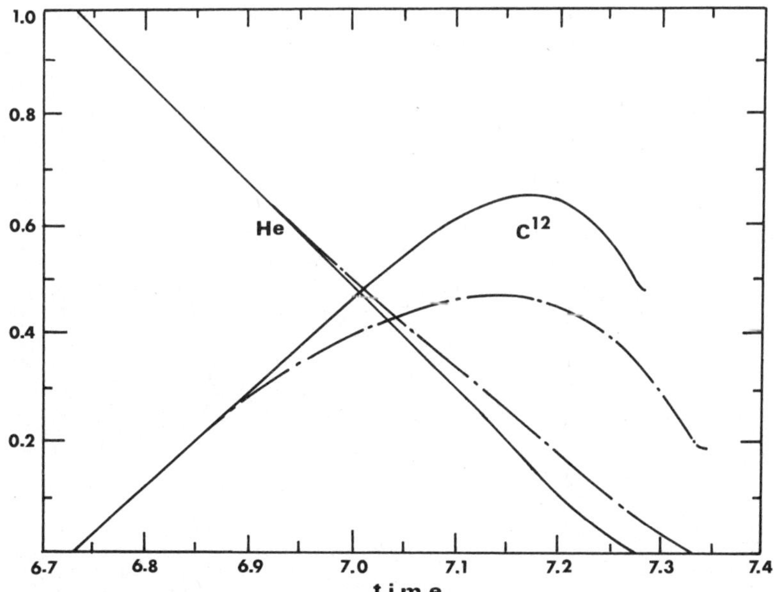

Fig.2. Behaviour of heliun and carbon vs time in the convective core of a M = 25 M⊙ Y = 0.25 Z = 0.003 with old carbon-α rate (solid lines) and with new carbon-α rate (dotted dashed lines).

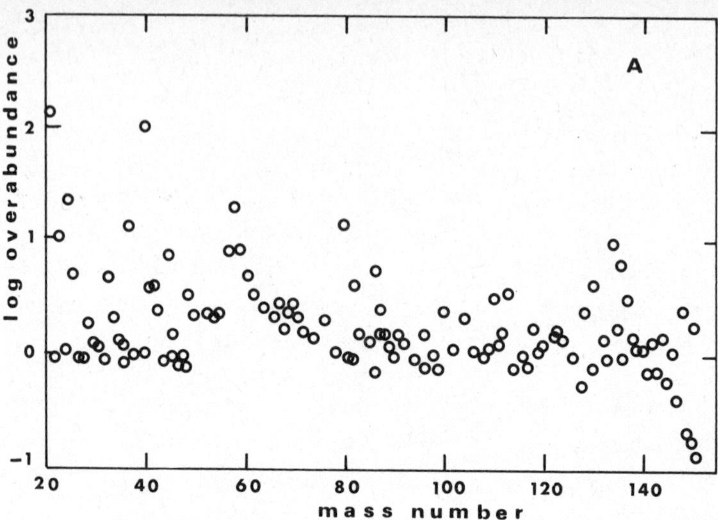

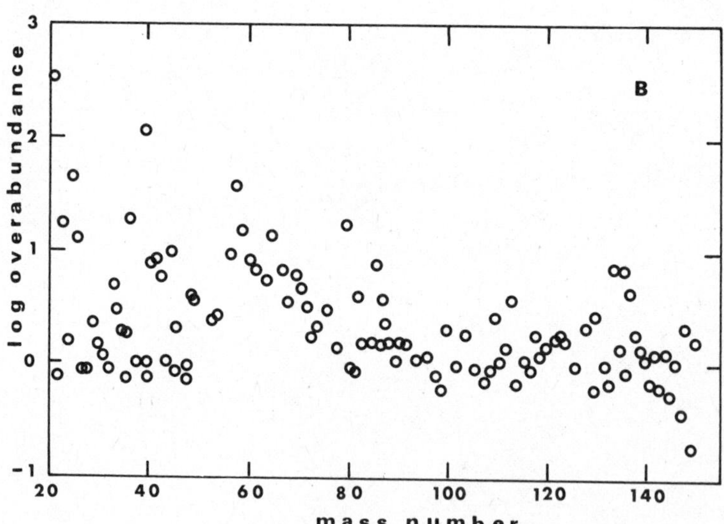

Fig.3. Overabundances in the convective core of a $M = 25 M_{\odot}$ $Y = 0.25$ $Z = 0.003$ at the end of core He-burning for the two cases: a) with old carbon-α rate, b) with new carbon-α rate. The s-production is slightly increased for the b) case in the region of mass number 60 ÷ 70.

IV. CONCLUSIONS

From these computations one gets some general indications:

1) If the hypothesis of the B.P. is true, they may increase the s-process production by a factor of 4 - 5 in the range of mass number 60 ÷ 70 (in the 15 M⊙ Y = 0.28 and Z = 0.02).

2) The effect of the new $C^{12}(\alpha,\gamma)$ rate is again to slightly increase the production of s-elements in the range of mass number 60 ÷ 70 by a factor of 2 (in the 25 M⊙ Y = 0.25 Z = 0.003).

We are now computing models of different masses and chemical composition to obtain more detailed information on the effect of this new physics on the production of s - elements.

References

Almeida,J.,Käppeler,F.:1983,Astrophys.J.265,417

Arnett,W.D.:1972,Astrophys.J.176,681

Arnett,W.D.,Thielemann,F.-K.:1985,Astrophys.J.295,589

Busso,M.,Gallino,R.:1985,Astrom.Astrophys.151,205

Castellani,V.,Chieffi,A.,Pulone,L.,Tornambe',A.:1987,Astrophys.J. in press.

Caughlan,G.R.,Fowler,W.A.,Harris,M.J.,and Zimmermann,B.A.1985 Atomic data Nucl. Data Tables,32,197.

Fowler,W.A.,Caughlan,G.R.,Zimmerman,B.A.:1975,Ann.Rev.Astron.Astrophys.13,69.

Harris,M.J.,Fowler,W.A.,Caughlan,G.R.,and Zimmerman,B.A.:1983,Ann.Rev.Astron. Astrophys.21,165.

Lamb,S.A.,,Howard,W.M.,Truran,J.W.,Iben,I.,Jr.:1977,Astrophys.J.217,213.

ON THE SYNTHESIS OF THE PROTON-RICH NUCLEI

M. Rayet
Institut d'Astronomie, d'Atrophysique et de Géophysique, CP165
Université Libre de Bruxelles, 1050 Bruxelles, Belgium

I. Introduction

Most heavy nuclei (with mass number $A \gtrsim 56$) are supposed to be produced during some stages of the stellar evolution through successive neutron captures followed by β-decays. Stellar conditions, such as temperature ranges, time scales and neutron fluxes, can naturally distinguish between slow(-s) and rapid(-r) nucleosynthetic processes. These produce respectively the bunch of nuclei which figure the so-called "nuclear stability valley" and the fewer (although isotopically fairly abundant) neutron rich isotopes, these two groups being called, for heuristic if not always natural reasons, s- and r- nuclei.

The remaining, proton-rich, stable nuclei cannot be synthesized by neutron capture processes since the β-unstable progenitors produced in these processes will always meet, on their way from the neutron-rich side to the bottom of the stability valley a s- or r- nucleus first: the proton-rich, or *p-nuclei* are thus said to be shielded by s- and r-nuclei or to be bypassed in s- or r-processes.

This "shielding" is easily observed in Figure 1, where 35 proton-rich nuclei are shown on a N-Z plane together with the s- and r- stable isotopes. Their properties and identification as p-nuclei will be discussed in §II.

The field of neutron capture processes in astrophysics has been extensively studied since almost forty years and is still actively developing now (see Mathews and Ward, 1985, for a recent historical survey). Its basic ideas have become relatively well understood and it can now rely on slightly more realistic astrophysical scenarios and detailed nuclear physics investigations, even if the r-process still suffers from considerable uncertainties despite the large amount of work (both from the astrophysical and nuclear physics communities) which has been devoted to the subject.

In comparison, the interest for the stellar synthesis of p-nuclei has remained marginal, at least quantitatively. The idea of a nucleo-synthetic mechanism driven by radiative proton captures (together with photo-dissociations) on s- and r-

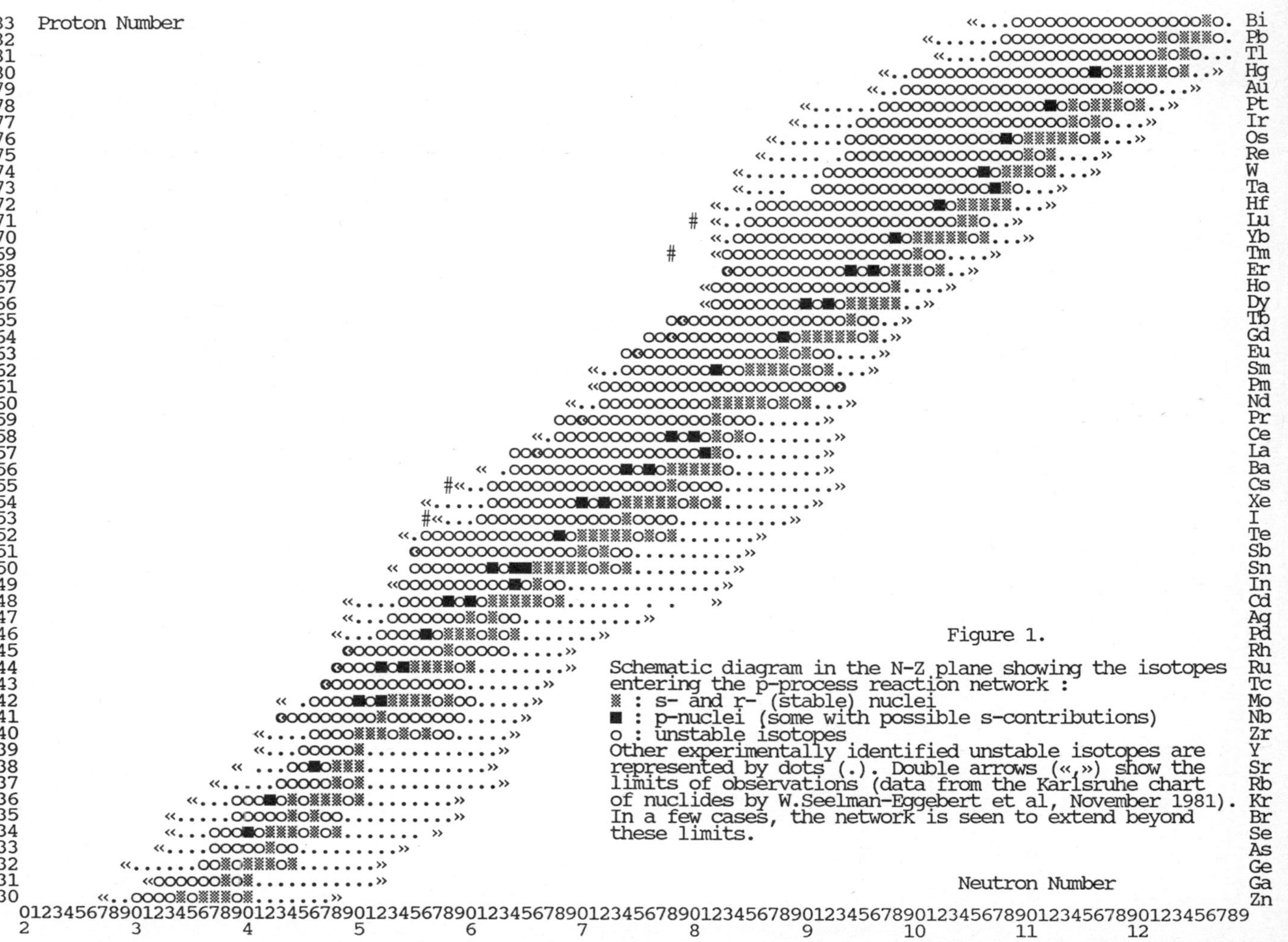

Figure 1.

Schematic diagram in the N-Z plane showing the isotopes entering the p-process reaction network :
▒ : s- and r- (stable) nuclei
■ : p-nuclei (some with possible s-contributions)
o : unstable isotopes
Other experimentally identified unstable isotopes are represented by dots (.). Double arrows («,») show the limits of observations (data from the Karlsruhe chart of nuclides by W.Seelman-Eggebert et al, November 1981). In a few cases, the network is seen to extend beyond these limits.

process seed nuclei was suggested in the now historical paper by Burbidge, Burbidge, Fowler and Hoyle (1957) (hereafter B^2FH). The synthesis of p-nuclei has since retained the conventional name of *p-process* regardless of the mechanism actually involved. The first quantitative attempts to produce all the p-nuclei in one astrophysical site appeared about 20 years later in only a few works, by Audouze and Truran (1975) (AT), Arnould (1976), Harrison (1978) and by Woosley and Howard (1978) (WH). These will be briefly reviewed in §III.

Different mechanisms were shown in a few earlier approaches to account for the enrichment in some particular p-elements, and will not be discussed in this paper. They included spallation reactions (Frank-Kamenetskii, 1961; Audouze, 1970) which can only efficiently produce the heaviest species, or weak interactions (photo-β reactions or positron captures; see e.g. Arnould and Brihaye, 1969).

It will be argued in this paper that the relative scarcity of works devoted to the p-process is not due to the fact that the latter concerns but a few "minor" isotopes: it should rather be ascribed to the multiplicity of the mechanisms entangled in one given scenario and to the discouraging sensitivity of the production yields to the relevant stellar conditions. These considerations plead in favour of revisiting the p-process taking advantage of recently developed stellar evolutionary (esp. explosive) models and using for the nucleo-synthesis calculation a reaction network which can cope with any conceivable p-process scenario (reaction fluxes, branchings, a.s.o.) and contains the best possible nuclear physics input. The tentative construction of an extended network meeting these requirements is presented in §IV with a few preliminary, semi-quantitative, conclusions.

II. The proton-rich nuclei in the solar system

Table 1 shows the abundances of the 35 proton-rich nuclei conventionally called p-nuclei, going from Se(Z=34) to Hg(Z=80), and describes the way they were calculated. Most p-nuclei are characterized by very small isotopic ratios, typically 1 to 0.1% (with increasing Z), no single element having a p-nucleus as its dominant isotope, which supports the view that p-nuclei are formed from s- and r- seed nuclei. All but four of them (to be discussed below) are even-even nuclei, a high proportion which, in comparison with the neutron richer elements, indicates a propensity to local nuclear stability.

As usual, exceptions are worth a particular attention:

1) the exceptional abundances of ^{92}Mo and ^{144}Sm can be readily related to a magic neutron number (N=50 and N=82 resp.),meaning better resistance to photo-desintegrations. It would be interesting, but not at all demonstrated, that the same shell effect might also explain the high abundances of ^{94}Mo and ^{96}Ru. However the fairly large isotopic ratios also observed for the p-isotopes of Pd, Cd and Sn could anyway not find such a straightforward explanation;

Table 1: Abundances of p-nuclei#)

Element	Isot.%+)	N/10^6Si	Element	Isot.%+)	N/10^6Si
^{74}Se	0.87	0.54	^{130}Ba	0.106	0.00462
^{78}Kr	0.339	0.154	^{132}Ba	0.101	0.0044
^{84}Sr	0.56	0.132	^{138}La	0.089	4.0 10^{-4}
^{92}Mo	14.8	0.373*)	^{136}Ce	0.190	0.0022
^{94}Mo	9.3	0.234*)	^{138}Ce	0.254	0.0029
^{96}Ru	5.5	0.102	^{144}Sm	3.1	0.008
^{98}Ru	1.86	0.0346	^{152}Gd	0.2	6.6 10^{-4}
^{102}Pd	1.0	0.0139	^{156}Dy	0.057	2.27 10^{-4}
^{106}Cd	1.25	0.0199	^{158}Dy	0.100	3.98 10^{-4}
^{108}Cd	0.89	0.0142	^{162}Er	0.14	3.54 10^{-4}
^{113}In	4.3	0.0079	^{164}Er	1.56	0.00395
^{112}Sn	1.01	0.0386	^{168}Yb	0.135	3.28 10^{-4}
^{114}Sn	0.67	0.0256	^{174}Hf	0.16	2.8 10^{-4}*)
^{115}Sn	0.38	0.0145	^{180}Ta	0.0123	2.78 10^{-6}
^{120}Te	0.091	0.0045*)	^{180}W	0.13	1.78 10^{-4}*)
^{124}Xe	0.114	0.00496*)	^{184}Os	0.018	1.29 10^{-4}
^{126}Xe	0.111	0.00483*)	^{190}Pt	0.013	1.78 10^{-4}
			^{196}Hg	0.15	7.8 10^{-4}*)

#) After Anders and Ebihara's (1982) compilation of C1 chondrite analyses. A recent discussion of solar abundances (N.Grevesse, 1984) demonstrates a nice convergence for almost all elements between both sources of information. The only exception, in the above table, is Tungsten, whose solar abundance seems to be 60% larger than its meteoritic abundance quoted by Anders and Ebihara. This latter quantity has however been subject to repeated controversy in the past.
+) Natural isotopic ratios taken from Lederer et al.(1978) as in Anders and Ebihara. Differences with IUPAC(1980) compilation are minor.
*) Values differing by more than 20% from Cameron(1973). Differences come essentially from discrepancies in elemental abundances.

2) several among presumed p-nuclei can, in certain circumstances, be produced in a s- (or even r-) process. This possibility has often been mentioned and in some cases investigated in more detail. Here are some brief remarks:

- ^{94}Mo, ^{108}Cd and ^{152}Gd may receive contributions from a s-process branch provided time is given to one of their (long lived) progenitor, resp. ^{93}Zr (1.5 10^6y), ^{107}Pd (6.5 10^6y) and ^{151}Sm (93y), to beta decay before neutron capture happens. This puts upper limits on the surrounding neutron density which are orders of magnitude less than that typically associated with the s-process (10^8 n cm^{-3}), while some s-process production of ^{152}Gd is less improbable. Even qualitatively, such conclusions should be taken with care, since terrestrial β-decay rates could be considerably enhanced in stellar conditions (by bound-state β-decays or thermalization of nuclear excited states);
- of all the Sn and In isotopes refered to as p-nuclei in table 1, only ^{112}Sn is completely shielded from any s- and r-process contributions. Ward and Beer (1981) have studied in detail the neutron capture paths leading to the tin and indium isotopes, considering the branching to the many isomeric states met in this complicated mass region. They concluded that a combination of s- and r- processes can account for a significant fraction of ^{113}In and ^{115}Sn (which would explain the

relatively large isotopic abundance of the former), while the neutron process contributions to ^{114}Sn appears to be negligible. As suggested before by Clayton (1978) the origin of the tin isotopes may hint at the dominant mechanism at work in the p-process (see §III);
3) both odd-odd nuclei ^{138}La and ^{180}Ta have exceptionnally low abundances in comparison with the other neighbouring nuclei, which again indicates that it is difficult to produce weakly bound nuclei in the p-process. Note that ^{180}Ta, by far the least abundant of all stable nuclei, might also be produced in the s-process.

In meteorites, isotopic anomalies involving p-nuclei have been reported on a few occasions, for krypton, strontium, tellurium, samarium. The most famous of them concern the xenon ,for which strongly correlated enhancements of its p- and r-components (Xe-HL) have been repeatedly observed in fine-grained carbon components of primitive meteorites. These were recently associated with the presence of diamonds which have most probably a presolar origin (Lewis et al., 1987). Other mineral fractions (spinels) of Allende and Murchison meteorites contained "mirror" anomalies (i.e. enriched s-xenon), while Allende's residues also showed similar correlations between p- and r-products of tellurium (Ballad et al., 1979). These findings have been put forward in support of the view that the anomalous components are nucleogenetic products from one single supernova event associated to the formation of the solar system and leaving records of successive nucleosynthetic processes (e.g. Heymann and Dziczkaniec, 1981).

Isotopic anomalies in tin have not been reported so far but would be of utmost interest in view of the above considerations on the possible origin of the tin isotopes. De Laeter et al. (1984) have recently emphasized the technical problems raised by the accurate measurement of the tin isotopic abundances and have proposed a standard for "normal" composition which, together with improved mass spectrometric techniques, should enable a search for tin isotopic anomalies in meteorites.

III.Mechanisms for the synthesis of p-nuclei

Two main mechanisms are naturally expected to be responsible for the production of proton enriched isotopes from already processed s- and r-seed nuclei: one is to add protons (radiative proton captures), the other is to remove neutrons (neutron photodissociations).

Examination of (γ,n) reaction rates reveals that temperatures in the range of 10^9 K are required in order to strip neutrons off stable nuclei, even if one considers long time scale stellar events (photodisintegration rates are already negligible for T_9=1 (T_9=T/10^9 K), at least for nuclei not too far from stability). Photo-nuclear rates are extremely sensitive functions of the temperature, which means also extreme sensitivity to the temperature profiles in the stellar environment where the p-process is taking place. Since they are equally dependent on n, p, and α-particle nuclear binding energies, photo-nuclear reactions will, at a given temperature, destroy the less-bound heavy nuclei after a much shorter time than they would with lighter species; for the same reason, nuclear structure details will also be of importance.

On the other hand, (p,γ) reactions are much less dependent on temperature and nuclear structure details but, because of the Coulomb barrier, they are progressively inhibited with increasing nuclear charge: astrophysical (maxwellian averaged) reaction rates are reduced by factors 10^6-10^9, at temperatures of a few 10^9 K, when going from Fe to Bi. As a consequence, proton radiative captures will only contribute significantly in the lower mass region (from Fe to Sn to give a range), except in exceptionnaly proton-rich environments.

With these observations in mind, we will briefly sketch here four works which have attempted to reproduce the whole solar abundance pattern of the p-nuclei in one single stellar scenario.

III.1. p-process in explosive hydrogen burning.

Audouze and Truran (1975) (AT) have explored quantitatively the original suggestion of B^2FH of a p-process in the hydrogen-rich envelope of a supernova heated briefly to temperatures of a few billion K by the passage of the shock wave. The achievement of temperatures $T_9>1$ implied, from hydrodynamical considerations, a high proton density $\rho\approx10^4$ g cm^{-3} with a time scale of about 10s. Calculated enhancements reproduced within factors of 10 the solar abundances of the lightest p-nuclei (up to samarium) in a relatively wide temperature range ($1.5\lesssim T_9\lesssim2.5$) while heavier species were produced only for $T_9\approx2$. These results are easily explained by the fact that photo-nuclear reactions must dominate the process in the high mass region while the very high proton densities used in this calculation are able to drive radiative captures fast enough to produce proton-rich isotopes up to Sm.

However the required temperature-density conditions rule out type II supernova envelopes as a plausible site for this scenario and the difficulty of finding such conditions in realistic stellar models remains the main objection to this calculation, which, on the other hand, easily succeeds in giving comparable enhancements for the low and high mass p-nuclei, from a solar composition of seed nuclei.

III.2 γ-process in explosive oxygen burning.

The strong anti-correlation observed between the solar abundances of p-nuclei and their total photodisintegration rates led Woosley and Howard (1978) (WH) to explain the formation of p-elements in explosive conditions from photo-nuclear reactions alone, without any interference from particle reactions.

Quantitatively, this schematic model meets some difficulties if one tries to get comparable overproduction factors over the whole mass range of p-nuclei. One single temperature profile, parametrized by a peak temperature and an exponential decrease with constant time scale, results in a marked underproduction of light p-nuclei. An appropriate combination of exposures with different peak temperatures (smaller for high-mass, larger for low-mass p-nuclei production) is necessary to cure this problem.

Explosive oxygen burning zones meet the physical conditions requirements for such a γ-process, although contrary to the canonical p-process of AT and B^2FH, the γ-process is relatively independent of the chemical composition of the site. However, large deviations in the nuclear seed abundances from solar values must be expected in zones which have experienced hydrostatic helium and probably carbon burning, and were shown to influence the resulting p-abundance pattern. Background particles (essentially neutrons and protons) present in largely unknown amounts in the different burning stages are also likely to perturb a pure γ-process if in sufficiently high concentrations.

III.3. p-process in hydrostatic oxygen burning.

Because of the extremely fast variation of photo-nuclear reaction rates with temperature, the physical conditions for p-processing in quiescent phases of stellar evolution, i.e. on much larger time scales, also require the high temperatures needed in explosive situations. Arnould (1976) treated this conjecture in a quantitative way and proposed hydrostatic oxygen burning zones as possible candidates for p-element synthesis. Characteristic conditions for these zones were extracted from detailed evolutionary models available at the time: $2 \lesssim T_9 \lesssim 2.2$, $\rho \approx 10^6$ g cm^{-3} and neutron, proton and α-particle densities in the ranges $10^{13} \lesssim n \lesssim 10^{15}$ cm^{-3}, $10^{16} \lesssim p \lesssim 10^{18}$ cm^{-3} and $10^{17} \lesssim \alpha \lesssim 10^{20}$ cm^{-3}.

It was shown that after times in the range $10^4 \lesssim t \lesssim 10^6$ s, depending on temperatures and mass ranges, thermal equilibrium is established between photons, neutrons and a fair number of isotopes (with the exception of the lightest elements A<92). Isotopic abundances for each element are then found to peak at one or two A values, but as in WH (although in quite different conditions), heavy and light p-elements cannot be produced in similar amounts for only one set of physical conditions. The heavier ones (above Sm) require lower temperatures (T_9=2.0) and $n \gtrsim 10^{14}$ cm^{-3} whereas when their mass goes down, p-elements can be produced in non-negligible amount only for increasing temperature (up to $T_9 \approx 2.2$) and decreasing neutron densities (down to $n \approx 10^{13}$ cm^{-3}). Furthermore, in the $Z \gtrsim 62$ range, the thermal equilibrium is followed, after times of the order of 10^4 s, by a phase of transformations through charge particle reactions and β^+-decays, which results in a progressive destruction of the p-elements into lower mass nuclei. With a combination of five different sets of T_9 and n values in the limited range described above, and assuming s-like seed composition, Arnould (1976) obtains abundance patterns of the same quality as in the explosive models, but after a time not much longer than 1.8 10^4 s. This corresponds to shell oxygen burning conditions. In other circumstances, it would be necessary to invoke an ad-hoc mechanism like an explosion to freeze-out the process. In this sense, Arnould's calculations do not contradict WH's model, but it would rather complement this model in a natural sequence of stellar events.

III.4. γ-process with non-thermalized photons.

Harrison (1978) has demonstrated the feasibility of a p-process driven by energetic photons of nuclear origin, at least if these photons are not thermalized, which is far from being proved. These are emitted in various reactions of the p-p and p-CNO cycles in hydrogen burning zones, but the more interesting possibility is the emission of the very energetic γ ($\approx$20 MeV) in the reaction $^{3}T(p,\gamma)^{4}He$ which closes the p-p cycle in electron degeneracy conditions. An interesting aspect of this model is that the emited photon, being above the $(\gamma,2n)$ threshold, can produce in a highly correlated way the 2 contiguous p-isotopes which are observed for some elements. Since it involves a totally different mechanism, this model cannot be compared to the other 3 and we will not discuss it further here, if only to remark that it calls for rather specialized stellar environments which are not expected, if they exist at all, to contribute significantly to the p-nuclei enrichment.

IV. Revisiting the p-process

IV.1. A few comments on the existing p-process models.

Beyond the global differences between the mechanisms involved in the 3 models outlined in §III, which were all able to approximately explain the abundance pattern of most p-nuclei, it is interesting to discuss their more detailed predictions concerning a few especially relevant nuclei:

1) the two odd-odd isotopes ^{138}La and ^{180}Ta and the odd ^{113}In are markedly underproduced in all calculations, which suggests different origins of these nuclides (either s-process, or Harrison's, or another peculiar mechanism mentioned in §I). The same happens for ^{152}Gd, which might have some contribution from the s-process (Prantzos et al, 1987);

2) the odd isotope ^{115}Sn is efficiently produced in AT's p-process which, as mentioned before, may appear to be in contradiction with its conjectured s-process origin (Ward and Beer, 1981), while, in AT's prefered conditions (T_9=2) the typical p-process nucleus ^{112}Sn is strikingly underproduced;

3) models 2 and 3 produce light p-nuclei only with some difficulty. The situation is particularly bad for the Mo and Ru isotopes (except for the fact that ^{92}Mo is well produced in Arnould's model). Comparison with AT shows that this region is

better enriched in p-nuclei by a radiative capture process, but ^{94}Mo, which is also underproduced in AT, remains a puzzling problem. The region between the two closed shells N=50 and Z=50 deserves special attention and it is not impossible that unexpected nuclear physics effects would lead to the re-evaluation of some of the reaction rates which have been used so far in the calculations.

These remarks and the considerations of §III suggest the following tentative conclusions:

- the various mechanisms proposed for the synthesis of the p-nuclei require a very fine tuning between time scales and temperatures. For not too unrealistic stellar conditions, it seems that reasonable solar abundances can only be obtained from a combination of exposures to slightly different temperatures. Contributions from both photon and particle reactions probably favour a balanced production of light and heavy p-elements;
- in so far as p-nuclei are produced in evolved stellar regions, there are major uncertainties concerning the abundance of seed-nuclei and the densities of neutron, proton and α-particles;
- explosive sites are most probably the source of the p-elements, not only because they correspond to high temperatures, but also because they allow the p-process to freeze out before most of the processed nuclei experience a rapid disruption into lighter nuclei. However, their synthesis may well have occured in a previous hydrostatic burning phase;
- reliable calculations must involve extremely large nuclear reaction networks which allow for all possible branching and include all possible reactions, calculated with the best available nuclear reaction model.

Existing calculations have not been satisfactory neither on the latter point, nor, for models 1 and 2, on the use of well-defined and realistic stellar environments. At the time, the reasons for these deficiencies lay, for the latter, in the lack of evolutionary models followed self-consistently up to late stages of stellar evolution and for the former in the fact that restricted networks and reaction rates calculated with schematic nuclear reaction models were prefered to avoid computer time consuming network and nuclear rate calculations.

IV.2. Towards a more satisfactory investigation of the p-process.

Progresses made in stellar evolution theory offer now the opportunity of using more reliable stellar conditions for hydrostatic neon and oxygen burning zones as well as for explosive evolution (see e.g. Woosley, 1986). Of particular interest for the p-process is the recent recognition of the possibility for very massive mass-losing stars (VMMS) (with $M_{ZAMS} \approx 100\ M_{\odot}$) to evolve towards a $e^{\pm}$-pair creation instability, eventually leading to a so-called pair-creation supernova. The evolution of a VMMS was recently followed by Langer and El Eid (1986) to He-core exhaustion where it becomes a Wolf-Rayet star; it was later shown by El Eid and Langer (1986) that during its further evolution, the star reached a region of pair instability propagating in an appreciable fraction of its mass and triggering a collapse which ended with the ignition of explosive oxygen burning at a temperature of $T_9 \approx 2.5$. Complete disruption seems energetically possible in case the Wolf-Rayet star has developed a core of about 60 $M_\odot$.

Detailed temperature and density profiles can be obtained from these calculations for different mass radii. With a 61 $M_\odot$ core (model A of El Eid and Langer, 1986), peak temperatures at the time of explosion vary from $T_9=4$ at the center to $T_9 \gtrsim 2$ in the outer region and temperature is maintained above $T_9=1$ for a time of the order of 80 s (El Eid, 1987). These physical conditions cover therefore a range of interest for a p-process scenario, and pair creation supernovae constitute good candidates for detailed p-process calculations, especially if neutron and proton fluxes can be reliably followed.

The other essential ingredient for such calculations is a nuclear reaction network which is large enough to avoid artificial border effects on the reaction fluxes and which reduces as much as possible the uncertainties in the theoretically estimated reaction rates. Such a complete network has been constructed, starting from the neutron rich stable elements and going deeply into the proton-rich side of the nuclear stability valley, extending even in some occasions beyond the last observed isotope. The limits in atomic mass numbers are shown in Figure 1, where for comparison are also seen the present observational limits for unstable nuclides. The network limits roughly correspond to those used in WH. Here however all possible reactions on a given isotope, viz. (n,p), (n,α), (n,γ), (p,α), (p,γ), (α,γ) and their inverse, have been included in the network.

The lowest element shown in Figure 1 is Zn, which is certainly low enough for a p-process calculation, but the network has been extended down to magnesium and its lower part can be used to complement existing networks for oxygen burning. As it stands, the entire network contains almost 1000 nuclei and over 10200 reactions, among which 10% concern the low mass region (Mg to Ni) of interest for oxygen burning.

Particle induced reaction cross-sections have all been calculated in the framework of the Hauser-Feshbach statistical theory. Photo-nuclear reactions were calculated using the reversibility theorem (which is formally exact here since the targets have been thermalized). Use has been made for the calculations of the code SMOKER developed lately by F.-K. Thielemann. A full description of the underlying physics can be found in Thielemann, Arnould, Truran(1986) and J.W. Truran (1987) (see these papers for additional references). Let us only remind here some of its salient characteristics:

- experimental nuclear levels are used wherever possible. Where not, level densities are calculated with an improved version of the back-shifted Fermi gas model, which cures its known deficiencies at very low energy. This is important when dealing with very unstable nuclei for which no levels are known;

- transmission coefficients for neutron and proton reactions are calculated with the "theoretical" optical potential of Jeukenne et al. (1977), which has been shown to give an excellent overall agreement to the s-wave neutron strength functions. Such a "theoretical" approach is certainly safer, when applied to very unstable nuclei, than the hazardous extrapolation of phenomenological potentials.

Astrophysical reaction rates obtained from the above prescriptions are claimed by Thielemann et al. (1986) to be predicted safely within a factor of 2, although their application to very exotic nuclei should still, in our mind, deserve more investigation, particularly regarding the adequateness of the level density formulae. Comparison of the calculated cross-sections with available measurements for medium-weight nuclei are indisputably satisfactory (see e.g. Thielemann et al., 1986). A quite recent measurement of the (p,γ) reaction on a heavier nucleus, namely ^{90}Zr (Laird et al, 1987) almost perfectly agrees, between 2 and 6 MeV, with our prediction calculated with SMOKER. This comparison is obviously encouraging, if probably not quite representative, especially in view of the above mentioned problems linked with the molybdenum region.

The huge amount of reaction rates generated in function of the temperature in the range 10^8-10^{10} K have been parametrized by a seven parameter formula of the usual exponential type. The resulting set of parameters has been automatically inserted in a self-generating network routine by N. Prantzos (1987). A first preliminary test has been performed in a schematic hydrostatic oxygen burning calculation at constant temperature, during which enhancements and eventual decreases of p-nuclei abundances have been observed.

V. Conclusions.

From a critical examination of the few models devoted to the p-process, we conclude that the synthesis of the p-elements in stars is difficult to explain by one single or simple mechanism. The alluring regular appearance of these elements in the whole nuclear landscape must not hide the fact that this lanscape is made of regions of widely different nuclear characteristics. The very narrow range of very high temperatures which seem to be required to produce p-nuclei (nothing happens below, everything is destroyed above) makes the p-process, whatever it exactly is, a very critical and unforeseeable phenomenon. In view of this situation we think that a complete reaction network, calculated in a general framework likely to be valid in the whole atomic mass range considered, and an explosive stellar environment resulting from a self-consistent evolutionary model are good prerequisites to face a new p-process calculation founded on firmer grounds than its predecessors.

REFERENCES

Anders, E., and Ebihara, M. 1982, *Geochim. Cosmochim. Acta* **46**, 2363
Arnould, M. 1976, *Astron.Astrophys. 46, 117*
Arnould, M., and Brihaye, C. 1969, *Astron. Astrophys.* **1**, 193
Audouze, J. 1970, *Astron. Astrophys.* **8**, 436
Audouze, J., and Truran, J.W. 1975, *Ap. J.* **202**, 204
Ballad, R.V., Oliver, L.L., Downing, R.G., Manuel, O.K. 1979, *Nature* **277**, 615
Burbidge, E.M., Burbidge, G.R., Fowler, W.A., Hoyle, F. 1957, *Rev.Mod.Phys.***29**, 547
Cameron, A.G.W. 1973, in *Explosive Nucleosynthesis*, eds. D.N. Schramm and W.D.Arnett (Austin: University of Texas Press)
Clayton, D.D. 1978, *Astrophys. J. Lett.* **224**, L93
De Laeter, J.R., Rosman, K.J.R., Loss, R.D. 1984, *Astrophys. J. 279, 814*
El Eid, M.F., Langer, N. 1986, *Astron. Astrophys. 167, 274*
El Eid, M.F. 1987, private communication
Frank-Kamenetskii, D.A. 1961, *Soviet Astr.* **5**, 66.
Jeukenne, J.-P., Lejeune, A., Mahaux, C. 1977, *Phys. Rev.***C16**, 80
Grevesse, N. 1984, *Phys. Scripta* **8**, 49
Harrison, T. 1978, *Astrophys. J. Suppl. 36, 199*
Heymann, D., Dziczkaniec, M. 1981, *Geochim. Cosmochim. Acta 45, 1829*
Laird, C.E., Flynn, D., Hershberger, R.L., Gabbard, F. 1987, *Phys. Rev. C 35, 1265*
Langer, N., El Eid, M.F. 1986, *Astron. Astrophys.* **167**, 265
Lederer, C.M., Shirley, V.S., Browne, E., Dairiki, J.M., and Doebler, R.E. 1978, *Table of Isotopes*, 7th. edition (New York: J.Wiley)
Lewis, R.S., Tang Ming, Wacker, J.F., Steel, E. 1987, *Lunar Planet. Sci.* **18**, (submitted)
Mathews, G.J., Ward, R.A. 1985, Rep.Prog.Phys. **48**, 1371
Prantzos, N. 1987, private communication
Prantzos, N., Arnould, M., and Arcoragi, J.-P. 1987, *Astrophys. J.*, **315**, 209
Thielemann, F.-K., Arnould, M., Truran, J.W. 1986, in *Advances in Nuclear Astrophysics*, eds. E. Vangioni-Flam, J. Audouze, M. Cassé, J.-P. Chièze, J. Tran Thanh Van (Gif-sur-Yvette: Editions Frontières), p525
Truran, J.W. 1987, this volume
Ward, R.A., Beer, H. 1981, *Astron. Astrophys.* **103**, 189
Woosley, S.E. 1986, in *Saas-Fe Lectures: Nucleosynthesis and chemical evolution*, Eds. B. Hauck, A. Maeder and G. Meynet(Geneva observatory), pl
Woosley, S.E., and Howard, W.M. 1978, *Astrophys. J. Suppl.*, **36**, 285

Studies of non-local and time-dependent convection

Rudolf Kuhfuß

Max-Planck-Institut für Astrophysik

Karl-Schwarzschild-Str. 1

8046 Garching bei München

Abstract: Using the convection model of Kuhfuß (1986) [1] the timedependence as well as the nonlocal character of the model are investigated in detail. The timedependence is studied within a one-zone model for variable stars. Nonlocality leads to a consistent description of overshooting including the influences on the temperature gradient.

1. Introduction to the Model

The model for convection used for the studies presented is described in Kuhfuß 1986 [1]. It can be derived from the full set of hydrodynamic equations via averaging over spherical surfaces (Ω denotes the solid angle):

$$\langle\rho\rangle(r) := \int_{S_2} \rho(r,\Omega)\,\frac{d\Omega}{4\pi} \tag{1a}$$

$$\langle\vec{v}\rangle(\vec{r}) := \langle\vec{v}\cdot\vec{e}_r\rangle\vec{e}_r \tag{1b}$$

The definition of the fluctuations, denoted as ρ' or $\vec{v}'$, is obvious. Very often one is lead to terms which are averages of products of fluctuating quantities. The problem is to close the resulting equations. First the anelastic approximation [2] is used. Unfortunately lateron at some places the pressure fluctuations have to be dropped in addition, thus destroying the accuracy of this approximation. The corrections are of minor importance for low Mach numbers. Moreover a first order approximation for the Reynolds tensor is used. The second major assumption is the diffusion approximation. There often occur fluxes of the form $\langle\vec{v}'(\langle\rho\rangle\varphi)'\rangle$. These are modelled like usual diffusive fluxes with the help of a typical scale for the velocity which one gets from the specific ('convective') kinetic energy inherent in the convective motion $\vec{v}'$:

$$\langle\vec{v}'(\langle\rho\rangle\varphi)'\rangle \approx -\,\alpha_\varphi\Lambda\sqrt{\varpi}\langle\rho\rangle\,\mathrm{grad}\,\langle\varphi\rangle \tag{2a}$$

$$\varpi := \left\langle v'^2/2\right\rangle \tag{2b}$$

From now on only averaged quantities will occur, therefore I drop the brackets ($\langle\rho\rangle \to \rho$). The main result of the whole derivation is an additional equation for the 'convective' kinetic energy ϖ :

$$d_t\varpi = \mathbf{S}_\varpi - c_{\mathrm{D}}\frac{\varpi^{3/2}}{\Lambda} - \frac{1}{\rho}\operatorname{div}\vec{j}_t$$
$$- \frac{p_t}{\rho}\operatorname{div}\vec{v} + (\textit{Reynolds shear}) \tag{3}$$

c_{D} denotes a free parameter which governs the dissipation of kinetic energy into smaller and therefore in the sense of transport less efficient scales of the turbulence cascade. This equation looks like the usual thermal energy equation. There are source and sink terms, there is a flux of energy $\vec{j}_t$, a pressure work term due to the turbulence pressure and a term resulting from Reynolds shear forces. As already mentioned the flux of 'convective' kinetic energy is modelled via diffusion

$$\vec{j}_t = -\alpha_t\Lambda\sqrt{\varpi}\rho\operatorname{grad}\varpi \tag{4}$$

with a dimensionless free parameter α_t and the turbulence pressure results from the ansatz for the Reynolds tensor (isotropic turbulence assumed):

$$p_t = \frac{2}{3}\rho\varpi \tag{5}$$

As a length scale for convective motion the pressure scale height is used:

$$\Lambda := \alpha_{\mathrm{ML}}H_p \tag{6}$$

The Navier-Stokes equation looks familiar only being modified to include the turbulence pressure and Reynolds shear. In addition in the energy equation very naturally the convective flux of thermal energy appears, also being modelled as a diffusive flux (s denotes specific entropy)

$$\vec{j}_w = -\alpha_s\Lambda\sqrt{\varpi}\rho T\operatorname{grad}s \tag{6}$$

with a dimensionless free parameter α_s. For simplicity gradients in the chemical composition have been dropped. The full expressions can be found in [1]. The source term for 'convective' kinetic energy is closely related to this flux:

$$\mathbf{S}_\varpi = \frac{\nabla_{ad}}{\rho H_p}\vec{j}_w\cdot\vec{e}_r \tag{7}$$

Taking into account gradients of chemical composition one also obtains diffusion equations for all species. The free parameter for the diffusive transport is called α_m. Another one is used in the first order model for the Reynolds shear terms.

2. 'Determination' of free parameters

There are many free parameters in the model:

Mixing length parameter	:	α_{ML}
dissipation of kinetic energy	:	c_{D}
transport of kinetic energy	:	α_t
transport of thermal energy	:	α_s
transport of species	:	α_m
Reynolds shear terms	:	α_μ

Because the last five parameters occur always together with the scale height Λ the mixing length parameter could be included, thus only five parameters are independent.

The criterion for the onset of convection resulting from the model should be the Ledoux criterion because this is a result from the general hydrodynamic equations. The Ledoux criterion results from the model iff $\alpha_s = \alpha_m$.

As a first estimate the mixing length theory convective velocity and flux should result from the model. From this one can determine α_s and c_{D}.

In the static models there is no contribution from the Reynolds shear tensor because of absence of radial motion. In the one zone model the Reynolds shear is neglected. Therefore α_μ can be dropped in this discussion.

The last parameter which governs the flux of kinetic energy and therefore determines the extend of overshooting is modelled according to the following picture. Suppose one has a turbulence element moving with no temperature excess only due to an initial velocity through an adiabatically stratified layer. After a certain distance it dissolves and sets free all its entropy. But there have been kinetic energy losses already before dissolving and only the rest of it is deposited at the end of the path. The way kinetic energy is lost during the motion is governed by the equation of motion. This equation is motivated by the equation for the kinetic energy ϖ to be

$$\frac{dv}{dt} = -kv^2$$

and the constant k can be derived under the assumption that the scale height does not change. From this one gets $\alpha_t = f(\alpha_s, c_{\mathrm{D}})$ and thus has determined the parameter α_t. If one adopts the values for the constants α_s and c_{D} as derived above one gets for totally isotropic turbulence about 61% efficiency of transport of kinetic energy compared to transport of entropy, in the totally anisotropic case about 74%.

Thus one is left with the parameter α_{ML} or with the determination of the scale height in general. This is the major problem with almost all models for convection.

3. Timedependence: A one-zone model for variable stars

In order to study the changes of convection in time caused by changes of the structure of a convective region a one-zone model for variable stars was constructed. It is derived similar to the investigations of Stellingwerf (1972 and 1986) [3,4] with the major difference that the convection model was replaced. One arrives at a set of one second order (Navier-Stokes) and two first order (energy equation and convection equation) coupled ordinary differential equations. A linear stability analysis leads to criteria for stability which are well known in the convection free case from other one-zone investigations, e.g. Baker (1966) [5]. With convection present the criteria look slightly different to those of Stellingwerf [4] because of the different convection model.

Convection seems to have a damping influence. First the stability criteria seem to indicate that, but also the nonlinear calculations lead to this statement. I found three classes of solutions:

1. 'stable' solutions, in which the perturbations of the stationary state are damped out

2. periodic solutions

3. 'runaway' solutions, which look somehow periodic but have more and more growing amplitudes and periods.

For a parameter set which leads to a 'runaway' solution in the radiative case one may get a periodic solution ore even a 'stable' one if a certain fraction of the luminosity is assumed to be carried by convection (this is one of the free parameters). This illustrates the damping effect of convection. A parameter choice which leads to a periodic solution is the 'strip' case in Stellingwerf (1986) [4]. The first plot shows the outer shell radius normalized to the equilibrium shell radius (solid line), the velocity normalized to equilibrium shell radii per free fall time (dashed line), the luminosity normalized to the equilibrium luminosity (dotted line) and the fraction of the luminosity which is carried by convection (dash-dotted line), all as functions of time in units of the free fall time.

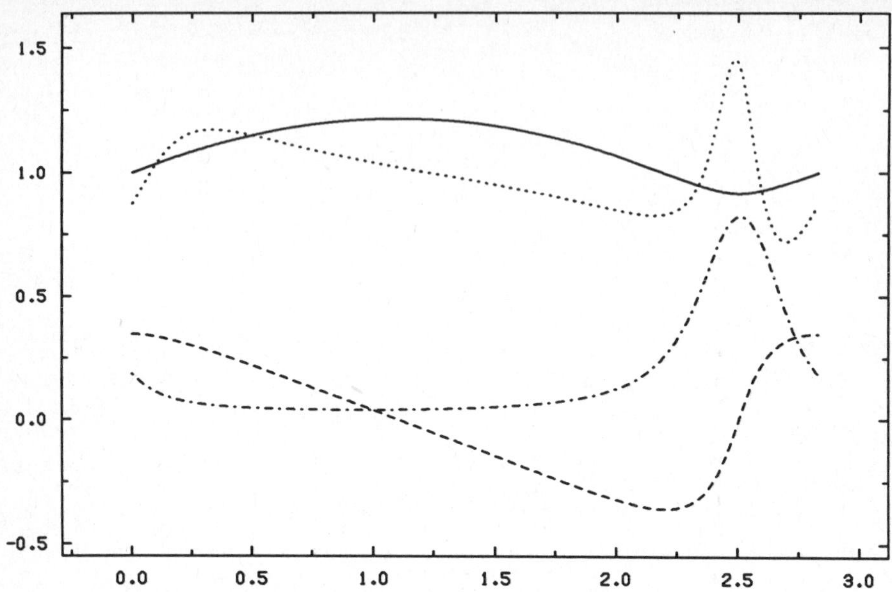

fig. 3.1 : radius (solid line), velocity (dashed), luminosity (dotted) and fraction due to convection (dash-dotted) versus time in units of the free-fall time for the one-zone model with the parameters of Stellingwerf [4] in the 'strip' case

The radius and velocity curves look very natural, but the spike in the luminosity at the time of maximal compression of the shell is surely unphysical. It is due to the fact that one cannot treat the inner core (inner boundary condition of the shell under consideration) as a hard core. Instead one knows from many other linear as well as nonlinear extended calculations that immediately beneath the main driving regions there is a 'principal dissipation zone' (Baker, Kippenhahn 1962, Christy 1966) [6, 7]. Especially this problem is not due to the convection treatment because one can find those spikes also for purely radiative models if one insists on parameter choices which lead to a real periodic solution. Most of the solutions given in literature seem to be not periodic. Also the curves in [4] corresponding to those shown above seem to be not completely relaxed to the fully periodic solution with damped out initial conditions. This is ensured in the models presented because I use a different method which treats the whole periodic solution as a nonlinear eigenvalue problem with the period as the eigenvalue.

It might be possible to have convection also acting as a driving mechanism. This is indicated by another model with constant opacity and 90% of the luminosity being carried by convection. Thus one may ask whether there is another mechanism responsible for the pulsation rather than the κ−effect [5, 6]. The model yields very long periods in this case (about 25 free fall times). For better optical resolution the dashed line stands for velocity + 1.

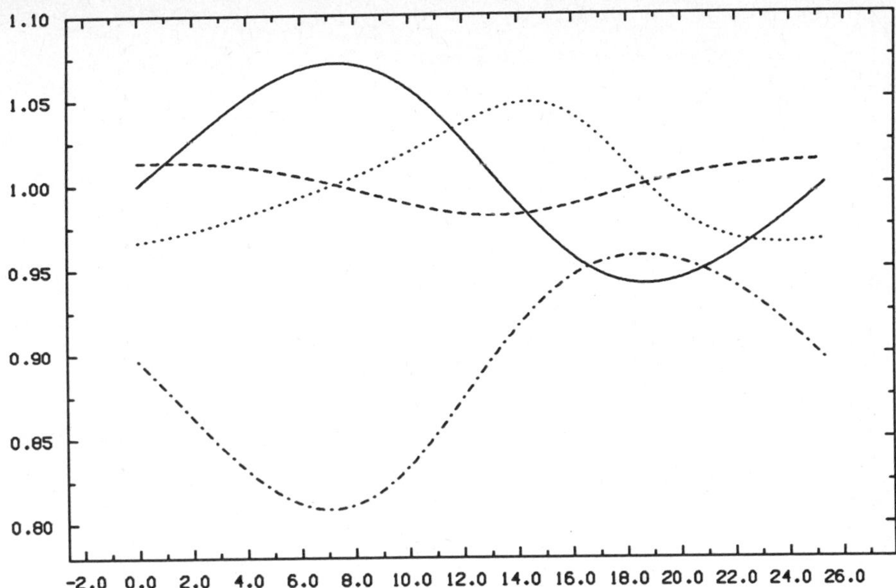

fig. 3.2 : radius (solid line), velocity+1 (dashed), luminosity (dotted) and fraction due to convection (dash-dotted) versus time in units of the free-fall time for the one-zone model with constant opacity and 90% convective flux in the equilibrium model

4. Nonlocality: Overshooting in ZAMS stars

The special nonlocal behaviour is studied in the stationary case at the example of chemically homogenous ZAMS stars. It should be mentioned that one could handle the effects of composition gradients in the same selfconsistent way [1]. For the energy generation a fit formula for p-p and CNO burning was used [8]. The composition is $X = 0.685, Y = 0.294$. The MLT parameter α_{ML} was chosen to be 1.5. The opacity is taken from a fit formula cited in [9]. Due to this very crude estimate the $1\text{M}_\odot$ star already has a slightly convective core. Nonetheless the qualitative aspects of the overshooting should be represented to sufficient accuracy.

First a $20\text{M}_\odot$ star is studied in detail. As in all other cases a MLT model was calculated together with a nonlocal one for comparison. In figure 4.1 the function

$$\text{sign}(\nabla - \nabla_{\text{ad}})\big|\log(|\nabla - \nabla_{\text{ad}}|)\big|$$

is plotted as a function of the mass coordinate. This is so to say the logarithm of the nonadiabaticity with the additional information whether convection is driven (positive sign) or damped. At the Schwarzschild boundary $\nabla = \nabla_{\text{ad}}$ the function has a pole with change of sign and this causes the interpolation routine of the plot program to mark this boundary by a vertical line. The nonlocal model is represented by the solid line, the MLT model is indicated by the dashed line.

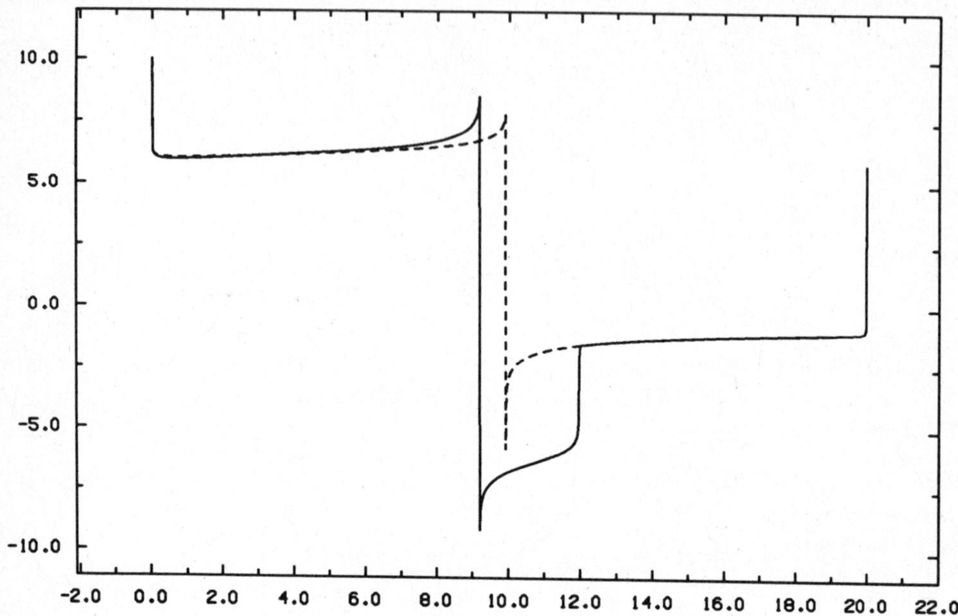

fig. 4.1 : logarithmic nonadiabaticity as a function of mass for the nonlocal (solid line) and the MLT model (dashed line)

There are two effects to be seen:

1. There is a certain region where in the nonlocal model the nonadiabaticity is small, though damping (antibuoancy). This is the overshooting region, which becomes almost adiabatic because of a large convective backward flux (see figure 4.2).

2. The nonlocal Schwarzschild 'boundary' $\nabla = \nabla_{ad}$ is shifted a bit towards the center of the star. Therefore one is left with a reduced 'effective' overshooting compared to the MLT model.

There are three different luminosities for the energy transport in the nonlocal model. These are shown in figure 4.2. The solid line indicates the total luminosity, the dotted line stands for convective thermal luminosity, the radiative luminosity is marked by the dashed line, and the luminosity due to transport of kinetic energy, which causes overshooting but is small compared to the others, is indicated by the dash-dotted line. The unit for the luminosity is $L_O = 2.15597 \cdot 10^{11} L_\odot$. The convective backward flux and the corresponding local maximum of the radiative flux is well to be seen. Figure 4.3 shows the same for the MLT model.

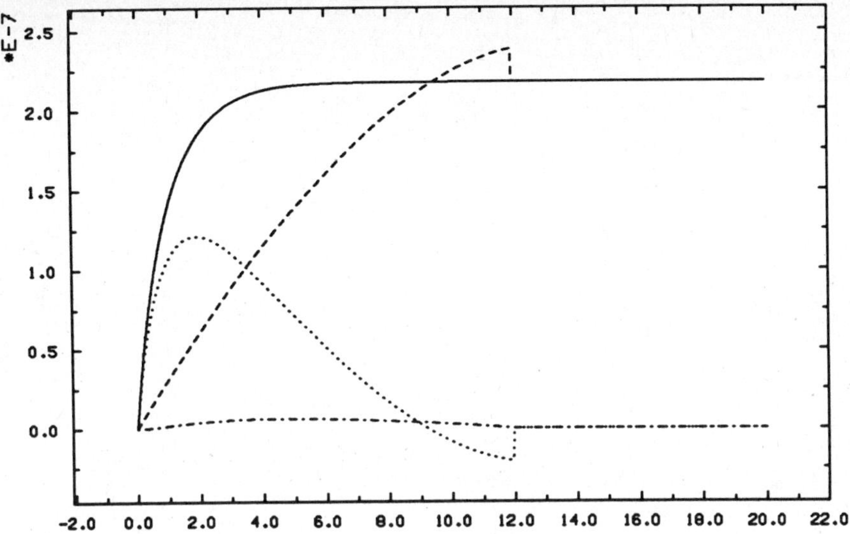

fig. 4.2 : comparison of all luminosities as functions of the mass coordinate for the nonlocal model: total (solid line), radiative (dashed), convective (dotted) and luminosity due to kinetic energy flux (dash-dotted)

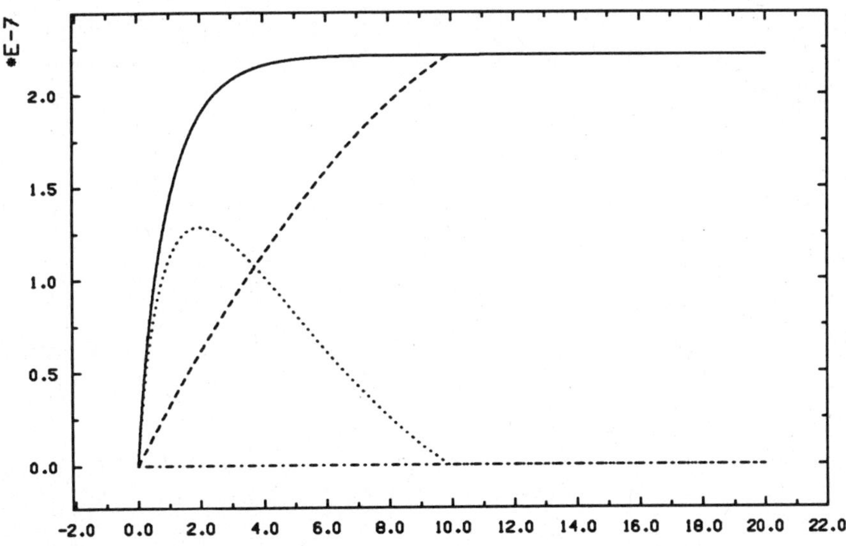

fig. 4.3 : the same as 4.2, but for the MLT model

It is furthermore interesting to study the dependence of overshooting on the mass of a star. In the following the results of a series of stellar models from $1M_\odot$ to $100M_\odot$ are plotted. In figure 4.4 the core masses in % are plotted for the nonlocal model (dots) and for the MLT model (triangles). The relative increase, which is defined as the difference of the core masses normalized to the MLT core mass in %, is given by the crosses.

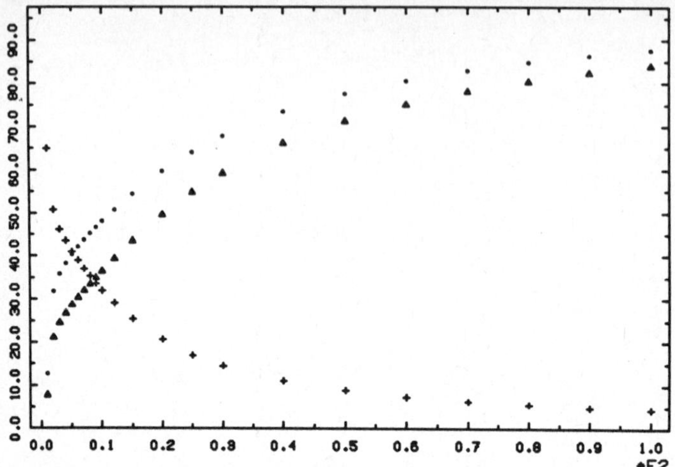

fig. 4.4 : relative core masses for the nonlocal model (dots), the MLT model (triangles) and the relative increase of the core of the nonlocal model compared to the MLT core (crosses), all in %, as functions of the total mass

Another common way to give the extent of overshooting is the so called overshooting parameter. This is defined to be the extent of the overshooting core minus the extent of the MLT core in units of the pressure scale height taken at the boundary of the MLT core. In figure 4.5 the crosses indicate the shift of the Schwarzschild boundary, the triangles stand for the overshooting and the dots for the resulting effective overshooting, all normalized to the pressure scale height taken at the MLT core boundary.

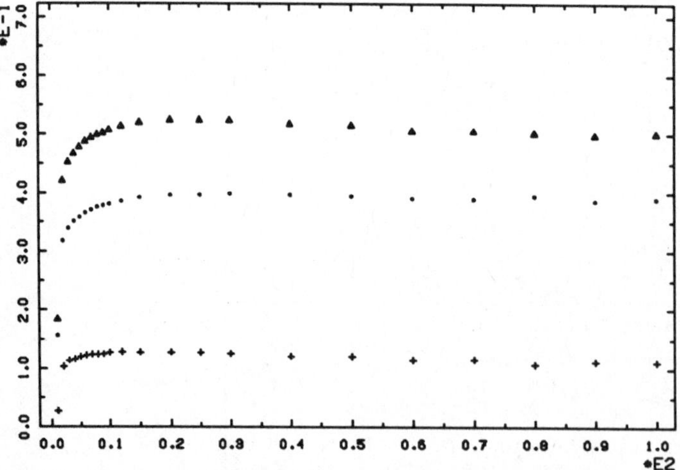

fig. 4.5 : shift of Schwarzschild boundary (crosses), total overshooting (triangles) and corresponding effective overshooting (dots) as functions of the total mass and all normalized to the pressure scale height at the boundary of the MLT core.

The overall structure of the nonlocal models, that means central density, central temperature, radius and luminosity, does not differ very much from that of the MLT models. For masses say larger than about $15M_{\odot}$ the overshooting parameter seems to be approximately constant and about 0.4. But this depends on the MLT parameter α_{ML}. To investigate this dependency a parameter study was performed. For a $20M_{\odot}$ star a series of models was calculated for α_{ML} varying between 1.0 and 3.0. Figure 4.6 gives the core mass increase as a function of the MLT parameter α_{ML} in the same way as in figure 4.4, figure 4.7 correponds to figure 4.5 and shows the 'overshooting' parameter as a function of α_{ML}.

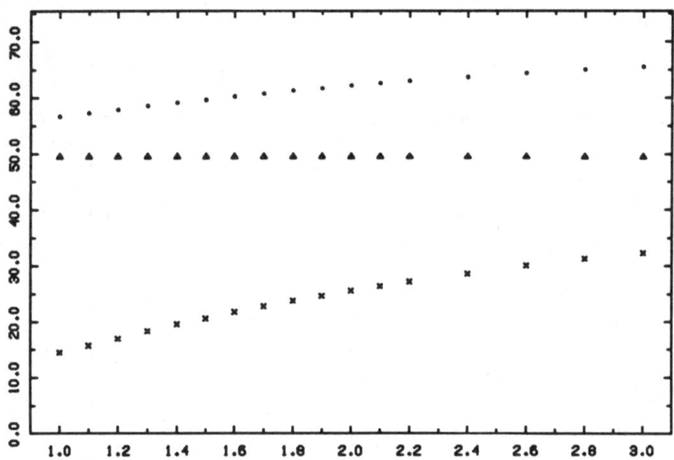

fig. 4.6 : relative core masses for the nonlocal model (dots), the MLT model (triangles) and the relative increase of the core of the nonlocal model compared to the MLT core (crosses), all in %, as functions of α_{ML}.

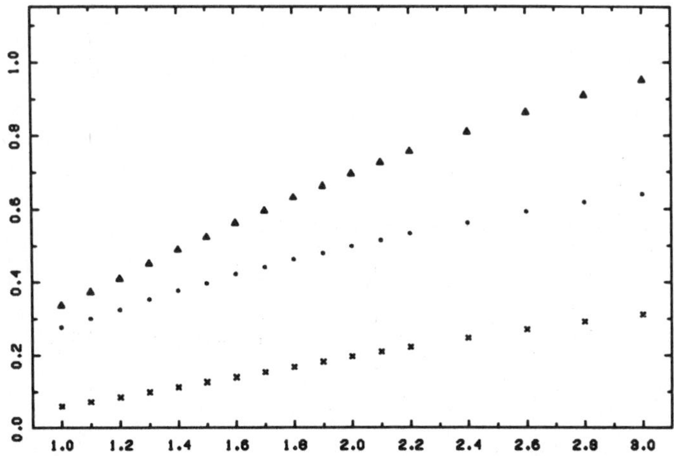

fig. 4.7 : shift of Schwarzschild boundary (crosses), total overshooting (triangles) and corresponding effective overshooting (dots) as functions of α_{ML}, all normalized to the pressure scale height at the boundary of the MLT core.

There is a certain dependence of the extent of overshooting on the MLT parameter α_{ML}. Maeder and Meynet [10] claim that the overshooting parameter should be about 0.25 to get consistency with observations. This could be achieved using $\alpha_{ML} \approx 0.9$. Again the overall structure of the nonlocal models is not changed much compared to the MLT model. The most severe differences occur for the highest α_{ML}, with almost no change in central density and temperature, but slight changes of about 2% in radius and 5% in luminosity. For α_{ML} less than 1.5, which seems to be more natural, the luminosity changes reduce to less than 2%.

5. Conclusions

Some of the special effects of a nonlocal timedependent convection model for stellar evolution have been investigated to get a feeling for the differences to the usual MLT treatment. The qualitative aspects should be well represented, quantitative statements however should be checked very carefully because of the variety of free parameters in the model. Observations of different kind, e.g. variable stars as well as HR diagrams for clusters, could tell us how to choose these parameters to yield a consistent description of convection for many applications in stellar structure and evolution calculations.

6. References

1 **R. Kuhfuß**: 1986, *Astron. Astrophys.* **160**, 116

2 **D.O. Gough**: 1969, *J. Atmos. Sci.* **26**, 448

3 **R.F. Stellingwerf**: 1972, *Astron. Astrophys.* **21**, 91

4 **R.F. Stellingwerf**: 1986, *Astrophys. J.* **303**, 119

5 **N.H. Baker**: 1966, in**Stellar Evolution**, ed. by R.F. Stein and A.G.W. Cameron, 333

6 **N.H. Baker, R. Kippenhahn**: 1962, *Ztschr. Astrophys.* **54**, 114

7 **R.F. Christy**: 1966, *I.A.U. Symposium* **28**, 105

8 **E. Hofmeister, R. Kippenhahn, A. Weigert**: 1964, *Ztschr. Astrophys.* **59**, 215

9 **G.M.H.J. Habets**: 1985, PhD thesis, Amsterdam

10 **G. Meynet**: 1987, these proceedings

NUCLEOSYNTHESIS IN EXPLOSIONS OF HIGH METALLICITY SUPERMASSIVE OBJECTS

Wolfgang Hillebrandt[1], Friedrich-Karl Thielemann[2], and Norbert Langer[3]

[1] Max-Planck-Institut für Physik und Astrophysik, Institut für Astrophysik, Garching, FRG
[2] Department of Astronomy, Harvard University, Cambridge, USA
[3] Universitätssternwarte, Göttingen, FRG

Abstract

We investigate the hydrodynamic evolution of a high metallicity (Z=0.04) supermassive object ($M=5x10^5 M_\odot$) through contraction and explosion due to hot hydrogen burning. It is demonstrated that these objects, if they occasionally should form in the dense cores of galaxies, can produce interesting amounts of ^{13}C, ^{17}O, and ^7Li, together with radioactive ^{26}Al. In fact, we find that the presence of about 5 $M_\odot$ of ^{26}Al at or near the center of our own galaxy can easily be explained if such an object exploded there within the last 2 million years.

1. Introduction and Motivations

Originally, supermassive star-like objects (SMOs, $M \gtrsim 5x10^4 M_\odot$) have been invented by Hoyle and Fowler (1963) in order to account for the high luminosity of quasars. Since, however, those stars are radiation dominated (adiabatic index $\gamma \simeq 4/3$) they become gravitationally unstable during hydrogen burning due to post-Newtonian effects. Therefore their lifetime is roughly equal to the quasistatic contraction phase prior to hydrogen ignition and thus too short to be associated with quasars. SMOs may, however, be candidates to explain the activity of certain Seyfert galaxies (Stoner and Ptak, 1984).

Motivated by the apparent need for some pre-galactic nucleosynthesis several authors have investigated low-metallicity (or even zero metallicity) SMOs as possible candidates (Appenzeller and Fricke, 1972; Fricke, 1973, 1974; Fricke and Ober, 1980). Recently Fuller, Woosley and Weaver (1986) have reinvestigated this problem by using up-to-date nuclear reaction rates for hot hydrogen burning (Wallace and Woosley, 1981), an implicit hydrodynamic code, a realistic equation of state, and neutrino losses. Their main conclusion was that SMOs will explode due to hot hydrogen burning provided the initial metallicity is at least $Z \simeq 0.005$. They also found that the metallicity required for explosions increased with increasing mass of the SMO. In all cases where they obtained explosions the explosion energy, mostly in form of kinetic energy of the ejecta, exceeded 10^{56} erg. An explosion was never accompanied by the formation of a compact (black hole) remnant.

We have investigated the hydrodynamic evolution of a SMO of $5 \times 10^5 M_\odot$ of high metallicity (z=0.04). The choice of roughly twice the solar metal abundances was motivated by the discovery of the γ-ray line flux from the decay of ^{26}Al in the interstellar medium by Mahoney et al. (1982, 1984) and its later confirmation by Share et al. (1985). The idea was that a single SMO exploding at the galactic center might produce a sufficient amount of ^{26}Al to explain these observations (see also Hillebrandt, Mair and Ziegert (1986), and Hillebrandt, Thielemann and Langer (1986)). Such an event some 10^6 years ago would easily explain the observations of Ballmoos, Diehl and Schönfelder (1986) who found that the flux distribution is strongly peaked towards the galactic center and is fitted best by a "point source" of $(5 \pm 2) M_\odot$ of ^{26}Al at the galactic center, the 1σ angular resolution of the MPI Compton telescope being 3.5^0.

A scenario which may explain that SMOs form occasionally from dense star clusters in galactic cores has been given by Begelman and Rees (1978) (see also Sanders (1970)). They discuss the possibility that a dense cluster of low mass main sequence stars undergoes coalescence and disruptive collisions leading to the formation of a massive gas cloud which then contracts and passes through a phase of hydrostatic equilibrium. It is interesting to note that the relaxation time for a dense cluster of 10^7 stars of 1 $M_\odot$ in a volume of 1 pc^3 is of the order of a few 10^8 years only and therefore much shorter than the evolution times of typical low mass stars.

Given the possibility that once in a while even metal-rich SMOs may form in nature one should ask the question whether or not there is additional evidence or if our hypothesis is in conflict with observations. Firstly, large radio lobes extending from several 10^2 pc to a few kpc out of the centers of some galaxies, including our own galaxy, are observed. They have been interpreted as beeing due to gigantic explosions releasing about 10^{57} erg (Sofue, 1984; Hayakawa, 1986). The energetics of the galactic center also seems to indicate that there might have been a very active phase a few 10^6 years ago (Oort, 1984). Again, the explosion of a SMO would be a straight forward explanation.

Thus we are left with the question whether the explosion of metal-rich SMOs would pollute the gas near the galactic center with matter of strange chemical and isotopic composition. It is the main aim of this paper to show that this is not the case. On the contrary, it will be demonstrated that our model may even be able to explain non-solar isotopic ratios of certain elements found in dense clouds near the galactic center.

2. Outline of the Computations

We have not intended to perform a series of evolutionary calculations for various masses and initial metallicities but rather restricted ourselves to what seems to be a typical case, namely a $5\times10^5\ M_\odot$ object composed of 70% H, 26% He and 4% "metals" which seems to be reasonable for galactic center material. Since from the previous works we could expect that for the rather high metallicities under consideration the temperature inside the star would never exceed roughly 2×10^8 K we have chosen a simplified nuclear reaction network in order to obtain the energy generation rate in hot hydrogen burning. According to Wallace and Woosley (1981) it is sufficient to use a 9 nuclei network and the following set of reactions:

$$^{12}C(p,\gamma)\ ^{13}N(p,\gamma)\ ^{14}O(\beta^+)\ ^{14}N(p,\gamma)\,^{15}O(\beta^+)\ ^{15}N(p,\alpha)\ ^{12}C$$

and

$$^{13}N(\beta^+)\ ^{13}C(p,\gamma)\ \ ^{14}N\ .$$

The initial stellar model was chosen to be a n = 3-polytrope with a central density sufficiently low to allow for the establishment of a thermal equilibrium during the quasi-static contraction phase. The evolution was then computed by means of a implicit hydro code gravity being included in post-Newtonian approximation.

At the onset of collapse, the central values of density and temperature were $2.3 \times 10^{-2} gcm^{-3}$ and $5.1 \times 10^{7} K$, respectively, and nuclear energy generation ($1.4 \times 10^{6} erg\ g^{-1} s^{-1}$) was too small to stabilize the star. The peak values of density and temperature were reached about 2×10^{6}s later (see fig.1), and more than $5 \times 10^{51} erg\ s^{-1}$ were liberated by hot hydrogen burning for roughly 2×10^{4}s (see fig.2), leading to an explosion with a total kinetic energy of 10^{56}erg. During the whole evolution only about $10^{4} M_{\odot}$ of hydrogen have been converted into helium. In agreement with previous computations no compact remnant was formed and the star was completely disrupted.

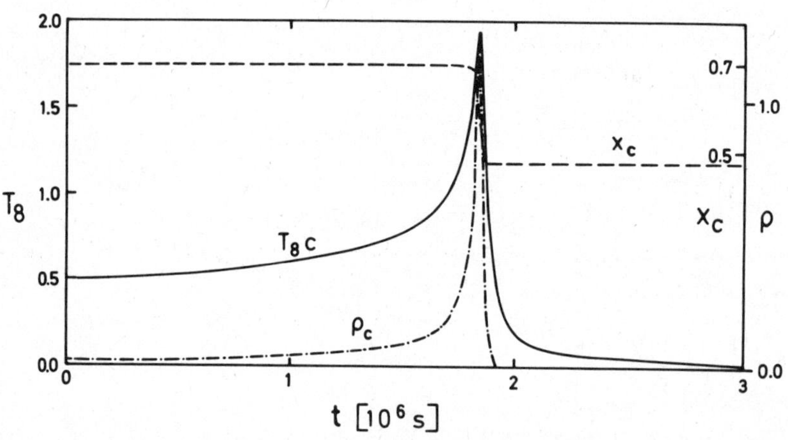

Figure 1: Temperature T_c (in units of 10^8K), density ρ_c (in gcm^{-3}), and hydrogen mass fraction X_c of the central zone versus time (in 10^6s).

For the inner $2.7 \times 10^{5} M_{\odot}$ of the stellar model described above we have computed the nuclear abundances as functions of time by applying the reaction network discussed in detail in the work of Wiescher et al. (1986). The temperatures in the outer layers of the model stayed sufficiently low so that we did not have to include them (see fig.3).

The initial abundances of all heavy elements in the network were assumed to be twice their solar system values.

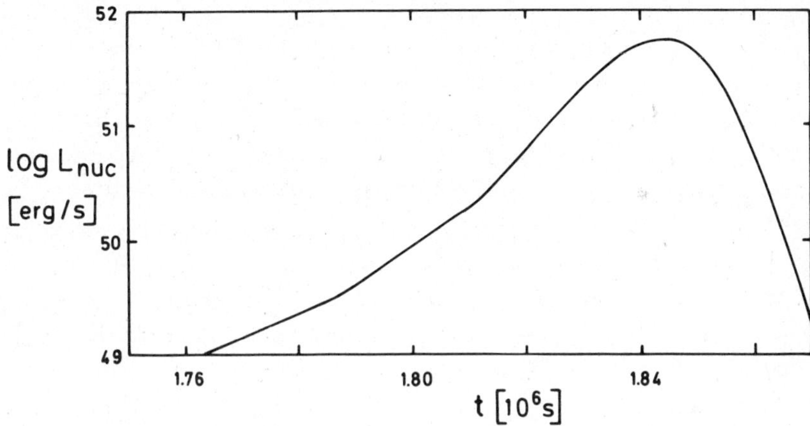

Figure 2: Nuclear energy generation integrated over the star versus time. The time is measured from the onset of the computations.

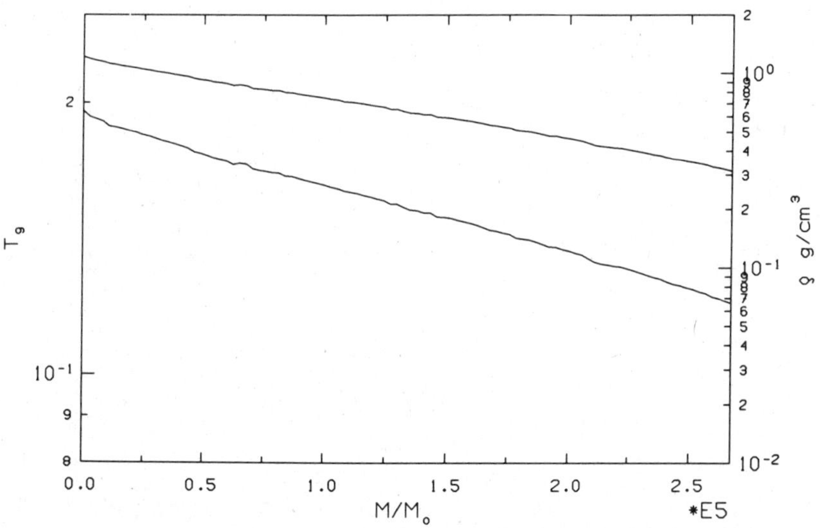

Figure 3: Peak values of temperature (in 10^9K, lower curve) and density (in gcm^{-3}, upper curve) as a function of mass (in 10^5 $M_\odot$).

Our main results are displayed in figures 3 to 6. Figure 3 shows the peak values of temperature and density reached in various mass zones of the star during the explosion. It becomes obvious that only in the inner $10^5 M_\odot$ the conditions are appropriate for the formation of ^{26}Al. Figure 4 gives the mass fractions of various nuclei at freeze-out for different positions inside the star. It can be seen that in the inner mass zones ^{24}Mg is destroyed and transformed into ^{25}Mg and ^{26}Al, but that heavier elements such as Si and S are not much affected because the peak temperatures were too low. The total amount of ^{26}Al produced in our model is $48.7 M_\odot$, about one order of magnitude more than in the previous computations of Hillebrandt, Mair and Ziegert (1986), mainly because of the lower temperatures of the present model. The abundance patterns of the CNO-isotopes agree with those expected form the operation of the hot CNO-cycle (see Wiescher et al. (1986) for an extended discussion).

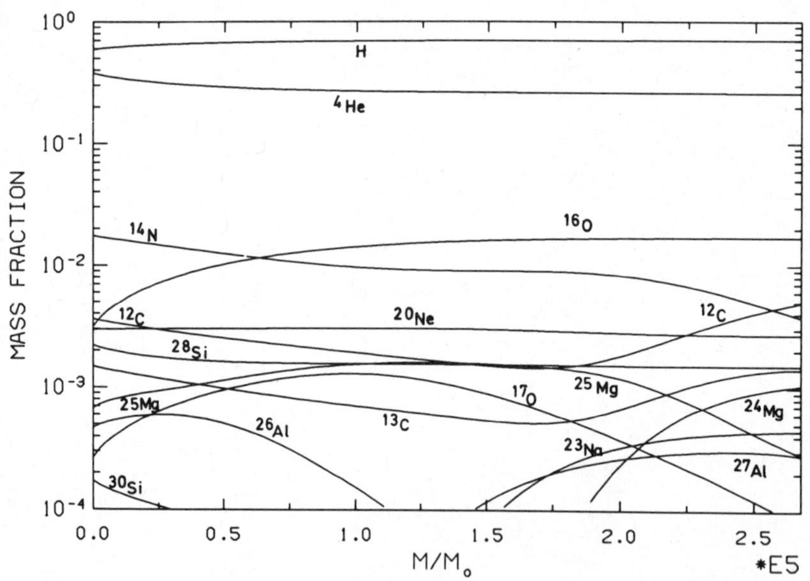

Figure 4: Mass fractions versus mass of various isotopes after the explosion of the SMO. It can be seen that ^{26}Al is produced in the inner 10^5 $M_\odot$ only.

Figures 5 and 6 show the abundances of elements and isotopes, respectively, relative to their solar system values after averaging

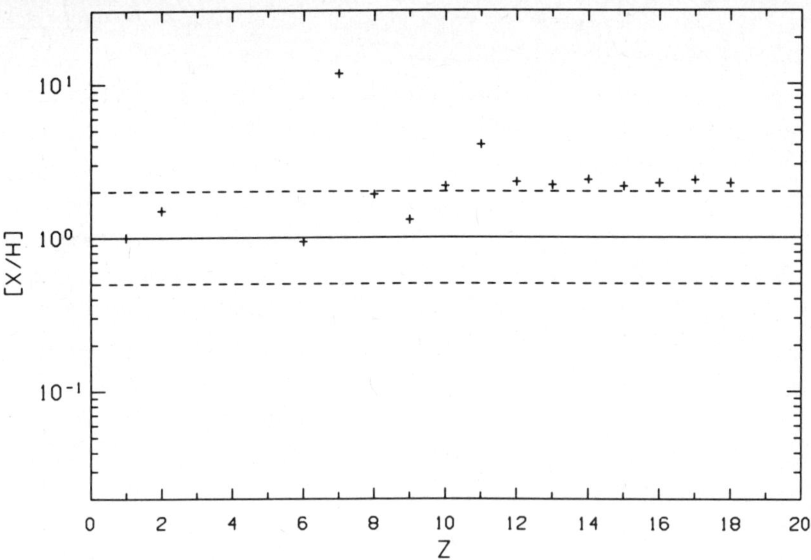

Figure 5: Mass fractions of the elements normalized to hydrogen relative to solar values after the explosion. Li is not shown because it is off-scale.

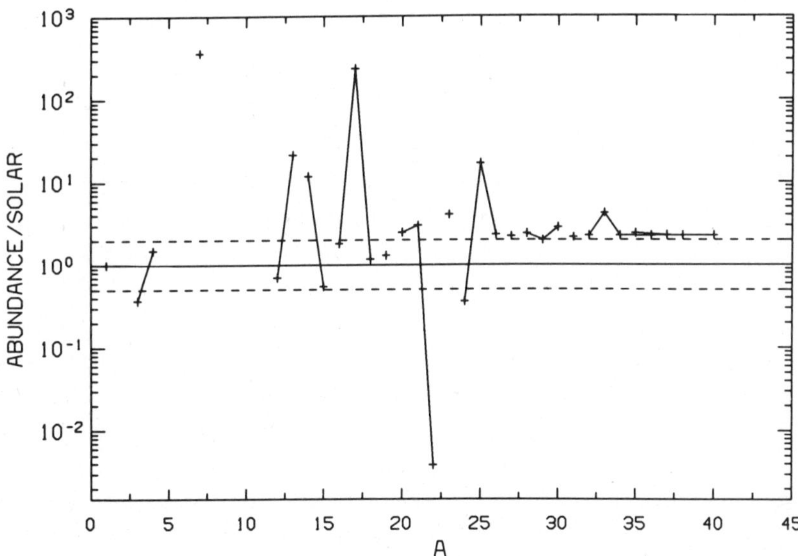

Figure 6: Isotopic abundances relative to solar system abundances. Isotopes of a given element are connected by solid lines.

over the inner $2.7x10^5 M_\odot$ of the star. With the exception of Li (off-scale in figure 5; see, however, figure 6) and N and, to a lesser extend, Na all elemental abundance ratios are roughly solar. The overall enhancement by a factor of about 2 is due to the initial metal enrichment of the model. As a consequence of the operation of the hot CNO-cycle carbon is somewhat depleted. The isotopic abundances in figure 6 show more clearly the effects of hot hydrogen burning. ^7Li as well as ^{17}O are enhanced by more than a factor of 200, ^{13}C and ^{25}Mg are enhanced by roughly a factor of 20, and ^{22}Ne is almost completely destroyed. The nuclear mass 22 is predominantly formed as ^{22}Na.

3. Implications and Conclusions

We have demonstrated that in the explosion of a single high-metallicity (z=0.04) supermassive object of $5x10^5 M_\odot$ approximately 50 $M_\odot$ of ^{26}Al are produced. Although we have not performed a systematic parameter study this value seems to be a typical one (Hillebrandt, Thielemann and Langer, 1986). It seems, therefore, that the explosion of a single SMO near the galactic center within the last 1 to 2 million years could easily explain the presence of several solar masses of ^{26}Al there. Of course, the statistical accuracy of the MPI measurements (Ballmoos, Diebl and Schönfelder, 1986) was not sufficient to exclude a diffuse origin of the γ-ray line flux from the decay of ^{26}Al, but if it should be confirmed that the flux is indeed concentrated in a region of a few hundred pc close to or at the galactic center, our model would give a possible explanation (see also Prantzos (this volume)).

Proceeding on the assumption that a SMO exploded at the galactic center about a million years ago we have to show that this is not in conflict with other observations. Firstly, such an explosion should not be singular event. In this respect the predicted huge overproduction of ^{17}O seems to cause a problem for our model. If we assume that events as the one discussed here mix their ejecta with the matter of the inner, say, $10^9 M_\odot$ of the galaxy the total amount of ^{17}O should not exceed about $10^4 M_\odot$. This limits the number of events to about 50 over the lifetime of the galaxy, and the average frequency to about one event every $3x10^8$ years, which seems to be uncomfortably low. One should keep in mind, however, that the $^{17}O/^{18}O$ production ratio is strongly affected by rather uncertain reaction rates involving unstable target nuclei. Here the reaction $^{17}F(p,\gamma)$ ^{18}Ne may serve as an example of a rate for which theoretical predictions vary by orders of magnitude (Wiescher and Kettner, 1982). A somewhat higher rate than was used in our computations will lower the $^{17}O/^{18}O$ ratio and thus will weaken the constraint obtained for the event rate. Moreover, the production of ^{17}O is strongly dependent upon the peak temperatures and will strongly decrease for somewhat larger stellar masses. The huge overabundance of ^{17}O found in our model, therefore, may not be typical. If, on the other hand, we use the overproduction of ^{13}C in order to limit the possible event rate we can allow for a frequency that is almost a factor of 10 higher without encountering serious overproduction problems.

The remnant of an explosion of a supermassive object near the galactic center some 10^6 years ago would by now have swept up a few times $10^8 M_\odot$ of interstellar matter (Hillebrandt, Mair and Ziegert, 1986). If the processed matter of the exploded star has been mixed homogeneously with that of presumably normal composition no significant effects on elemental and isotopic abundance ratios are expected. If, however, mixing was incomplete we expect to find large non-solar ratios of $^{17}O/^{16}O$, $^{13}C/^{12}C$ and $^{14}N/^{15}N$. $^{13}C/^{12}C$ and $^{14}N/^{15}N$ ratios of the order of 4 to 5 times their solar values have indeed been found in dense clouds near the galactic center (Güsten, Henkel and Batrala, 1985; Güsten and Ungerechts, 1985), but the roughly solar ratio of $^{18}O/^{16}O$ obtained from our model does not agree with the observation of those clouds. However, as we have discussed before, one can overcome this problem by either increasing the $^{17}F(p,\gamma)$ ^{18}Ne reaction rate or by increasing the mass of the SMO.

Finally, we want to note that explosions of SMOs may provide an interesting source of ^7Le. In the model presented here about $1M_\odot$ of ^7Li is ejected. A few hundred explosions, therefore, could have produced as much ^7Li as is attributed to big bang nucleosynthesis.

References

1) Appenzeller, I, and Fricke, K. 1972, Astron. Astrophys. 21, 285.
2) Ballmoos, P.v., Diehl, R., and Schönfelder, V. 1986, Ap.J., in press.
3) Begelmann, M.C., and Rees, M.J. 1978, M.N.R.A.S. 185, 847.

4) Fricke, K.J. 1973, Ap.J. 183, 941.
5) Fricke, K.J. 1974, Ap.J. 189, 535.
6) Fricke, K.J., and Ober, W. 1980, Ann. N.Y. Acad. Sci. 336, 399.
7) Fuller, G.M., Woosley, S.E., and Weaver, T.A. 1986, Ap.J. 307, 675.
8) Güsten, R., and Ungerechts, H. 1985, Astron. Astrophys. 145, 241.
9) Güsten, R., Henkel, C., and Batrala, W. 1985, Astron. Astrophys. 149, 195.
10) Hayakawa, S. 1986, preprint, Prog. Theor. Phys.
11) Hillebrandt, W., Mair, G., and Ziegert, W. 1986, Proc. 2nd IAP Rencontre on Nuclear Astrophysics, J. Audouze et al., eds., in press.
12) Hillebrandt, W., Thielemann, F.-K., and Langer, N. 1986, Ap.J., in press.
13) Hoyle, F., and Fowler, W.A. 1963, M.N.R.A.S. 125, 169.
14) Mahoney, W.A., Ling, J.C., Jacobson, A.S., and Lingenfelter, R.E. 1982, Ap.J. 262, 742.
15) Mahoney, W.A., Ling, J.C., Wheaton, W.A., and Jacobson, A.S. 1984, Ap.J. 286, 578.
16) Oort, J.H. 1984, IAU-Symp. 106, v. Woerden et al., eds., p. 349.
17) Sanders, R.H. 1970, Ap.J. 162, 791.
18) Share, G.H., Kinzer, R.L., Kurfess, J.D., Forrest, D.UJ., Chupp, E.L., and Rieger, E. 1985, Ap.J. 292, L61.
19) Sofue, Y. 1984, Publ. Astron. Soc. Japan 36, 539.
20) Stoner, R. and Ptak, R. 1985, Ap.J. 297, 611.
21) Wallace, R.K., and Woosley, S.E. 1981, Ap.J. Suppl. 45, 389.
22) Wiescher, M., and Kettner, K.U. 1982, Ap.J. 263, 891.
23) Wiescher, M., Görres, J., Thielemann, F.-K., and Ritter, H. 1986, Astron. Astrophys. 160, 56.

ISOTOPIC ANOMALIES AND WOLF-RAYET STARS

J.B. Blake
Space Sciences Laboratory
The Aerospace Corporation
P.O. Box 92957
Los Angeles, CA 90009

D.S.P. Dearborn
Lawrence Livermore National Laboratory
P.O. Box 808
Livermore, CA 94550

ABSTRACT

Wolf-Rayet stars can generate some of the isotopic anomalies seen in meteoritic material. The cases of ^{41}Ca and ^{107}Pd are discussed.

INTRODUCTION

Wolf-Rayet stars have attracted the attention of nuclear astrophysicists for several reasons, one of which is the fact that large amounts of newly synthesized nuclei are injected into the ISM by the very strong stellar winds of such stars, cf. [1]. Wolf-Rayet stars are believed to result in general from very massive progenitors (ZAMS mass $\gtrsim$ 40 M_o) which evolve with large mass loss. The stellar sequence expected under this hypothesis is O→Of→WN→WC [2, 3, 4]. The nuclei ejected during the Of and WN star phases have been subjected to hydrogen burning via the CNO, NaNe, and MgAl cycles, resulting in material which is N rich and includes ^{26}Al [5, 6, 7]. Material ejected during the subsequent WC stage will contain the products of core helium burning which include nuclei resulting from neutron capture (a weak s-process).

Because Wolf-Rayet stars in general are massive objects and thus relatively rare, they are expected to be minor contributors to the overall nucleosynthesis in the galaxy, and would affect the bulk composition of the solar system only if such a star had happened to be near the protosolar nebula. The ejecta from some Wolf-Rayet stars is observed to contain grains. Thus some of the exotic isotopic and elemental abundances produced in these stars should be trapped in grains. Because of the conditions surrounding a Wolf-Rayet star, such grains must be as durable as any produced in other sites such as Novae and Red Giants. If some of these grains were injected in the solar nebula they could provide an exotic component for incorporation into meteoritic material.

Because the mass-loss rates in Wolf-Rayet stars are high (> few x 10^{-5} $M_{\odot}$/yr), the time between production of an isotope and its injection into the ISM can be relatively short. We have made extensive calculations of the production ofextinct radioactivities in Wolf-Rayet stars [8]. They show that some short-lived radioactivities of interest are produced in high relative abundance in WR stars and must subsequently be ejected in the stellar wind. In this report the isotopes ^{41}Ca and ^{107}Pd are discussed.

MODEL CALCULATIONS

Stellar models with masses of 50, 100, and 150 $M_{\odot}$ were evolved to represent the range of ZAMS masses which are believed to become Wolf-Rayet stars. The models all had a standard composition of X = .7 and Z = .02. For reasons discussed in [9] we did not include any ad hoc prescriptions for convective overshoot. While such an effect may occur, it is unclear how much actual material mixing that cause, and more importantly, it is not an important effect for the discussion here, leading perhaps to a larger core for a given ZAMS mass and a slight enhancement in the overall yield of the s-process abundances in question.

The nucleosynthesis network used was described in [5] and includes hydrogen burning via the CNO, NaNe, MgAl cycles, helium burning, and neutron production from ^{13}C, ^{17}O, and ^{22}Ne. Mixing due to convection and semiconvection is handled simultaneously with nuclear burning by a diffusion approximation.

While most of the physics pertain to models of massive stars during hydrogen and helium burning is relatively well understood, the physics of mass loss is an exception. Fortunately, the main effect of differing mass loss formalisms is the relationship between the initial and core masses. A constant mass-loss rate leads to a lower core mass (smaller Wolf-Rayet star) than a mass-loss rate which is initially low and increases with luminosity (and perhaps has episodes of very high mass loss as seems to have occurred in η Carinae). While the mass-loss rate is a principle uncertainty in these models, the results that we discuss here will not be qualitatively affected by this uncertainty. We are interested in the relative production ratio of various extinct nuclei, and the mass-loss rate affects only the total yield of these species. Isotopic ratios of material of interest from WC stars will go through a wide range of values as helium burning products from deeper layers are ejected, but the ratios of interest in the present context rapidly approach a steady-state value. These calculations are discussed in detail in [8].

In a recent review, Wasserberg [10] gave a detailed discussion of short-lived nuclei which might be found in the primitive solar system. He provided the data in Table 1 for ^{41}Ca and ^{107}Pd:

Table 1

Parent Nucleus	Daughter Nucleus	Mean Life (yrs)	Solar System Abundance at ~ 4.6 AE
^{41}Ca	^{41}K	1.44×10^5	^{41}Ca/^{40}Ca $(8 \pm 4) \times 10^{-9}$
^{107}Pd	^{107}Ag	9.38×10^6	^{107}Pd/^{108}Pd ~ 2.0×10^{-5}

Figure 1 shows the calculated ^{41}Ca/^{40}Ca ratio as a function of the residual mass of our stellar model. One may note that this ratio has reached a maximum for the 50 M_o and 150 M_o ZAMS model but not for the 100 M_o model. This difference depends upon the mass-loss profile used in the calculations. If a somewhat larger mass-loss rate were used in the 100 M_o case, the curve would have flattened out as the case for the other two models. Figure 2 shows the evolution of the ^{107}Pd/^{108}Pd for the same stellar models. It can be seen that there is a substantial production of ^{107}Pd as well as ^{41}Ca.

DISCUSSION

Our calculations show that both ^{41}Ca and ^{107}Pd are produced by WC stars in a relative abundance well above the solar system observations given in Table 1. A period of free decay and substantial dilution with "normal" material in the ISM can occur without reducing the relative abundance below the observed values and that the relative abundances are similar for the three different ZAM masses.

The time available for free decay to reach the observed relative abundance values given in Table 1 can be estimated. For ^{107}Pd/^{108}Pd, Figure 2 indicates a ratio > 0.1 obtained during much of the WC phase. A ^{107}Pd/^{108}Pd ratio of ~ 2×10^{-5} would allow a free decay of t ~ 8×10^7 years. In the case of ^{41}Ca, the corresponding time period is t ~ 2×10^6 years. Decay can reduce the ^{41}Ca/^{40}Ca ratio to the observed upper limit and still have a ^{107}Pd/^{108}Pd ratio well above the observed value.

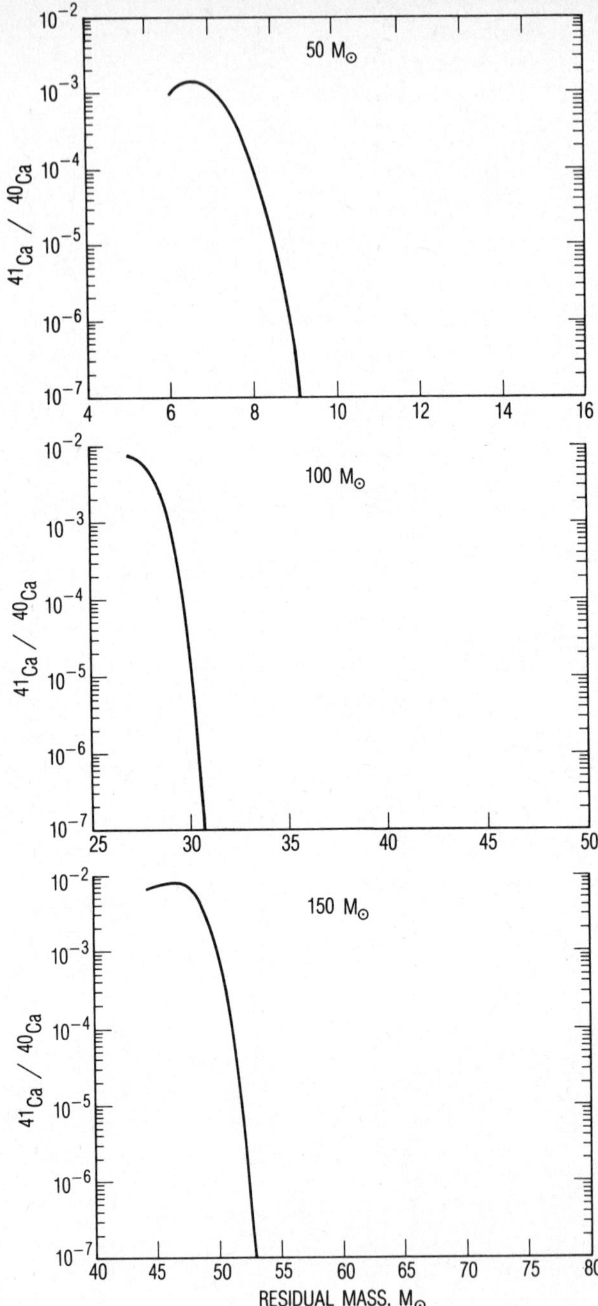

Figure 1. The evolution of the $^{41}Ca/^{40}Ca$ ratio as a function of residual mass is shown for stellar models of 50, 100, and 150 M_o.

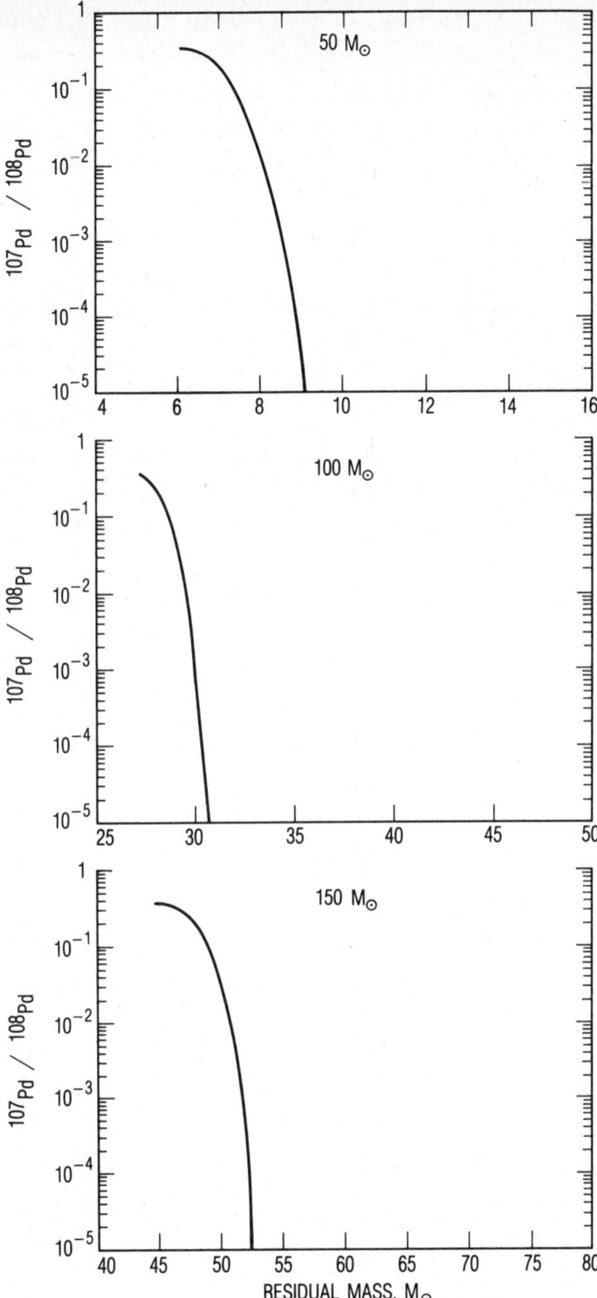

Figure 2. The evolution of the $^{107}Pd/^{108}Pd$ ratio as a function of residual mass is shown for stellar models of 50, 100, and 150 M_o.

The stellar wind of Wolf-Rayet stars contains relative abundances characteristic of the time when that parcel of gas was left outside the retreating stellar convective core. Thus, as the star evolves, the isotopic composition of the gas and dust in the wind continually changes. A collection of dust grains formed over time in the stellar wind of a Wolf-Rayet star would exhibit a complex, changing array of isotopic "anomalies". The ways in which the dilution of this exotic composition occurs is a crucial issue. Observations have shown the condensation of dust around WC stars [11, 12, 13]. This suggests that some of the relatively large abundance ratios of short-lived radioactivities which occur in the WC phase (cf. Figures 1 and 2) can be preserved from immediate dilution with the surrounding ISM. Margolis [14] has studied the injection of grains into the protosolar nebula. He found that grains are stopped within a column density equal to the grain mass. Thus grains do not penetrate deeply. If grains from a nearly WC star peppered the periphery of the protosolar nebula, a non-uniform nebula composition would arise. Presumably refractory elements are preferentially selected in the formation of grains. The time-varying abundance ratios in various elements in the stellar winds which result as the neutron exposure increases leads to some potentially interesting correlations. Figure 3 shows the evolution of the

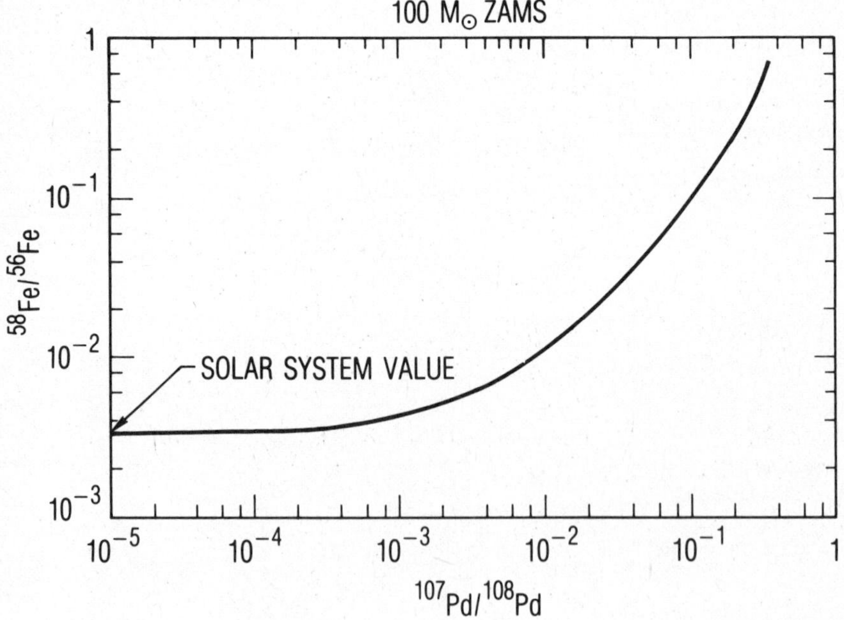

Figure 3. The correlation of the $^{58}Fe/^{56}Fe$ ratio with the $^{107}Pd/^{108}Pd$ ratio is shown for the 100 $M_\odot$ model. The other two models yielded similar results.

$^{58}Fe/^{56}Fe$ ratio as a function of the $^{107}Pd/^{108}Pd$ ratio. The Pd-Ag anomaly is seen in iron meteorites. If some of this iron originated with the ^{107}Pd in a WC star, an enhancement in ^{58}Fe would be expected. In general, if Wolf-Rayet stars contributed to the isotopic anomalies seen in early solar-system material there would be expected to be correlations between the anomalies seen in various elements.

The separation of the anomalies produced in WC stars into gas and dust portions effectively decouples the dilution of the two components. In this way a single WC star near the protosolar nebula could contribute significantly to certain isotopic anomalies, and to the solar system low fluence s-process. This issue is discussed by Dearborn and Blake [8].

ACKNOWLEDGEMENTS

This work was supported at the Aerospace Corporation by the company-sponsored research program, and at the Lawrence Livermore National Laboratory under the auspices of the U.S. Department of Energy.

REFERENCES

[1] D.C. Abbott, 1982, Ap.J., 263, 723.
[2] A. Maeder, 1983, Astr. Ap., 120, 113.
[3] C. Chiosi and A. Maeder, 1986, in Annual Rev. Astr. and Ap., ed. G. Burbridge, D. Lazzer and J.G. Phillips (Palo Alto: Ann. Rev., Inc), p. 329.
[4] N. Prantzos, M. Arnold and J.P. Arcoragi, 1987, Ap. J., 315, 209.
[5] D.S.P. Dearborn and J.B. Blake, 1984, Ap. J., 277, 783.
[6] D.S.P. Dearborn and J.B. Blake, 1985, Ap. J. Rett., 288, L21.
[7] N. Prantzos and M. Casse, 1986, Ap. J., 307, 324.
[8] D.S.P. Dearborn and J.B. Blake, 1987, in preparation.
[9] J.B. Blake and D.S.P. Dearborn, 1986, in press.
[10] G.J. Wasserberg, 1985, in Protostars and Planets II, ed. by D.C. Black and M.S. Matthews (Tucson: University of Arizona Press), p. 703.
[11] Hackwell, J.A., Gehrz, R.D., Smith, J.R. and Strecker, D.W., 1976, Ap. J., 210, 137.
[12] Hackwell, J.A., Gehrz, R.D. and Grasdalen, G.L., 1979, Ap. J., 234, 133.
[13] P.M. Williams, A.J. Longmore, K.A. van der Hucht, A. Talenera, W.M. Wamsteher, D.C. Albott and C.M. Telescon, 1985, Mor. Nat. R. Ast. Soc., 215, 23 p.
[14] S.H. Margolis, 1979, Ap. J., 231, 236.

THE ^{26}Al γ-RAY LINE : A STATUS REPORT

Nikos PRANTZOS

Institut d' Astrophysique de Paris
98bis, Bd. Arago, 75014 Paris, FRANCE

and

Service d' Astrophysique, Institut de Recherche Fondamentale
CEN Saclay, 91191 Gif sur Yvette, FRANCE

Introduction

The detection of the 1.8 MeV γ-ray line from the galactic center (GC) direction (Mahoney *et al.* 1982,1984), attributed to the decay of $\sim$3 $M_\odot$ of ^{26}Al in the interstellar medium (ISM), is a discovery of paramount importance: indeed, the presence in the ISM of a nucleus with an *intermediate* half-life of a million years or so (that is, very short compared to the time-scale of galactic chemical evolution) is a confirmation of current nucleosynthesis in the Galaxy and offers an opportunity of direct comparison between nucleosynthesis theories and observational data (Fowler 1984; Clayton 1984). That is why ^{26}Al, which can be produced in substantial amounts in various astrophysical sites, has been proposed as early as 1977 (Ramaty and Lingenfelter 1977), as an interesting candidate for γ-ray line astronomy. Its subsequent detection in the ISM came as no surprise, but the origin of the emission discovered by Mahoney *et al.* (1982) is not clear as yet.

In this paper we summarize the observational data concerning the 1.8 MeV emission in the galactic plane (sec. I), and we review the various astrophysical sources of ^{26}Al that have been proposed up to now (sec. II), with an emphasis on our "best-candidate": massive, mass losing stars.

I. The observations

Four different experiments, up to now, have detected the 1.801 MeV line due to the decay of ^{26}Al in the ISM: the HEAO satellite (Mahoney *et al.* 1982, 1984), the SMM satellite (Share *et al.* 1985), and 2 recent balloon experiments (Balmoos *et al.* 1986; Leventhal *et al.* 1987).

The γ-ray experiment on board of the HEAO satellite consists of four pure Ge detectors with an energy resolution of 3.3 keV (FWHM) and a field of view of 42° at 1.8 MeV. The angular response of the instrument is very broad above a few hundred keV. The measured width of the 1.8 MeV line (FWHM $<$ 3 keV; Fig. 1) implies velocity dispersions of the emitting ^{26}Al atoms less than 250 km s^{-1}, a limit compatible with emission from the ISM. For various technical reasons (poor statistics and angular resolution, among others), a direct mapping of the 1.8 MeV line emission in the galactic plane was not possible. Assuming that the line emission has the same longitudinal profile as the high energy (50 MeV $<$ E $<$ 5000 MeV) γ-rays detected by the COS-B satellite (Mayer-Hasselwander *et al.* 1982; Fig. 2), Mahoney *et al.* (1984) derived a γ-ray flux F

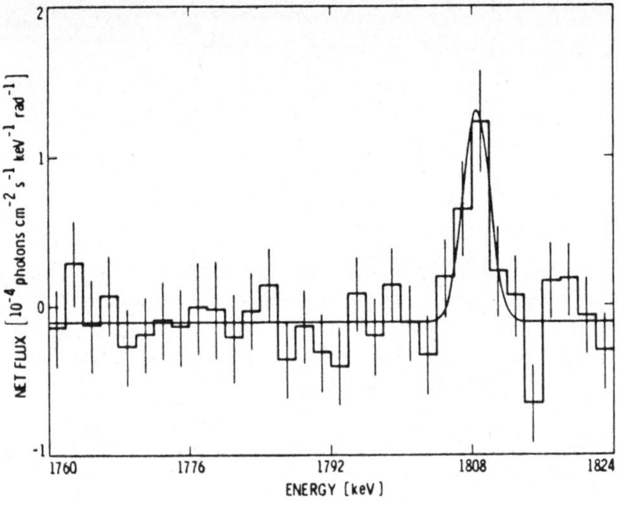

Figure 1

The γ-ray line of ^{26}Al at 1.8 MeV, detected by the HEAO satellite in the galactic plane (*Mahoney et al.* 1984).

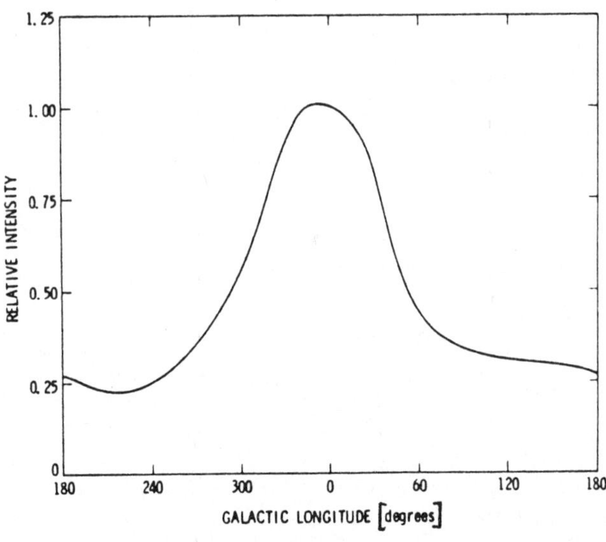

Figure 2

Longitude profile in the galactic plane of the high energy gamma-rays (50 MeV – 5000 MeV), as detected by the COS-B satellite. This profile was used for the original data analysis of *Mahoney et al.* (1984).

$= 4.8 \pm 0.7\ 10^{-4}$ photons $cm^{-2}\ s^{-2}\ rad^{-1}$ toward the GC region. In that case the coresponding galactic γ-ray emissivity Q (in photons s^{-1}) is related to the flux by : $F = 1\ 10^{-46}$ Q (Higdon and Lingenfelter 1976). This implies that $N_{26} \sim 10^{56}$ nuclei or $M_{26} \sim 3\ M_{\odot}$ of ^{26}Al are currently present in the ISM, assuming equilibrium between its production and decay rates (e.g. $Q \sim dN_{26}/dt \sim N_{26}/\tau_{26}$).

The SMM experiment (7 cylindrical 7.5 cm x 7.5 cm NaI scintillator detectors) detected the line at the 10σ confidence level (Share *et al.* 1985) and confirmed its observed parameters (e.g. energy and width, as well as flux). The line flux ($F = 4.\pm 0.4\ 10^{-4}$ photons $cm^{-2}\ s^{-1}\ rad^{-1}$, if the COS-B distribution is again assumed) did not vary from year to year since 1980, when observations started. The data is consistent with a source centered at the GC and a galactic center/anticenter intensity ratio > 2.5, but the longitudinal flux distribution cannot be derrived because of the poor angular resolution ($\sim 130^{o}$) of the experiment.

As recently reported (Balmoos *et al.* 1986, Weber *et al.* 1986), a balloon borne experiment of the Max Planck Institute (MPI) equipped with a Compton telescope detected the 1.8 MeV line at the 4.5σ level and localized it within 10° of the GC. According to the authors the data does not exclude a diffuse galactic emission, but it is more consistent with a point source located at the GC. In that case, the detected flux ($F = 6.7 \pm 2.7\ 10^{-4}$ photons $cm^{-2}\ s^{-1}$ for a point source) implies the existence of $5 \pm 2\ M_{\odot}$ of ^{26}Al in the GC region. However, in the case of an extended source following the COS-B profile, the derived flux ($F = 21 \pm 8.4\ 10^{-4}$ photons $cm^{-2}\ s^{-1}\ rad^{-1}$) is higher than both the HEAO and SMM fluxes by a factor of 5.

This balloon observation is clearly in contradiction with both satellite experiments, as well as with the data obtained from four balloon observations between 1977 and 1984 with NaI scintillators, as reported by Leventhal *et al.* (1987). The combined result for all four experiments gives a flux of (3.9 +2.0/-1.7) $10^{-4}\ cm^{-2}\ s^{-1}\ rad^{-1}$, quite consistent with the satellite experiments. The equivalent point source flux from the GC direction is $(1.6 \pm 0.9)\ 10^{-4}\ cm^{-2}\ s^{-1}\ rad^{-1}$, much lower than the MPI results, but the data favor a distributed source over a point source at the 90% confidence level.

Obviously, the precise determination of the flux profile is of crucial importance for the determination of the nature of the source (see below).

II. The nucleosynthesis of ^{26}Al

^{26}Al can be synthesized in astrophysical sites, either during quiet burning phases of stellar evolution (e.g. massive main sequence stars, red giants), or in explosive burning (e.g. supernovae, novae). The different temperature and density regimes of the corresponding astrophysical sites imply different modes for the production and destruction of this radioactive nucleus:

— at low temperatures ($T \sim 3$ - $5\ 10^{7}$ K for central H burning in massive stars, $T \sim 7$ - $9\ 10^{7}$ K for shell H burning in red giants) ^{26}Al is produced through the operation of the Mg-Al cycle, essentially by $^{25}Mg(p,\gamma)^{26}Al$. It is destroyed by $^{26}Al(\beta^{+})^{26}Mg$ (predominant at $T < 4\ 10^{7}$ K) and/or by $^{26}Al(p,\gamma)^{27}Si$. During the subsequent phase of He burning ($T > 2\ 10^{8}$ K and no protons present) any ^{26}Al that survived H exaustion is rapidly destroyed through (n,α) and (n,p) reactions, the neutrons being produced through (α,n) reactions on ^{13}C, ^{17}O and ^{22}Ne.

— at intermediate temperatures (T $\sim$ 2 - 4 10^8 K and a proton-rich environment in typical novae) the situation becomes a little more complicated: ^{26}Al may also be produced by two successive proton captures on ^{24}Mg (^{24}Mg(p,γ)^{25}Al(β^+)^{25}Mg(p,γ)^{26}Al), which greatly favours its production, since ^{24}Mg is more abundant than ^{25}Mg. But, on the other hand, some leakage out of the Mg-Al cycle starts occuring as the temperature increases, through ^{26}Al(p,γ)^{27}Si(p,γ)^{28}P etc., instead of ^{27}Si(β^+)^{27}Al(p,α)^{24}Mg (Wiescher *et al.* 1986).

— at high temperatures (T $\sim$ 2 - 3 10^9 K, corresponding to explosive C or Ne burning in Type II supernovae) the very rapid proton reactions may transform not only ^{24}Mg and ^{25}Mg into ^{26}Al, but also some Ne (that is, the nuclear flow goes *through* the Mg-Al region instead of being recycled there). On the other hand, neutrons liberated through ^{13}C(α,n) and ^{22}Ne(α,n) reactions are now the main agent of destruction of ^{26}Al, instead of (p,γ) reactions.

The following remarks can be made on the nucleosynthesis of ^{26}Al:

* in general, the reaction rates of stable nuclei are better known in the high temperature regime (where extrapolation from experimental data is relatively easy) than in the low temperature one (where uknown, or difficult to evaluate, resonances may interfere). However, the higher the temperature, the more unstable (proton rich) nuclei are involved in the nuclear flow. Since no experimental information is available for such nuclei, one has to rely on theoretical (and quite uncertain) estimates for the relevant reaction rates. Thus, the nuclear physics is better treated in the case of quiet nucleosynthesis of ^{26}Al than in the case of explosive nucleosynthesis.

* for temperatures T < 4 10^8 K the short-lived ($\tau \sim$ 7 s) isomeric state ^{26}Al^m (E=226 keV) is not thermalized (e.g. its population with respect to the ground state ^{26}Al^g is not given by the equations of statistical equilibrium) and it should be treated as a separate species in detailed nucleosynthesis computations (e.g. Ward and Fowler 1980).

* the production rate of ^{26}Al is proportional to the metallicity of the concerned site, whereas its destruction rate is proportional to metallicity only in the high temperature regime (because neutron production through (α,n) reactions is proportional to metallicity). As a result, the *net* production rate of ^{26}Al may be considered as (roughly) proportional to metallicity in the case of massive stars, red giants and novae, but not in the case of supernovae (this holds, of course, under the assumption that the physical conditions of the site do not depend sensitively on metallicity -e.g. through the energy production, the opacity etc.- which can not be exluded).

The explosive nucleosynthesis of ^{26}Al in the case of *supernovae* has beeen studied by many authors. Truran and Cameron (1978) and Arnett and Wefel (1978) studied the nucleosynthesis during the adiabatic expansion of a C-O shell brought to peak temperatures $T_{peak}\sim$2 10^9 K and peak densities ρ_{peak} $\sim$ 10^5 gr cm^{-3}. Morgan (1980) and Woosley and Weaver (1980) considered somewhat higher temperatures (T_{peak} $\sim$3 10^9 K, corresponding to explosive Ne burning), using more realistic models of the explosion site. They all came essentialy to the same conclusion, namely that an abundance ratio $(X_{26}/X_{27})_{SN}$ $\sim$ 4 10^{-4} – 2 10^{-3} could be considered as typical of those sites. More recently, Woosley and Weaver (1986) find $(X_{26}/X_{27})_{SN}$ $\sim$ 6 10^{-3}, with improved modelisation of the explosion and revised reaction rates. Supposing that supernovae produced all the ^{27}Al ($X_{27} \sim 6\ 10^{-5}$) in the ISM (that is, in a mass $M_{ISM} \sim 4\ 10^{9}\ M_\odot$) during the past $T_G \sim 10^{10}$ years (the age of the Galaxy), the quantity of ^{26}Al produced by supernovae with $(X_{26}/X_{27})_{SN} \sim 6\ 10^{-3}$ during the last $\tau_{26} \sim 10^6$ years should be:

$$M_{26} \sim (X_{26}/X_{27})_{SN}\ (\tau_{26}/T_G)\ X_{27}\ M_{ISM} \sim 0.15\ M_\odot$$

if *nucleosynthesis at constant rate* is assumed all over the galactic history. Thus, it seeems that supernovae fall short of producing the observed quantity of ^{26}Al by a factor of ~20 (stated in a different way, if supernovae were at the origin of ~3 $M_\odot$ of ^{26}Al in the ISM, they should have overproduced ^{27}Al by a factor of ~20, as Clayton (1984) remarked). However, if the assumption of constant rate nucleosynthesis is dropped, and *galactic chemical evolution* effects are taken into account (as they should), the supernovae contribution may be slightly modified. In particular, if *infall* on the galactic disk is assumed, the ^{27}Al difficulty is alleviated, since ^{27}Al synthesized by supernovae is diluted in the infalling metal-poor gas. Using simple chemical evolution models Clayton and Leysing (1987) find that the supernova contribution in the ISM may be as high as 0.4 $M_\odot$ of ^{26}Al in the past 10^6 years.

The nucleosynthesis of ^{26}Al in the case of *novae* has been studied in a parametrised, one zone, approximation by Arnould *et al.* (1980) and Hillebrandt and Thielemann (1982). They adopted thermodynamic conditions corresponding to the novae models of Starrfield *et al.* (1978) and found a production ratio $X_{26}/X_{27} \sim 0.1$ - 1. The rates of many relevant reactions - concerning essentially unstable nuclei - have recently been revised (Wiescher *et al.* 1986), in some cases by many orders of magnitude. Because of the resulting leakage out of the Mg-Al cycle (through ^{27}Si(p,γ)), the production of ^{26}Al is found to be considerably reduced with respect to previous estimates : only $2\ 10^{-7}$ (by mass fraction) for *hot* novae and $7\ 10^{-5}$ for *cold* ones, instead of a few 10^{-4} previously obtained. Taking as an (optimistic) average $X_{26} \sim 5\ 10^{-5}$, $M_{NOV} \sim 10^{-4}\ M_\odot$ for the mass ejected by a typical nova explosion, and a frequency of $n_{NOV} \sim 40$ novae year^{-1} in the Galaxy (rather an upper limit), we obtain:

$$M_{26} \sim X_{26}\ M_{NOV}\ \tau_{26}\ n_{NOV} \sim 0.2\ M_\odot$$

that is ~15 times less than the value derived from observations.

O-Ne-Mg rich novae could also contribute to the ^{26}Al production, their yield being proportional to their metallicity (e.g. Delbourgo-Salvador *et al.* 1985). Recent computations suggest, however, that the mass ejected by that kind of novae (which may constitute up to 25% of the total) should be much less than the "canonical" value of $5\ 10^{-5}\ M_\odot$ (Starrfield et al. 1986). It should be stressed, however, that the production of ^{26}Al in novae is particularly sensitive to the modelisation of the site: the basic difficulty comes from the treatment of *convection* (the time-scale of which is comparable to the nuclear one), but also from the treatment of the mass loss and the *dredging – up* mechanisms. Woosley (1986) reported recently results of parametrised, two-zone, computations (a lower and hotter one, where ^{26}Al is produced but can also be destroyed because of the high temperatures prevailing there; and an upper and colder one, where it can be preserved before been ejected to the ISM). In some cases (depending on the adopted physical conditions) he finds spectacular enhancements to the production of ^{26}Al, a fact which clearly illustrates the uncertainties affecting all current (parametrized) nucleosynthesis computations in novae and the need for completely self-consistent computations (e.g. coupling hydrodynamics and nucleosynthesis).

The nucleosynthesis of ^{26}Al in the envelopes of *asymptotic red giants* has been considered be Norgaard (1980), on the basis of previous estimates (Iben and Truran 1978) concerning the thermodynamic conditions, the *dredging – up* and the mass loss of stars with M $\sim 5-7$ $M_\odot$, during this evolutionary phase. He found that nucleosynthesis at the bottom of the convective H envelope may produce abundance ratios $X_{26}/X_{27} \sim 0.5-1$. On the basis of those computations Cameron (1984) and Truran (1986) argued (in a very qualitative way) that red giants could significantly contribute to the production of ^{26}Al at the Galactic level. However, no reliable quantitative estimates for that nucleosynthetic site exist yet.

After the recent suggestion of Balmoos *et al.* (1986) that the 1.8 MeV line emission is (possibly) due to a point source at the GC, Hillebrandt *et al.* (1986a,b) suggested that the explosion of a supermassive star (M $\sim$ 5 10^5 $M_\odot$) in the GC a few million years ago could be at the origin of the inferred quantity of ^{26}Al ($M_{26} \sim 5\pm2$ $M_\odot$). However, beyond the problems of the the formation and existence of such "monster" stars (see Fuller *et al.* 1986, for a discussion), the probability of such a special event at the "right moment" (e.g. a few million years ago) seems quite low. Moreover, it should be stressed that the data of that experiment (if confirmed) is not, in any case, inconsistent with an extended source distribution, strongly peaked towards the GC region.

Dearborn and Blake (1984, 1985) suggested that *massive, mass losing stars* could produce ^{26}Al during their main-sequence phase and eject it in the ISM through their intense stellar winds. They found that such stars might have a rather marginal contribution (up to 0.2 $M_\odot$) to the quantity of ^{26}Al in the Galaxy. These ideas have been followed and substantially extended by our group in Saclay (Cassé and Prantzos 1986; Prantzos and Cassé 1985, 1986), with more "realistic" stellar models for Wolf-Rayet (WR) stars and more recent nuclear data.

WR stars are massive, (presumably) core He burning stars, losing mass at high rates (3 10^{-5} $M_\odot$/year). According to models of massive star evolution (Prantzos *et al.* 1986) they produce ^{26}Al during core H burning (at central temperatures higher than 35 10^6 K) and eject it in the ISM when the former convective core appears at the stellar surface. We checked that the recent update of the relevant nuclear reaction rates (Caughlan and Fowler 1987) does not alter the ^{26}Al yield predicted by the above models. Thus, the average yield of WR stars with progenitor masses $M_{ZAMS} > 50$ $M_\odot$ in the solar neighborhood is 4 10^{-5} $M_\odot$ and their average 1.8 MeV line luminosity 5 10^{37} s^{-1}. (In a recent preprint Timmermann *et al.* (1987) suggest spectacular modifications for the ^{27}Al(p,γ) and ^{27}Al(p,α) reaction rates in the $20 < T_6 < 60$ temperature range, which could substantially modify the flow in the Mg-Al cycle and alter the above quoted values. Work is currently in progress to check the effect of that suggestion).

The contribution of WR stars to the total quantity of ^{26}Al in the Galaxy and the corresponding γ-ray line flux depend sensitively on their galactic distribution (largely unknown, because the catalogues are complete only within 2.5 kpc from the Sun). It seems, however, that there is a radial gradient in their surface density distribution, implying a strong WR concentration in the inner Galaxy. Moreover, the ratio of WR/O stars seems also to increase in the inner Galaxy (see Prantzos and Cassé 1986 for a discussion of these points). This could be due to the effect of the higher metallicity (z) of these regions on the stellar winds (Kudritzki *et al.* 1987), which could make the former stellar core to appear earlier at the surface. We adopted for this effect the parametrization of Maeder (1984), i.e. $N_{WR}/N_O \propto z^{1.7}$ for all (single + binary) WR stars, based on data for the Small and Large Magellanic Clouds, as well as for the outer Galaxy. Finally, the ratio of binary to single WR stars seems to increase in regions with metallicity lower than the solar one (Hidayat *et al.* 1984), which seems to imply that in these regions the single WR formation mechanism (through strong stellar winds) becomes less effective than the binary one (through Roche lobe overflow). We take this effect into account (see also Prantzos, Cassé and Arnould 1987) by extrapolating the $N_{WR,single}$ / $N_{WR,binary}$ data of Hidayat *et al.* (1984) to the high metallicity regions of the inner Galaxy: in the solar neighborhood (i.e. z=0.02) 60% of WR stars are single ones, but at z=0.03 the corresponding fraction is 84%.

The distributions of single WR stars adopted in this work are slightly different from the ones used in our previous work (Prantzos and Cassé 1986), and are presented in Fig. 3. Case A is a conservative estimate, based on the new distribution of H_2 given by Scoville and Sanders (1986), which should be a good tracer of massive star formation and, therefore, of WR distribution. Moreover, it is based on the recently revised galactocentric distance of the Sun (R_O = 8.5 kpc, instead of 10 previously assumed). This distribution, folded with the $z^{1.7}$ effect leads to our case B (a rather extreme one). In both cases, the resulting radial *single* WR distribution is normalized to 0.9 kpc^{-2} in the solar neighborhood (that is, we neglect the possible contribution of binary WR stars, which should be very small, since their low central temperatures should not allow a substantial nucleosynthesis of ^{26}Al). The total number of WR stars in the Galaxy (N_G) and in the Galactic Center region (N_C) is: N_G = 1800 and 5500, and N_C = 300 and 2000, in case A and B respectively. The corresponding quantity of ^{26}Al ejected by WR stars in the last ^{26}Al lifetime ($\sim 10^6$ years) is 0.12 and 0.50 $M_\odot$, respectively.

Fig. 4 presents the longitude distribution of the resulting 1.8 MeV line flux in the galactic plane. In the longitude range $\pm 20^o$ from the galactic center the flux is F = 5.2 10^{-5} and 2.5 10^{-4} photons cm^{-2} s^{-1} rad^{-1}, for cases A and B, respectively. Even in case B the flux obtained is lower by a factor of 2 than the observed one. Note that the relationship F = α Q, betweeen F and the total galactic 1.8 MeV luminosity Q (or, the total ^{26}Al mass) depends sensitively on the adopted distribution: the COSB profile, used in the HEAO data analysis, leads to α = 1. 10^{-46}, but in our cases A and B we obtain α = 2. 10^{-46} and 3. 10^{-46}, respectively. These values imply a smaller quantity of ^{26}Al than previously thought (1 or 1.5 $M_\odot$, instead of $\sim$3 $M_\odot$), and we think that the experimental data should be reexamined in the light of the new profiles. This difference with respect to our previous work is due to the sharper WR distribution and the reduced R_O we adopted. We should note, however, that in case B the theoretically obtained ^{26}Al yield and γ-ray line luminosity of WR stars are (roughly) half the corresponding quantities derrived observationally (*if* the distribution B is adopted). This self-consistency of the model makes WR stars a serious candidate as a source of (a great part of) interstellar ^{26}Al. In any case, it is clear that none of the experiments performed up to now is able to distinguish between a sharply peaked flux distribution and a point source in the GC region.

The number of WR stars in the Galaxy obtained in case B seems to be high (compared to the usually quoted number of 1000). There are some indications, however, that this case may not be so unrealistic (Cassé 1986; Prantzos 1986): it seems indeed that substantial quantities of molecular gas are present in the central regions of the Galaxy (Sanders *et al.* 1984), and that the star formation rate (especially for massive stars) may be particularly enhanced there (Ho *et al.* 1985). On the other hand, high-resolution, far infrared surveys of the galactic center suggest the

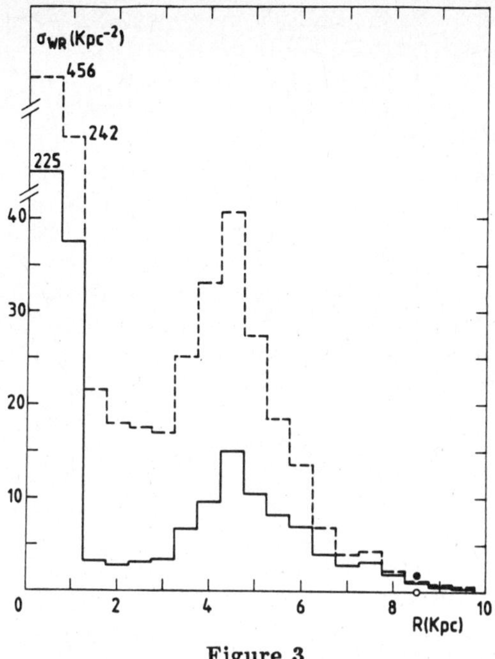

Figure 3

Plausible galactocentric WR distributions adopted in this work and in *Prantzos, Cassé and Arnould* (1987). The *single* WR surface density in the galactic plane (normalized to the *single* WR density in the solar neighborhood : 0.9 kpc^{-2}) is plotted as a function of the galactocentric distance (see text for details). The solar galactocentric distance is $R_{\odot}$=8.5 kpc.

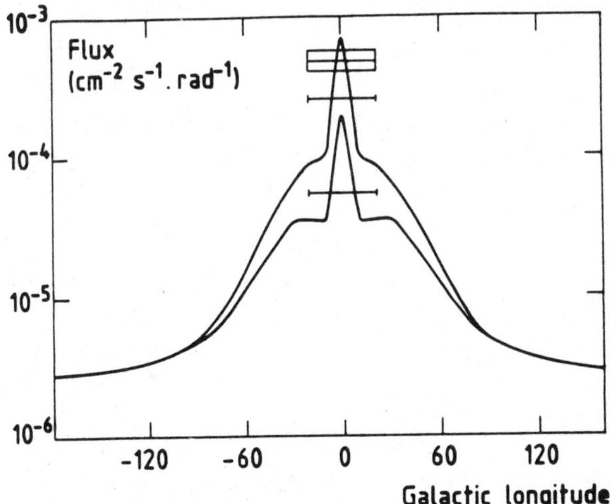

Figure 4

Theoretically derrived γ-ray flux profiles from the collective emission of WR stars in our Galaxy (see also *Prantzos, Cassé and Arnould* 1987). They correspond to the 2 WR distributions of Fig. 3. The flux detected by HEAO is indicated, together with the corresponding theoretical average flux (e.g. in the l=$\pm 20°$ range) for each case.

existence of a few 10^4 BO stars in that region (Odenwald and Fazzio 1984) that could give rise to WR stars, if the mass loss rate is enhanced there due to high metallicity.

Another, rather indirect, argument for a large number of WR stars in the Galaxy came recently from X-ray Astronomy (Montmerle 1986; Matsuoka 1987). Recent X-ray observations with the EXOSAT (Warwick *et al.* 1986) and TENMA (Koyama *et al.* 1986) satellites found evidence for a diffuse galactic emission in the hard X-ray region (kT $\sim$ 6 +14/-2 keV for EXOSAT, which had a good angular but poor energy resolution; kT $\sim$ 2-16 keV for TENMA, with a good energy but poor angular resolution). The ubiquitous presence of Fe line emission suggests a thermal bremmstrahlung from an optically thin plasma. The width of the ridge is about 2°, which requires a scale-height of $\sim$100 pc, while the rapid fall-off in the longitude distribution for galactic longitudes $l > 40^o$ implies a disk radius less than $\sim$6 kpc. The total luminosity of the ridge is $\sim 10^{38}$ ergs s^{-1}, too high to be explained by the interaction of supernovae or novae with the ISM ($\sim$ 0.10 SN/y or 10^5 n/y are required).

Recently, Dorland and Montmerle (1987) modelized the energy dissipation in (and heating of) surrounding dense clouds by intense stellar winds. On the basis of that model, Montmerle (1986) proposed that WR stars, the most "mechanically luminous" stars (with an average wind energy of $L_w \sim 10^{38}$ ergs s^{-1}) could indeed be responsible for the galactic X-ray ridge. If the wind energy is converted to X-ray emission with an efficiency of 0.0002, as the modelisation of Dorland and Montmerle (1987) suggests, $\sim$ 5000 WR stars are needed in order to explain the observed galactic X-ray emission. That number is comparable to the one needed to explain (half) the 1.8 MeV line flux, according to our case B. Thus, Montmerle (1986) and Matsuoka (1987) suggest that interstellar ^{26}Al and the Galactic X-ray ridge may have a common origin, namely galactic WR stars.

Conclusion

The recent detection in the GC direction of a γ-ray line at 1.8 MeV due to the decay of ^{26}Al, boosted theoretical as well as experimental and observational work on the origin of the $\sim$3 $M_\odot$ of ^{26}Al currently present in the ISM.

Because of the poor angular resolution and/or the insufficient sensitivity of the experiments, uncertainties persist about the precise source distribution (e.g. point-like in the GC region, or extended and peaked towards the GC direction?). Obviously, further observations, with better angular resolution ($<5^o$) and sensitivity ($\sim 10^{-6}$ photons cm^{-2} s^{-1}) are badly needed to clarify the situation. The GRO (Gamma Ray Observatory) γ-ray line experiment (initially scheduled to fly in 1988) and the GRASP experiment, planned to fly on the European space platform EUREKA, are qualified to fulfill that mission (Durouchoux 1986).

All the basic ingredients (nuclear physics, source models, and galactic distribution) of the proposed sources (supernovae, novae, red giants, Wolf-Rayet stars, and supermassive stars) suffer from (several) serious drawbacks. Physical conditions and nucleosynthesis in explosive sites are difficult to model; on the other hand, their galactic distribution is somewhat better known than the WR one (but less well than the location of the proposed supermassive star).

Among the proposed astrophysical sites, the WR model is the best developed up to now : it makes clear-cut predictions about the quantity of ^{26}Al ejected in the ISM. If those stars are really at the origin of the detected emission, the observed γ-ray line distribution could allow to trace their galactic distribution, unaccessible to other wavelengths.

In our opinion, none of the above proposals can be completely discarded at the present time: much more refined models for the corresponding sources are needed before any definite conclusions can be drawn. On the other hand, the detection of correlated (e.g. with the same flux profile as the ^{26}Al emission) γ-ray lines could be of crucial interest: for example, the detection of a correlated ^{22}Na line should strongly favour the novae hypothesis, whereas a correlated ^{60}Fe line (detectable at the 10^{-7} photons cm^{-2} s^{-1} level, according to the estimates of Prantzos, Arnould and Arcoragi 1987, for WC stars) would give strong support to the WR senario.

Clearly, much work is needed, both from theoreticians and observers. But nuclear physicists might hold the key of the problem, after all: a carefull estimate of the relevant reaction rates, especially at low energies, might reveal some surprises, as recently demonstrated (e.g. Wiescher *et al.* 1986; Schmalbrock *et al.* 1986; Champagne *et al.* 1986; Timermann *et al.* 1987). On the same ground, the measuring of some key reaction rates involving unstable targets should be of the uttermost importance (see Arnould 1986).

AKNOWLEDGEMENTS: I am grateful to M. Arnould and M. Cassé for their help in this work. I especially thank F. Mateucci for a very useful discussion.

REFERENCES

Arnett D. W., Wefel J. P. 1978, *Ap. J.*, **224**, L139

Arnould M. 1986, *Proc. of the Accelerated Radioactive Beams Workshop*, Vankouver (to appear)

Arnould M., Norgaard H., Thielemann F.-K., Hillebrandt W. 1980, *Ap. J.*, **237**, 931

Ballmoos P., Diehl R., Schonfelder V. 1987 *Ap. J.*, to appear

Cameron A. G. W. 1984, *Icarus*, **60**, 416

Cassé M. 1986, Proc. of the COSPAR meeting (Toulouse, France), to appear in *Adv. Space Res.*

Cassé M., Prantzos N. 1986, in *Nucleosynthesis and its implications on Nuclear and Particle Physics*, eds. J. Audouze and N. Mathieu, (Dordrecht : Reidel), p. 339

Champagne A. E., Mc Donald A. B., Wang T. F., Howard A. J., Magnus P. V., Parker P. D. 1986, *Nucl. Phys.*, **A451**, 498

Clayton D. D. 1982, in *Essays in Nuclear Astrophysics*, eds. Barnes C., Clayton D., and Schramm D., (Cambridge: Cambridge University Press), p. 401

Clayton D. D. 1984, *Ap. J.*, **280**, 144

Clayton D. D., Leysing M. D. 1987 *Phys. Rep.*, **144**, 1

Dearborn D. S. P., Blake J. B. 1984, *Ap. J.*, **277**, 783

Dearborn D. S. P., Blake J. B. 1985, *Ap. J.*, **288**, L21

Delbourgo-Salvador P., Mochkovitch R., Vangioni-Flam E. 1985, *Proceedings of the ESA Workshop : Recent results on Cataclysmic Variables*, Bamberg, p. 229

Dorland H., Montmerle T. 1987, *Astr. Ap.*, in press

Durouchoux Ph. 1986, *Advances in Nuclear Astrophysics*, eds. E. Vangioni-Flam, J. Audouze, M. Cassé, J. P. Chièze, T. T. Van, Editions Frontières, France, p. 309

Fowler W. A. 1984, *Rev. Mod. Phys.*, **56**, 149

Fuller G. M., Woosley S. E., Weaver T. A. 1986, *Ap. J.*, **307**, 675

Hidayat B., Admiranto A. G., van der Hucht K. A. 1984, *Ap. Sp. Sc.*, **99**, 175

Higdon J. C., Lingenfelter R. E. 1976, *Ap. J.*, **208**, L107

Hillebrandt W., Thielemann F.-K. 1982, *Ap. J.*, **255**, 657

Hillebrandt W., Mair G., Ziegert W. 1986a, Advances in Nuclear Astrophysics, eds. E. Vangioni-Flam, J. Audouze, M. Cassé, J.P. Chieze, T.T. Van, Editions Frontières, France, p. 343

Hillebrandt W., Thielemann F.-K., Langer N. 1986b, *preprint*

Ho P. T., Jackson J. M., Barrett A. H., Armstrong J. T. 1985, *Ap.J.*, **288**, 575

Koyama K., Ikeuchi S., Tomisaka 1986, *Publ. Astron. Soc. Japan*, **38**, 503

Kudritzki R.P, Pauldrach A., Pulls J. 1987, *Astr. Ap.*, **173**, 293

Iben I., Truran J. 1978, *Ap. J.*, **220**, 980

Leising M. D., Clayton D. D. 1985, *Ap. J.*, **294**, 591

Leventhal M., McCallum C.J., Huters A.F., Stang P.D. 1987, *Ap. J.*, to appear

Maeder A. 1984, *Adv. Space Res.*, **4**, 55

Mahoney W. A., Ling J. C., Jacobson A. S., Lingenfelter R. E. 1982, *Ap. J.*, **262**, 742

Mahoney W. A., Ling J. C., Wheaton W. A., Jacobson A. S. 1984, *Ap. J.*, **286**, 278

Mahoney W. A., Higdon J. C., Ling J. C., Wheaton W. A., Jacobson A. S. 1985, *Proceedings of the 19th ICRC*, La Jolla, USA, OG 3.2-3 (p. 357)

Matsuoka M. 1987, *Proceedings of the 20th ICRC*, Moscow, O.G.2.3-3

Mayer-Hasselwander et al. 1982, *9th Texas Symp. Relat. Ap.*, 211

Montmerle T. 1986, *Advances in Nuclear Astrophysics*, eds. E. Vangioni-Flam, J. Audouze, M. Cassé, J.P.Chièze, T.T. Van, Editions Frontières, France, p. 335.

Morgan J. A. 1980, *Ap. J.*, **238**, 674

Norgaard H. 1980, *Ap. J.*, **236**, 95

Odenwald S. F., Fazzio G. G. 1984, *Ap. J.*, **283**, 601

Prantzos N. 1986, *Advances in Nuclear Astrophysics*, eds. E. Vangioni-Flam, J. Audouze, M. Cassé, J.P. Chièze, T.T. Van, Editions Frontières, France, p. 321

Prantzos N., Cassé M. 1985, *Proc. of the 3d Workshop on Nuclear Astrophysics*, ed. W. Hillebrandt, W. Germany, p. 107

Prantzos N., Cassé M. 1986 *Ap. J.*, **307**, 324

Prantzos N., Doom C., Arnould M., de Loore C. 1986, *Ap. J.*, **304**, 695

Prantzos N, Cassé M., Arnould M. 1987, *Proceedings of the 20th ICRC*, Moscow

Prantzos N., Arnould M., Arcoragi J. P. 1987, *Ap. J.*, **315**,209

Ramaty R., Lingenfelter R. E. 1977, *Ap. J.*, **213**, L5

Scoville N. Z., Sanders D. B. 1987, in *Interstellar Processes*, eds. H. Thronson and D. Hollenbach, to appear

Share G. H., Kinzer R. L., Kurfess J. D., Forrest J. D., Chupp F. L., Rieger E. 1985 *Ap. J.*, **292**, L61

Schmalbrock P., Donoghue T. R., Hausman H. J., Wiescher M., Wijekumar V., Brown C. P., Rollefson A. A., Rolfs C. 1985, *Capture Gamma-ray spectroscopy and related topics*, AIP Conference Proceedings **125**, 785

Shaver P. A., Mc Gee R. X., Newton L. M., Danks A. C., Pottasch S. R. 1983, *M. N. R. A. S.*, **204**, 53

Starrfield S., Truran J. W., Sparks K. 1978, *Ap. J.*, **226**, 186

Starrfield S., Sparks K., Truran J. W. 1986, *Ap. J.*, **303**, L5

Truran J. W. 1986, *preprint*

Truran J. W., Cameron A. G. W. 1978, *Ap. J.*, **219**, 236

Ward R. A., Fowler W. A. 1980, *Ap. J.*, **238**, 266

Warwick R.S., Turner M.J., Watson M.G., Willingale R. 1985, *Nature*, **317**, 218

Wiescher M., Gorres, J., Thielemann, F.-K., Ritter, H. 1986, *A. A.*, **160**, 56

Weber W. R., Schonfelder V., Diehl R. 1986, *Nature*, **323**, 692

Woosley S. E. 1986, in *Saas-Fe Lectures : Nucleosynthesis and chemical evolution*, Eds. B. Hauck, A. Maeder and G. Meynet (Geneva Observatory), p.1

Woosley S. E., Weaver T. A. 1980, *Ap. J.*, **238**, 1017

Woosley S. E., Weaver T. A. 1986, *Nucleosynthesis and its implications on Nuclear and Particle Physics*, eds J. Audouze et N. Mathieu, (Dordrecht : Reidel), p. 145

A POSSIBLE RELATIONSHIP BETWEEN EXTINCT ^{26}Al AND ^{53}Mn IN METEORITES AND EARLY SOLAR ACTIVITY

G.J. Wasserburg
The Lunatic Asylum of the Charles Arms Laboratory
Division of Geological and Planetary Sciences
California Institute of Technology, Pasadena, CA 91125, USA
and

M. Arnould
Institut d'Astronomie, d'Astrophysique et de Géophysique, CP 165
Université Libre de Bruxelles, B-1050 Bruxelles, Belgium

1. Introduction

Much excitement in the meteoritic and astrophysical communities has been raised by the discovery in a small fraction of the meteoritic material of isotopic anomalies ascribable to nuclear effects. A special class of those anomalies relates to the in-situ decay of now-extinct short-lived ($t_{1/2} \lesssim 10^8$ yr) radionuclides. One of the most important of those cases is ^{26}Al [($t_{1/2}$ = 7.05x10^5 yr, following Norris et al. (1983)]. The short time scale for ^{26}Al has been used as the measure of time between the injection of freshly synthesized nuclei into a dense molecular cloud and its collapse. As reviewed by e.g. Wasserburg and Papanastassiou (1982), Wasserburg (1985) or Papanastassiou (1985), the $(^{26}\mathrm{Al}/^{27}\mathrm{Al})_0$ abundance ratio at the time T_0 of crystallization is 5x10^{-5}. Lower values have been found and are attributed to time differences or to heterogeneities in the solar nebula. There is some difficulty in maintaining large primary heterogeneities on the scale of 100 km in the solar nebula after collapse. There are also problems in relating the inferred abundances of short-lived nuclei with a simple or coherent set of nuclear astrophysical processes and sources. Strong evidence also exists for the presence of ^{107}Pd ($t_{1/2}$ = 6.5x10^6 yr) with $(^{107}\mathrm{Pd}/^{108}\mathrm{Pd})_0 \approx$ 2x10^{-5} in the early solar system (Kelly and Wasserburg 1978; Kaiser and Wasserburg 1983; Chen and Wasserburg 1983). As evidence of ^{107}Pd is found in iron meteorites, the time scale for this nuclide is related to the formation of small planetary objects.

The discovery of extinct ^{26}Al and ^{107}Pd in the solar system immediately suggests the possible presence of nuclei with comparable lifetimes. Birck and

Allègre (1985) have reported the discovery of a ^{53}Cr excess correlated with ^{55}Mn and attributed to the in-situ decay of ^{53}Mn ($t_{1/2}$ = 3.7x10^{6} yr) in Allende non-FUN inclusions. Those inclusions are thought to have condensed before significant decay of ^{53}Mn from a material somewhat depleted in Cr relative to Mn with respect to the bulk solar system composition. Birck and Allègre (1985) infer values of $(^{53}Mn/^{55}Mn)_0$ ranging from ≈2.5x10^{-5} to ≈8.9x10^{-5} (see, however, Lee, 1986). In the average solar system material with a normal Mn/Cr value, the ^{53}Cr deficit discovered in the above mentioned Allende inclusions would be offset by the total decay of ^{53}Mn if ^{53}Mn/^{55}Mn ≈ 3x10^{-5}. On the other hand, the FUN inclusion EK1-4-1 exhibits a ^{53}Cr excess which, if interpreted entirely in terms of ^{53}Mn decay, would require ^{53}Mn/^{55}Mn ≈ 4x10^{-4}. Instead, these data on the FUN samples have been interpreted in terms of substantial heterogeneities in the early solar system ^{53}Cr/^{52}Cr ratio, which could complement possible ^{53}Mn decay effects (Papanastassiou 1986). The same type of interpretation has been applied to the non-FUN inclusions as well (Lee 1986).

It should be noted that the radiogenic anomalies are part of a quite rich family of diverse anomalies in O, Ca, Ti, Cr, Ba, Sr, Mg, Sm, Nd, H and C isotopic patterns discovered in a variety of FUN and non-FUN inclusions.

The discovery of the isotopic anomalies in general, and of those related to short-lived radionuclides in particular, contradicts the canonical model of a homogeneous and gaseous protosolar nebula, and suggests instead that the solid bodies started to condensate out of imperfectly mixed reservoirs (made of gas and/or dust) with different compositions. Several models have been proposed for the origin of those reservoirs:

(1) the contamination from a single or several massive mass losing stars (MMLS) (of the Of or WR type) (Arnould and Prantzos 1986) or supernovae (SN) (e.g. Cameron and Truran 1977; Reeves 1978), or from a highly evolved low mass (≈$1M_\odot$) star during its asymptotic-giant-branch (AGB), or post AGB phase (Cameron 1984). One or another of those contaminating objects might also have triggered the solar system formation. In this type of model, the exotic material may be in the form of gas and/or grains. In the latter case, it is assumed that the grains at their entrance in the solar system still carried at least certain of the short-lived species in *live* form;

(2) the contamination by different grain components (referred to as stardust or sunocons) which might have been created around red giants, novae, SN, or MMLS such a long time before the solar system birth that all the short-lived species possibly carried by those grains at formation had time to decay in-situ before their inclusion into solar system solid bodies (e.g. Clayton 1982). In those views, the short-lived radionuclides have not been included live in the solar system;

(3) the differential irradiation of portions of the pristine solar system material by an intense flux of energetic particles from the Sun (e.g. Lee 1978, and references therein). This model considers a possible *local* origin for the observed isotopic anomalies.

While there is still a debate between the proponent of model (2) (e.g. Clayton 1982) and the proponents of various forms of model (1), it is generally considered that model (3) faces severe difficulties if a solar mass must be irradiated, due to scattering losses and hydrogen. Irradiation of dust grains after the removal of most of the gas is much more efficient and could produce many exotic nuclei. This has been discussed by Heymann and Dziczkaniec (1976) and by Lee (1978). In this paper, we will reexamine this local production by energetic particles. In particular, we will analyze the production of ^{26}Al and ^{53}Mn, and show that they are correlated. The thermonuclear and spallation models both have their virtues and it is not necessary that they be considered as entirely orthogonal. Our revisitation is motivated by the following considerations:

(1) a spallogenic production of various nuclides by present-day solar energetic particles (SEPs) is well known to be unavoidable. Phenomena like the 0.511 MeV annihilation line observed in solar flares also appear to be best interpreted in terms of the production of short-lived β^+-emitters by SEPs. Of course, this does not prove at all that such a spallative production has taken place $\approx 4.5 \times 10^9$ yr ago, when the early Sun possibly passed through a T-Tauri or X-ray star phase. However, such a possibility appears to be quite likely in view of the high activity exhibited by the observed T-Tauri stars (for a review, see e.g. Montmerle 1984);

(2) it appears to us desirable to reinvestigate the necessary input nuclear physics. In particular, some experimentally available spallation cross sections have not been used in previous investigations of a spallative origin of ^{26}Al and ^{53}Mn. This is especially the case for the work of Clayton et al. (1977), in which no use is made of any experimental data. These have also been overlooked by Lee (1978) for the production of ^{53}Mn. In addition, those two studies, as well as the one by Heymann et al. (1978), have neglected the possible production of ^{26}Al and ^{53}Mn by α-induced reactions;

(3) some of the models presented in the literature may be too specific or attempt to explain a wide variety of effects in a single complex model. In particular, they are often concerned with a blend of mechanisms involving chemical separation, mixing and irradiation effects in order to explain a particular anomaly or a set of anomalies. Previous discussions have relied on a power-law spectrum in kinetic energy for the SEPs. There is no *a priori* reason to limit the investigations to such a spectral shape. In fact, present-day SEPs are preferably described by e.g. a rigidity spectrum (e.g. McGuire and von Rosenvinge 1984). The most appropriate spectrum of a T-Tauri or X-ray star phase of the Sun is unknown. In the long run, it would also be worthwhile to extend spallation models in relation to the production of isotopic anomalies to other locations than the solar system, and in particular to the surroundings of non-exploding (AGB and WR) stars, or exploding objects (novae or SN);

(4) several arguments presented at various times have placed a spallation origin hypothesis of isotopic anomalies in disfavor or at least a weakened position. From energy considerations, Ryter et al. (1970) put drastic limitations on the amount of Li, Be and B which could have been produced by spallation in the early solar system. In the same way, Reeves (1978) concludes that the canonical $(^{26}Al/^{27}Al)_0$ ratio cannot be accounted for by such a mechanism, at least if the irradiation is extended to the *whole* solar system. Of course, this same argument does not apply to the localized irradiation envisioned by Lee (1978). The absence of certain anomalies has also been widely held against the spallation model. In particular, the assumed absence of effects due to the decay of ^{53}Mn was considered as a disproof of the spallation model. However, the ^{53}Mn anomaly is now with us! It has also been argued by Reeves (1978) that Li isotopic anomalies are unavoidable if ^{26}Al has to be of spallative origin. Taking the ^{53}Mn case as an example, it may be a bit dangerous to claim that Li isotopic anomalies are indeed not present in the solar system. In addition, gas/solid fractionation might blur the picture (e.g. Clayton et al. 1977);
(5) the production of ^{26}Al and ^{53}Mn is quite strictly related in spallation models. In our opinion, it is of great interest to try characterizing such a correlation and to provide *predictions* which might be confronted with the experiment. In contrast, the synthesis of those two radionuclides is expected to be largely uncorrelated in thermonuclear models. They are indeed predicted to occur either in different stellar objects, or at least in completely different zones of a given star. This is made clear in the next section.

2. A brief review of the nucleosynthesis processes for ^{26}Al and ^{53}Mn

In the framework of models (1) and (2), various mechanisms for ^{26}Al production have been proposed, namely:
(i) non-explosive H burning in the core of MMLS, followed by stellar wind ejection. As discussed in detail elsewhere (Arnould and Prantzos 1986; Prantzos and Cassé 1986), the ^{26}Al mass ejected in the ISM by such stars between the zero-age main sequence (ZAMS) and the end of the WR stage (identified in practice with the end of central He-burning) is predicted to vary from $\approx 10^{-5}$ to 10^{-4} $M_\odot$ for stars with initial masses M_{ZAMS} = 50 to 100 $M_\odot$. In such conditions, MMLS might be an important (even the dominant) ^{26}Al contaminant of the solar system (either in the form of gas or grains, which are known to form around certain WR stars; see e.g. Prantzos et al. 1986, for references). Such stars might also be of key importance for the feeding of the ISM with ^{26}Al at the level required by the 1.809 MeV γ-ray line observations (Prantzos and Cassé 1986);
(ii) explosive H-burning in novae. The ^{26}Al production in such a model has been the subject of several investigations (Arnould et al. 1980; Wallace and Woosley 1981; Hillebrandt and Thielemann 1982; Arnould 1985). From those calculations, it has been

estimated that novae could constitute a significant source of ^{26}Al in the ISM (Leising and Clayton 1985), and in meteorites;
(iii) C- or Ne-burning in Type II SNe. The various SN calculations (Arnett and Wefel 1978; Truran and Cameron 1978; Arnould et al. 1980; Woosley and Weaver 1986) predict ^{26}Al yields which are generally considered as insufficient to account entirely for the γ-ray line observations (Mahoney et al. 1984; Leising and Clayton 1985);
(iv) ^{26}Al production at the bottom of the convective envelope of AGB stars (Nørgaard 1980; Cameron 1984). No detailed calculations for such a mechanism have been performed up to now, so that such a scenario has still to be considered as speculative.

In all those scenarios, uncertainties of astrophysical as well as of nuclear physics nature still affect the ^{26}Al yield predictions (e.g. Arnould 1985). In that respect, it has to be remarked that the MMLS model (just relying on main sequence stars!) appears to be relatively more secure than models for highly complicated evolutionary phases (AGB or nova/SN explosions). On the other hand, the existence or absence of correlations between the ^{26}Al-related and other isotopic anomalies is of outermost importance in an attempt to assess the merits of the various proposed thermonuclear models. The minimum criterion for a good theory in this field is the ability to explain two (or more) isotopic effects which are observed.

As far as ^{53}Mn is concerned, it can be produced in Type I or II SN events through Si burning or in nuclear statistical equilibrium at temperatures in excess of $\approx 4 \times 10^9$ K. Some information concerning the ^{53}Mn production has been provided by Hainebach et al. (1974) in "normal freeze-out" nuclear statistical equilibrium conditions (as defined by Woosley et al. 1973). For given temperatures and densities, the amount of synthesized ^{53}Mn appears to depend critically upon the neutron excess parameter $\eta = \Sigma X_i (N_i - Z_i)/A_i$, where the sum is over free nucleons and over all nuclei with atomic number Z_i, mass number $A_i = N_i + Z_i$, and mass fraction X_i. In fact, a significant ^{53}Mn production, either as ^{53}Mn itself, or as ^{53}Fe, is expected only in partial Si burning or equilibrium burning zones with $\eta \lesssim 0.07$, while ^{53}Cr emerges from more neutron-rich zones ($\eta \gtrsim 0.08$). A simultaneous ^{53}Mn and ^{53}Cr production could be associated with $0.07 \lesssim \eta \lesssim 0.08$ zones. However, in such conditions, it appears difficult to avoid a concomitant and embarrassing ^{52}Cr overproduction. This can be seen from the results displayed by Hainebach et al. (1974) or Lee (1986).

It has also to be noted that the general pattern of Ca, Ti and Cr isotopic anomalies, and in particular the observed ^{48}Ca, ^{50}Ti and ^{54}Cr excesses, are generally interpreted in terms of Si-burning or equilibrium process in relatively more neutron-rich ($\eta \gtrsim 0.15$) zones than those required in the ^{53}Mn case. The precise blend of zones with different η values which have to be invoked depends largely upon the particular member of the vast family of isotopic anomalies one wants to account for. In such a framework, the ^{53}Mn-related anomaly would force adding a zip of relatively neutron-poor zones to the cocktail. In all of the cases where ^{53}Mn is the result of thermonuclear production, there will not be any ^{26}Al produced.

3. Spallation production of ^{26}Al and ^{53}Mn

3.1. Basic ingredients of the model. In order to evaluate the ^{26}Al and ^{53}Mn spallation yields, we limit ourselves in this work to a simple model the main ingredients and assumptions of which are as follows:

(i) only the action of SEPs is considered, the possible contribution of galactic cosmic rays being neglected. In such conditions, the production of secondary particles may be disregarded because of the relatively low energies of the SEPs (mainly $\leq$ 500 MeV/nucleon), and the resulting low multiplicity of the secondaries;

(ii) it is assumed that the abundances of the targets for the production of ^{26}Al, ^{53}Mn or ^{53}Fe [which decays to ^{53}Mn with a half-life of 8.5 min (ground state) and of 2.5 min (isomeric state)] do not change appreciably in time. That assumption is referred to as the *weak dosage* approximation;

(iii) the targets are assumed to be in dust grains (or embedded in a gas with high dust to gas ratio) for which the stopping of the SEPs can be neglected [e.g. Wasson (1963) for a detailed discussion of the validity of that approximation]. In such conditions, no depth profile of the SEP spectra has to be considered. That assumption is referred to as the *free space* approximation;

(iv) the SEP spectra are assumed to remain constant in time.

Two types of free space time invariant SEP spectra have been considered in the yield calculations. The first one

$$d\varphi/dE = A\cdot E^{-\gamma} \qquad (1)$$

is a power-law spectrum in kinetic energy/nucleon E (here and in the following, E is always expressed in the lab system), and is referred to in the following as the P-spectrum. The second one is an exponential spectrum in rigidity (momentum to charge ratio) R

$$d\varphi/dR = B\cdot \exp(-R/R_0), \qquad (2)$$

and is referred to as the R-spectrum. Both spectra are assumed to have non-zero values in the $E_0 \leq E \leq E_1$ range.

Calculations have been performed for values of the γ and R_0 parameters in the ranges $1 \leq \gamma \leq 6$ and $50 \leq R_0 \leq 250$ MV. Those ranges certainly encompass the values relevant to present-day SEPs (e.g. Lal 1972; McGuire and von Rosenvinge 1984). The same γ and R_0 values have been assumed for the proton and α-particle spectra. This appears to be quite well justified, at least to first approximation and for the present-day SEPs (e.g. McGuire and von Rosenvinge 1984) (the approximation might be a bit poorer, however, for the P-spectrum at $E \lesssim 20$ MeV). On the other hand, the cutoff energy E_1 has been set equal to 200 MeV. The adoption of a larger value would not affect the results to be described below, in view of the steep enough decrease of the considered spectra. In contrast, the value of the low cutoff energy E_0 may be influential. Calculations have been performed for some values in the $2 \leq E_0 \leq 8$ MeV range.

For the P- and R-spectra defined by Eqs. (1) and (2), one can define an integrated flux

$$\varphi = \int_{E_0}^{E_1} (d\varphi/dE)dE \tag{3}$$

and an effective spallation excitation function

$$\langle\sigma\rangle_{ij}(k) = \int_{E_0}^{E_1} \sigma_{ij}(E;k)(d\varphi/dE)dE/\varphi \tag{4}$$

for the production of nuclide k (in a given, ground or isomeric, state) by i-induced spallation reactions on target j, $\sigma_{ij}(E;k)$ being the corresponding E-dependent excitation function. From Eq. (4), one can also define the total excitation function

$$\langle\sigma\rangle_i(k) = \Sigma_j\ N_j\langle\sigma\rangle_{ij}(k)/\Sigma_j\ N_j\ , \tag{5}$$

where the sum extends over all possible targets with abundance numbers N_j.

3.2. Nuclear physics input data. Special care has been exercised in selecting the excitation functions $\sigma(E)$ appearing in Eq. (4), especially in view of the various shortcomings in that respect (see Sect. 1).

Table 1 lists the reactions considered in this work, and also provides information on the procedure adopted in order to select the corresponding $\sigma(E)$ in the relevant energy range. It has to be noted that only the production of the ground state of ^{26}Al is of relevance here. In contrast, both the ground and isomeric states of ^{53}Fe have to be considered.

Table 1. Procedure for selecting $\sigma(E)$. E is expressed in MeV. For the Hauser-Feshbach (HF) calculations, see Thielemann et al. (1986)

reaction	source and remarks
$^{23}Na(\alpha,n)^{26}Al$	E≤2.9 : HF calculations E>2.9 : estimate based on systematics
$Mg(p,X)^{26}Al$	E≤5.9 : experimental data from ref. (1) for $^{26}Mg(p,n)^{26}Al$ $^{25}Mg(p,\gamma)$ neglected E≥8 : ref. (2) 5<E<8 : interpolation
$^{27}Al(p,pn)^{26}Al$	values red on fig. from ref. (2)
$Mg(\alpha,X)^{26}Al$	values red on fig. from ref. (3)
$Al(\alpha,X)^{26}Al$	values red on fig. from ref. (3)
$Si(p,X)^{26}Al$	E≤16.4 : HF calculations for $^{29}Si(p,\alpha)^{26}Al$ E≥25.3 : ref. (2) 16.4<E<25.3 : interpolation
$Si(\alpha,X)^{26}Al$	values red on fig. from ref. (3)
$^{50}V(\alpha,n)^{53}Mn$	E≤1.8 : HF calculations E>1.8 : estimate based on comparison with other (α,n) reactions
$^{50}Cr(\alpha,n)^{53}Fe$	E≤4.5 : HF calculations E>4.5 : estimate based on comparison with other (α,n) reactions
$^{50}Cr(\alpha,p)^{53}Mn$	E≤7 : HF calculations E>7 : estimate based on comparison with other (α,p) reactions
$Cr(p,X)^{53}Mn$	E≤2.4 : experimental data for $^{52}Cr(p,\gamma)$ red on fig. from ref. (4) experimental data for $^{53}Cr(p,n)$ red on fig. from ref. (5) E≥14.3 : hybrid-model results red on fig. from ref. (6) 2.4<E<14.3 : interpolation based on HF calculations normalized to experimental data at E = 2.4
$^{55}Mn(p,X)^{53}Mn$	hybrid model red on fig. from ref. (6)
$Fe(p,X)^{53}Mn$	E≤10.8 : HF calculations for $^{56}Fe(p,\alpha)^{53}Mn$ E≥11.3 : red on fig. from ref. (7) 10.8<E<11.3 : interpolation
$Fe(p,X)^{53}Fe$	ref. (7)
$Fe(\alpha,X)^{53}Fe$	E≤9.45 : ref. (7) E>9.45 : estimate inspired by hybrid-model calculations
$Ni(p,X)^{53}Mn$	hybrid-model results red on fig. from ref. (6)
$Ni(p,X)^{53}Fe$	hybrid-model results red on fig. from ref. (6)

(1) Skelton et al. 1983, *Astrophys. J.* **271**, 404
(2) Tobailem et al. 1973, Rapport CEA - R - 4441, Saclay
(3) Tanaka et al. 1972, *J. Geophys. Res.* **77**, 4281
(4) Kennett et al. 1981, *Nucl. Phys.* **A363**, 233
(5) Gardner et al. 1981, *Aust. J. Phys.* **34**, 25
(6) Michel et al. 1980, *J. Radioanal. Chem.* **59**, 467
(7) Tobailem and de Lassus St-Genies 1975, Rapport CEA - N - 1466(3), Saclay

Whenever available, experimental data have been adopted. When necessary, this information has been complemented by theoretical estimates based on the Hauser-Feshbach (HF) model described by Thielemann et al. (1986) at relatively low energies, and on a "hybrid model" at energies for which the HF model is likely to fail. When neither experimental nor hybrid-model information were available in an energy range where the HF model breaks down, graphical interpolations were performed.

The situation concerning the α-induced reactions capable of producing ^{26}Al or ^{53}Mn is particularly unsatisfactory. No experimental data have been found, except in a limited energy range for $Fe(\alpha,X)^{53}Fe$. All the other relevant information derives from various estimates. Some calculations also exist for certain of the α-induced reactions considered (Keller et al. 1974). These predictions differ from our adopted estimates by factors which are well within the range of uncertainties of the various methods of evaluation. An experimental effort is urgently needed in order to reduce these uncertainties.

3.3. Effective excitation functions. Effective excitation functions relevant to the production of ^{26}Al (in its ground state), ^{53}Mn and ^{53}Fe have been calculated from Eqs. (4) and (5) in the framework of the model described in Sect. 3.1. They are displayed in Figs. 1 and 2. When possible, the values obtained in this work are compared with those derived by Clayton et al. (1977) and Lee (1978).

In the adopted range of values of the γ and R_0 parameters, the derived excitation functions are in general much more sensitive to the low cutoff energy E_0 in the P-spectrum than in the R-spectrum case. This is due to the fact that the P-spectrum overestimates the population of low-energy particles with respect to the R-spectrum.

3.4. Resulting abundances. Within the simple model defined in Sect. 3.1., the excitation functions exhibited in Figs. 1 and 2 can be used to evaluate the $^{26}Al/^{27}Al$ and $^{53}Mn/^{55}Mn$ ratios. Those abundance ratios can be expressed very simply either if the irradiation time t is much shorter than the lifetime τ of the relevant radionuclides, or in the opposite $t \gg \tau$ case. Those two situations are referred to as the short- and long-irradiation regimes, respectively.

In the short-irradiation regime, one obtains trivially

$$(^{26}Al/^{27}Al)(t) \approx t[(Mg+Al+Si)\varphi_p\langle\sigma\rangle_p(^{26}Al) + (Na+Mg+Al+Si)\varphi_\alpha\langle\sigma\rangle_\alpha(^{26}Al)]/^{27}Al, \quad (6a)$$

and

$$(^{53}Mn/^{55}Mn)(t) \approx t[(Cr+Mn+Fe+Ni)\varphi_p\langle\sigma\rangle_p(^{53}Mn) + (Fe+Ni)\varphi_p\langle\alpha\rangle_p(^{53}Fe) + (Cr+Fe)\varphi_\alpha\langle\sigma\rangle_\alpha(^{53}Fe) + Cr\varphi_\alpha\langle\sigma\rangle_\alpha(^{53}Mn)]/^{55}Mn, \quad (6b)$$

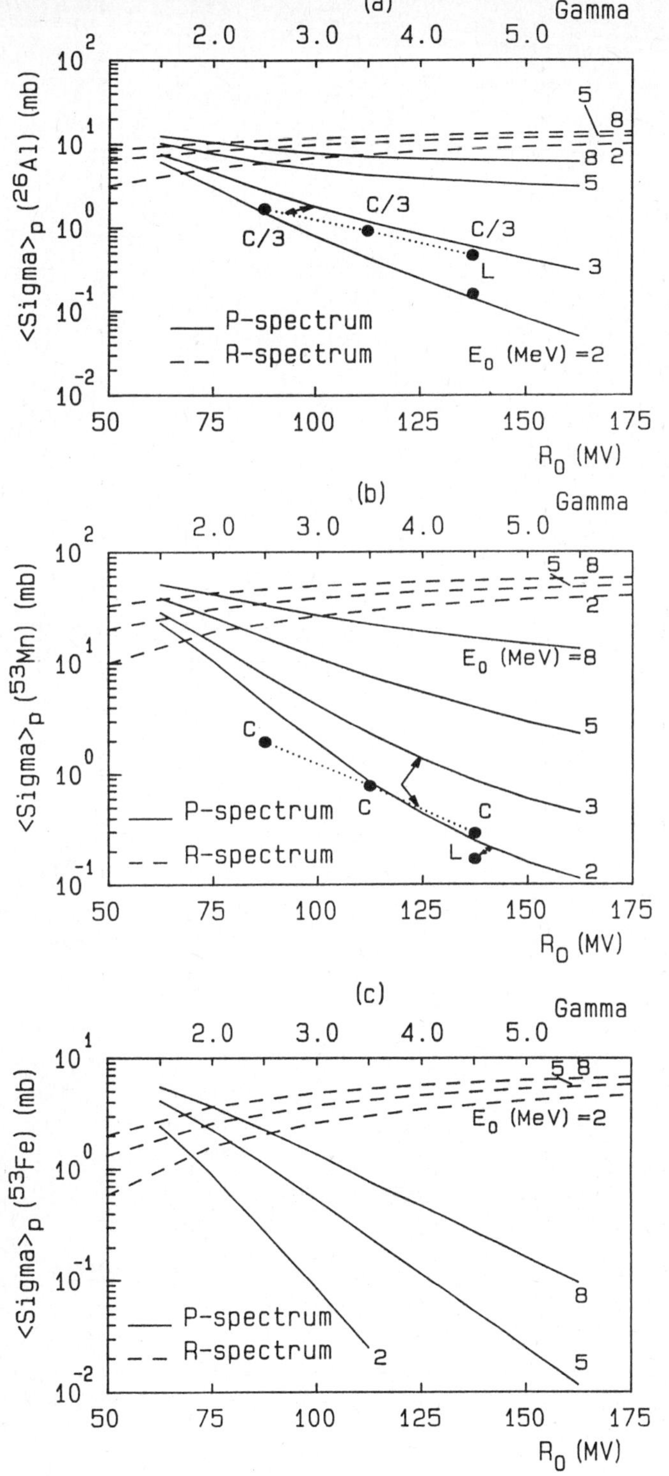

Figure 1:

(a) Total excitation function $\langle\sigma\rangle_p(^{26}Al)$ derived from Eq. (5) in the P- and R-spectrum cases, versus γ (upper abscissa) and R_0 (lower abscissa), respectively, and for various E_0 values. The considered targets are Mg, Al and Si (solar abundances). When possible, comparison is made with the $\langle\sigma\rangle$'s of Lee (1978) (point L), and Clayton et al. (1977) (point C; following their *Note added in proof*, the values given in their Table 1 are divided by 3 in order not to take into account the production of the ^{26}Al isomeric state). When necessary, arrows identify the data which have to be compared;

(b) same as **(a)**, but for $\langle\sigma\rangle_p(^{53}Mn)$. The considered targets are Cr, Mn, Fe, and Ni;

(c) same as **(a)**, but for $\langle\sigma\rangle_p(^{53}Fe)$. The considered targets are Fe and Ni.

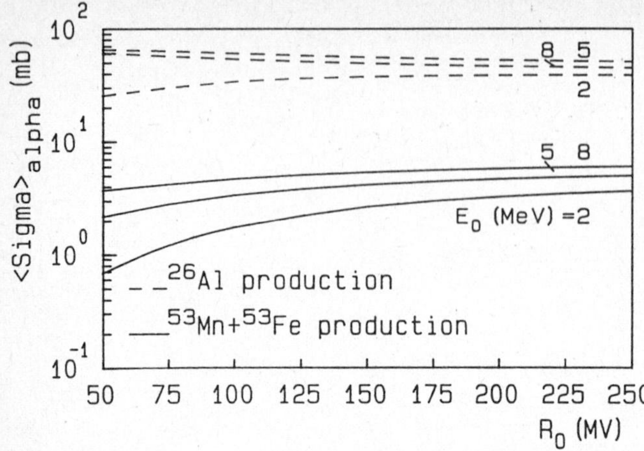

Figure 2: $\langle\sigma\rangle_\alpha(^{26}Al)$ and $\langle\sigma\rangle_\alpha(^{53}Mn+^{53}Fe)$ derived from Eq. (5) in the R-spectrum case, and for various E_0 values. The considered targets are Na, Mg, Al and Si for ^{26}Al production, and Cr and Fe for ^{53}Mn+^{53}Fe production.

where the chemical symbols designate the solar system number abundances of the corresponding elements or nuclides. We adopt Anders and Ebihara's (1982) abundances for the calculations. In the weak dosage limit, Eqs. (6) lead to the short-irradiation ratio

$$\frac{^{26}Al/^{27}Al}{^{53}Mn/^{55}Mn} \approx \left[\frac{\langle\sigma\rangle_\alpha(^{26}Al)}{3.985\langle\sigma\rangle_p(^{53}Mn)+3.889\langle\sigma\rangle_p(^{53}Fe)}\right]\frac{\varphi_\alpha}{\varphi_p} + \frac{\langle\sigma\rangle_p(^{26}Al)}{4.091\langle\sigma\rangle_p(^{53}Mn)+3.993\langle\sigma\rangle_p(^{53}Fe)}, \tag{7}$$

where only the numerically most important terms have been retained. In particular, the contribution of α-induced reactions to the ^{53}Mn and ^{53}Fe production turns out to be negligible with respect to the p-induced contribution, at least if the $^{50}Cr(\alpha,n)^{53}Fe$, $Fe(\alpha,X)^{53}Fe$ and $^{50}Cr(\alpha,p)^{53}Mn$ excitation functions discussed in Sect. 3.3 are accepted. It has to be stressed that Eq. (7) is independent of time, and also of the SEP flux if the α-induced reactions for ^{26}Al production are neglected. The ratio given by Eq. (7) has to be multiplied by $\tau(^{26}Al)/\tau(^{53}Mn) \approx 0.191$ in the long-irradiation regime.

Fig. 3 displays the short-irradiation values of the $(^{26}Al/^{27}Al)/(^{53}Mn/^{55}Mn)$ ratio obtained with the P- and R-spectra when the α-induced reactions are neglected [φ_α=0 in Eq. (7)]. In the R-spectrum case, typical values for that ratio cluster heavily around 0.05-0.07 for all the considered R_0 and E_0 values. They scatter much more in the P-spectrum case. For "reasonable" γ and E_0 values, it appears that $0.05 \leq (^{26}Al/^{27}Al)/(^{53}Mn/^{55}Mn) \leq 0.25$. In the study by Lee (1978), a low value of 0.016 was obtained, which would imply the overwhelming dominance of ^{53}Mn production. Fig. 4 generalizes Fig. 3 to the case where ^{26}Al production by α-induced reactions is taken into account. Results are presented in the R-spectrum case for φ_α/φ_p = 0.1 and 0.2. That range of values, complemented with the φ_α = 0 case of Fig. 3, most likely encompasses the actual SEP φ_α/φ_p values.

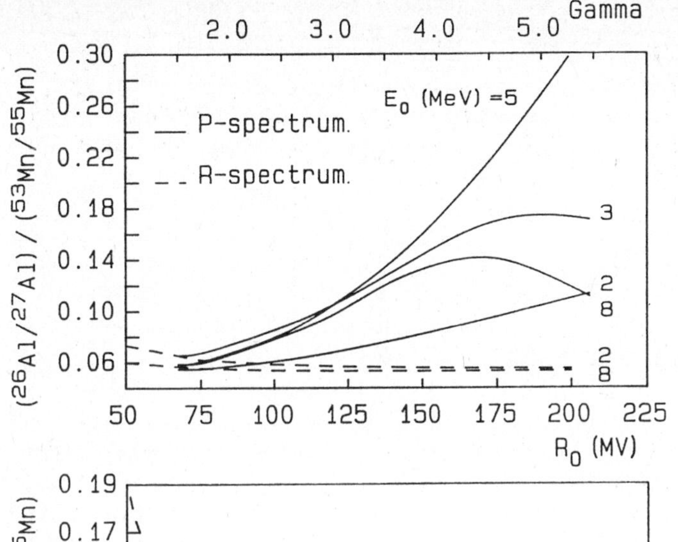

Figure 3:

Values of the indicated ratio obtained from Eq. (7) (weak dosage, short-irradiation limit) when $\varphi_\alpha=0$ (proton-induced reactions only), in the P-and R-spectrum cases (upper and lower abscissa, respectively), and for various E_0 values.

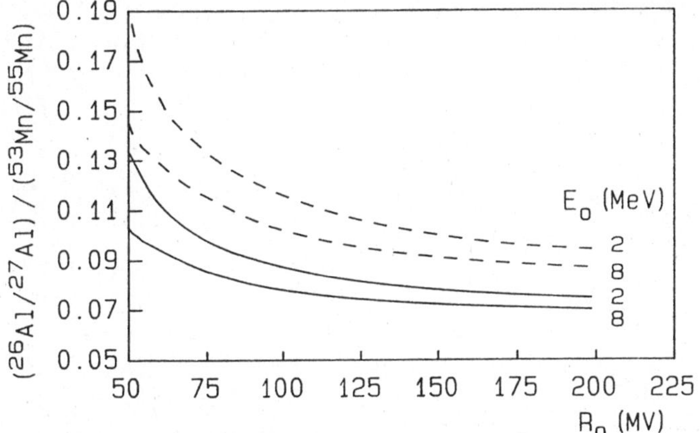

Figure 4:

same as Fig. 3, but when α-induced reactions are taken into account, and in the R-spectrum case only. The full and dashed lines correspond to φ_α/φ_p = 0.1 and 0.2, respectively.

A comparison between Figs. 3 and 4 indicates that α-induced reactions could indeed contribute significantly to the ^{26}Al production, especially in the $R_0 \lesssim$ 100-150 MV range, an upper limit $(^{26}Al/^{27}Al)/(^{53}Mn/^{55}Mn) \lesssim 0.15$ being reasonable, which is a factor $\approx$2 larger than typical values obtained without any contribution from the α-induced reactions. Of course, all those values have to be divided by a factor of $\approx$5 in the long-irradiation regime.

3.5. Fluences. From Eqs. (6), it is easy to evaluate the fluences $\varphi_p t$ (in the short-irradiation regime) or $\varphi_p \tau$ (in the long-irradiation regime) which are required in order to achieve given values of $^{26}Al/^{27}Al$ or $^{53}Mn/^{55}Mn$ at the end of the irradiation period. Table 2 summarizes some results obtained in the model already used for getting the results of Figs. 3 and 4. Those fluences are obtained without any consideration of dilution or chemical effects. In addition, no provision is made for possible time delays between the end of the irradiation and the condensation of Al and Mn into inclusions. In absence of such effects, the reported values represent

Table 2. Proton fluences $\varphi_p t$ or $\varphi_p \tau$ (in 10^{20} cm^{-2}) required for producing the indicated abundance ratios in the P- and R-spectrum cases, for two values of the cutoff energy E_0 (in MeV), and for extreme values of γ and R_0 (in MV). The possible contribution of α-induced reactions is neglected.

		$^{26}Al/^{27}Al=5x10^{-5}$		$^{53}Mn/^{55}Mn=5x10^{-5}$	
		γ=1.5	γ=5.5	γ=1.5	γ=5.5
P-spectrum	E_0=2	3	400	0.02	40
	E_0=8	2	3	0.1	0.4
		R_0=75	R_0=225	R_0=75	R_0=225
R-spectrum	E_0=2	4	2	0.2	0.001
	E_0=8	2	1	0.1	0.07

upper limits in the Al case, as α-induced reactions could contribute significantly to the ^{26}Al production.

The P-spectrum fluence values of Table 2 have to be compared with the values of $\approx 10^{21}$ to $\approx 10^{22}$ protons/cm^2 reported by Clayton et al. (1977) and Heymann et al. (1978). Taking dilution and chemical effects into account, Lee (1978) reports much higher values ($\approx 2x10^{25}$ protons/cm^2). It should be noted that our results for $^{26}Al/^{27}Al$ relative to $^{53}Mn/^{55}Mn$ are widely different from his calculations. We presume that this is due to his taking different proportions of irradiated and unirradiated materials with different chemical abundances in order to get such a large enhancement of $^{53}Mn/^{55}Mn$.

As made clear elsewhere (e.g. Clayton et al. 1977), certain of the proton fluences of Table 2 might induce other isotopic anomalies than the ^{26}Al- and ^{53}Mn-related ones. The examination of that question requires an extended compilation of spallation cross sections which has not been performed at the present stage of our study.

3.6. Can the spallation model account for negative anomalies? As already emphasized in Sect. 1, experimental data point toward a ^{53}Cr "negative" anomaly (i.e. deficiency with respect to solar) accompanying in certain cases the ^{53}Mn-related anomaly. A negative ^{26}Mg anomaly might also go along with the ^{26}Al-related anomaly in certain cases. Heymann et al. (1978) have provided an interpretation of that negative Mg anomaly in the framework of a spallation model which provides a sizable transformation of ^{26}Mg into ^{26}Al. Similar general arguments could probably be applied to the Mn-Cr case, the $^{53}Cr \rightarrow {}^{53}Mn$ transformation playing here the same key role as $^{26}Mg \rightarrow {}^{26}Al$ in the Mg-Al case.

4. Conclusion

We have shown that both ^{26}Al and ^{53}Mn can readily be produced by bombardment of dust grains with the composition of condensible solar matter by energetic protons and α-particles. These two nuclei would be strictly correlated by this model. The ratio $(^{26}Al/^{27}Al)/(^{53}Mn/^{55}Mn)$ is found to be restricted to a rather narrow range, 0.05-0.25 for short-irradiation regimes and for a wide choice of energy spectra and low energy cutoffs. For long-irradiation regimes, this range becomes 0.012 to 0.05. It follows that the values of the $(^{26}Al/^{27}Al)/(^{53}Mn/^{55}Mn)$ ratio derived from the particle bombardment model presented in this work are substantially lower than the estimates based on current observations. However, the results obtained here for most cases do not give $^{53}Mn/^{55}Mn$ values overwhelmingly greater than $^{26}Al/^{27}Al$, as was estimated earlier. A large α-particle flux relative to protons would increase the ^{26}Al yields.

Acknowledgments. This work has been supported in part by the Programme International de Collaboration Scientifique PICS nr.18. M. A. is Chercheur Qualifié F.N.R.S. (Belgium)

References

Anders, E., Ebihara, M. 1982, *Geochim. Cosmochim. Acta* **46**, 2363.
Arnett, W.D., Wefel, J.P. 1978, *Astrophys. J.* **224**, L139
Arnould, M. 1985, in *Proceedings of the Accelerated Radioactive Beams Workshop*, eds. L. Buchmann and J. D'Auria (TRIUMF TRI-85-1), p. 29
Arnould, M., Prantzos, N. 1986, in *Nucleosynthesis and its Implications on Nuclear and Particle Physics*, ed. J. Audouze and N. Mathieu (Dordrecht: Reidel), p. 363.
Arnould, M., Nørgaard, H., Thielemann, F.-K., Hillebrandt, W. 1980, *Astrophys. J.* **237**, 931
Birck, J.-L., Allègre, C.J. 1985, *J. Geophys. Res. Lett.* **12**, 745
Cameron, A.G.W. 1984, *Icarus* **60**, 416
Cameron, A.G.W., Truran, J.W. 1977, *Icarus* **30**, 447
Chen, J.H., Wasserburg, G.J. 1983, *Geochim. Cosmochim. Acta* **47**, 1725
Clayton, D.D. 1982, *Quat. J. Roy. Astr. Soc.* **23**, 174
Clayton, D.D., Dwek, E., Woosley, S.E. 1977, *Astrophys. J.* **214**, 300
Hainebach, K. L., Clayton, D. D., Arnett, W. D., Woosley, S. E. 1974, *Astrophys. J.* **193**, 157
Heymann, D., Dziczkaniec, M. 1976, *Science* **191**, 79
Heymann, D., Dziczkaniec, M., Walker, A., Huss, G., Morgan, J.A. 1978, *Astrophys. J.* **225**, 1030
Hillebrandt, W., Thielemann, F.-K. 1982, *Astrophys. J.* **255**, 617
Kaiser, T., Wasserburg, G.J. 1983, *Geochim. Cosmochim. Acta* **47**, 43

Keller, K.A., Lange, J., Munzel, A. 1974, *Landolt-Bornstein New Series* (Berlin: Springer-Verlag), vol. **5**, Part C
Kelly, W.R., Wasserburg, G.J. 1978, *Anal. Chem.* **50**, 1279
Lal, D. 1972, *Space Sci. Rev.* **14**, 3
Lee, T. 1978, *Astrophys. J.* **224**, 217
Lee, T. 1986, *Nature* **324**, 352
Leising, M.D., Clayton, D.D. 1985, *Astrophys. J.* **294**, 591
Mahoney, W. A., Ling, J. C., Wheaton, W. A., Jacobson, A. S. 1984, *Astrophys. J.* **286**, 278
McGuire, R.E., von Rosenvinge, T.T. 1984, *Adv. Space Res.* **4**, 117
Montmerle, T. 1984, *Adv. Space Res.* **4**, 357
Nørgaard, H. 1980, *Astrophys.J.* **236**, 895
Norris, T.L., Gansarz, A.J., Rokop, D.J., Thomas, K.W. 1983, *J. Geophys. Res.* **88**, B331
Papanastassiou, D.A. 1985, in *Isotopic Ratios in the Solar System*, ed. Centre National d'Etudes Spatiales (Toulouse: Cépadues-Editions), p. 77
Papanastassiou, D.A. 1986, *Astrophys. J. Lett.* **308**, L27
Prantzos, N., and Cassé, M. 1986, *Astrophys. J.* **307**, 324
Prantzos, N., Doom, C., Arnould, M., de Loore, C. 1986, *Astrophys. J.* **304**, 695
Reeves, H. 1978, in *Protostars and Planets*, ed. T. Gehrels (Tucson: University of Arizona Press), p. 399
Ryter, C., Reeves, H., Gradsztajn, E., Audouze, J. 1970, *Astron. Astrophys.* **8**, 389
Thielemann, F.-K., Arnould, M., Truran, J.W. 1986, in *Advances in Nuclear Astrophysics*, eds. E. Vangioni-Flam, J. Audouze, M. Cassé, J.-P. Chièze, J. Tran Thanh Van (Gif-sur-Yvette: Editions Frontières), p. 525
Truran, J.W., Cameron, A.G.W. 1978, *Astrophys. J.* **219**, 226
Wallace, R.K., Woosley, S.E. 1981, *Astrophys. J. Suppl.* **45**, 389
Wasserburg, G.J. 1985, in *Protostars and Planets II*, eds. D.C. Black and M.S. Matthews (Tucson: University of Arizona Press), p. 703
Wasserburg, G.J., Papanastassiou, D.A. 1982, in *Essays in Nuclear Astrophysics*, eds. C.A. Barnes, D.D. Clayton, and D.N. Schramm (Cambridge University Press), p. 77.
Wasson, J.T. 1963, *Icarus* **2**, 53
Woosley, S.E., Weaver, T.A. 1986, in *Nucleosynthesis and its Implications on Nuclear and Particle Physics*, eds. J. Audouze and N. Mathieu (Dordrecht: Reidel), p. 145
Woosley, S.E., Arnett, W.D., Clayton, D.D. 1973, *Astrophys. J. Suppl.* **26**, 231

THE CONTAMINATION OF COMETARY GLOBULES BY THE EJECTA OF NEARBY MASSIVE STARS

Jean-Pierre Arcoragi
Max-Planck-Institut für Astrophysik
Karl-Schwarzschild-Str. 1
8046 Garching bei München

ABSTRACT

It has been argued in the past that the ejecta of massive stars might explain the presence of short-lived nuclear species found in chondritic meteorites. The purpose of this paper is to provide new evidence in favor of such a picture. We will show that the Gum Nebula can be a good model for understanding the physical context in which our own solar system was born. This nebula is the sight of secondary low mass star formation induced by the massive stars at the center of the nebula. Recent theories of the propagation of stellar winds in the ISM (wind bubbles) and new models of massive stars, including detailed nucleosynthesis, allow one to follow in some detail the contamination process of the cometary globules by the ejecta of the massive stars located near the center of the nebula. We will show that the abundances of some short lived-species in chondritic meteorites are consistent with the amount that should be injected in the cometary globules of the Gum Nebula. The Gum Nebula is a region that offers the opportunity to understand stellar formation under simplified conditions.

I- INTRODUCTION

Not so long ago it was thought that the solar nebula was composed of one uniform chemical reservoir. The discovery of isotopic anomalies changed that picture and implied that there were more than one reservoir composing the primitive solar nebula. Many ideas were proposed in order to explain these anomalies. According to one line of thought the anomalies were produced by a very active young sun bombarding the nebula with high energy particles (e.g. Arnould, this volume). On the other hand, many authors advanced the idea that anomalies were the signature of the fact that the sun was born near massive stars (e.g. Cameron and Truran 1977; Reeves 1978; Olive and Schramm 1982; Blake, this volume).

In this paper, we will present a picture of the birth of our solar system that calls for the presence of nearby massive stars. The main contribution that this paper adds to the debate is to combine recent observations and new theoretical work in a self-consistent approach and use anomalies in order to check the plausibility of the scenario. More specifically, we will use the observations of the Gum nebula together with theoretical work on induced low mass star formation and wind bubbles, and recent models of massive stars in order to study the contamination of the uniform chemical reservoir of the cometary globules with short-lived nuclei ejected by massive stars. The degree of contamination will be shown to be compatible with solar system abundances of some short-lived nuclei. Section II will describe the Gum nebula and section III will concentrate on the details of the contamination process. Finally, section IV concludes by presenting a brief overview of observations that, taken together, give more credibility to the model presented in this paper.

II- THE GUM NEBULA

The Gum Nebula is a large region of ionized gas $\sim$ 125 pc in radius at the estimated distance of $\sim$ 450 pc (Brandt et al 1971; Beuermann 1973; Reynolds 1976 a, b; Reipurth 1983; Bruhweiler

1983). Seen projected upon the central regions are the Vela pulsar, the O4If star ς Puppis and γ^2 Velorum, a WC8+O9 binary system. The distance of the Vela pulsar has been estimated at 500 ± 100 pc on the basis of X-ray absorption measurements and that of γ^2 Velorum and ς Puppis at 450 ± 50 pc (Morton 1976; Morton and Wright 1978).

Further out from the center of the nebula one can see cometary globules. These structures appear projected on an annulus extending from ~ 51 to ~ 92 pc from the center of the nebula. They exhibit the following features (Zealey et al 1983):

(i) compact, dusty heads, sometimes containing embedded stars;

(ii) long faint luminous tails extending from one side of the head;

(iii) they are not physically attached to any neighbouring bright rims or nebulosity.

Most of the cometary globules in the Gum nebula have tails that point directly away from the central regions of the nebula and some show signs of low mass star formation. In order to explain the formation of cometary globules, Reipurth and Bouchet (1984) present a scenario in which (i) a more or less uniform molecular cloud contracts and (ii) creates a massive star which pushes its surroundings with its winds and radiation field leaving behind denser regions that (iii) eventually become cometary globules that (iv) finally evolve towards the Bok globule stage.

Recent observations have shown that the globules of the Gum Nebula are associated with low to intermediate mass pre-main-sequence stars. Pettersson (1987) found a large number of H_α emission line stars in the Gum Nebula. He selected 16 stars that appeared to be associated with dust lanes and 9 of them turned out to be T Tauri stars the others being background Be stars. These T Tauri stars are located in a part of the Gum Nebula rich in dark clouds near the cometary globule (CG) complex CG30-CG31. Reipurth (1983) had already identified 2 pre-main-sequence stars, Bernes 135 in CG1 and $PH_\alpha 92$ in CG22. He was able to estimate the mass of Bernes 135 at $\sim 2.5\, M_\odot$ and its age at $\sim 10^6$ years. Graham and Elias (1983) report a probable pre-main-sequence star associated with the dark cloud ESO210-6A. This cloud also contains 2 Herbig-Haro objects, HH46 and HH47. Thus, it seems likely that formation of stars of low to intermediate masses is going on on a large scale in the Gum Nebula. A systematic search for H_α-emitting stars might reveal more of these objects. The formation of stars (even low mass ones) in structures such as cometary globules is not unexpected from a theoretical point of view. Klein et al. (1985) have reviewed the situation of induced or secondary star formation and shown that the radiation pressure of massive stars acting on globule like structures enhances the density of such globules and thus favors the formation of stars.

III- THE CONTAMINATION PROCESS

In section II we have shown that the Gum Nebula is a site where low to intermediary mass stars are formed within cometary globules that owe their existence to the presence of massive stars near the center of the nebula. Using the wind bubble theory of Weaver et al. (1977), we will show how the Gum Nebula can be used as a model in order to understand the presence of some short-lived anomalies in the early solar system. Weaver et al. (1977) describe the structure of the wind bubble around a massive star (from the center going outwards) : region A is a region of free flowing stellar winds, region B is composed of high temperature ($\sim 10^6$ K) shocked winds, region C is shocked interstellar medium (ISM) and region D is the normal ISM. A rough model of the wind bubble in the Gum Nebula can be built by assuming that one star dominates in the center of the nebula. Using the numerical results of Reipurth (1983) in which the free flowing wind region is ~ 25 pc in radius, the external boundary (C-D) ~ 125 pc in radius and the region between the shocked winds and the shocked ISM (B-C) is ~ 115 pc in radius we can see that most, if not all, of the globules should be in the shocked wind region. The theory of Weaver et al. (1977) predicts expansion velocities of ~ 20 km/s in the shocked wind region (region B). Reynolds (1976a) claims to have observed such expansion velocities but Wallerstein et al. (1980)

attribute this velocity to dispersion. Zealey et al. (1983) maintain that expansion velocities of the order of 5 km/s are compatible with the observations. The situation is still far from clear on the observational side. On the theoretical side, however, it seems impossible not to have some kind of expansion occuring in the Gum Nebula, even if one allows for dissipation of energy in the form of X-rays at the free flowing wind— shocked wind interface (Dorland et al. 1986, Dorland and Montmerle 1987). In the following, we will assume that expansion velocities of $\sim$ 5 km/s are typical of the shocked wind region.

Let us now describe the contamination process in some detail. We start from a massive cloud which contracts and creates a few massive stars (the Vela progenitor, ς Puppis, and the binary system γ^2 Velorum being the most massive stars in the center of the Gum Nebula). These stars push the surrounding matter and evaporate and compress the higher density clumps of matter giving them the characteristic shapes of cometary globules. These globules form low mass stars that can be contaminated at some point by the winds and explosion products of massive stars. Recent models (Prantzos et al. 1986, 1987; Langer et al. 1987; Arcoragi et al. 1987) of the evolution of massive stars show that these stars can eject substantial amounts of short-lived nuclei during the late stages of their evolution in the form of stellar winds (Of, Wolf-Rayet stages) and supernova explosions. Only the most massive single stars ($\gtrsim 40\,M_\odot$ on the ZAMS) will eject newly synthesized nuclei in their winds. This is not true for close binary systems where mass transfer plays an imporatant role in the evolutionary process. In this case, stellar winds can eject short-lived nuclei for stars with M_{ZAMS} much less than $40\,M_\odot$.

Let us follow the stellar winds on their way to the cometary globules (supernova explosions might bring their share of short-lived nuclei but their yields are not known precisely and the propagation of the ejecta cannot be followed with the wind bubble theory): Once the matter is ejected in the winds it will travel in the free flowing region and perhaps form dust. Then it will arrive in a shocked region where the temperature is very high. Even the dust particles will be destroyed by collisional processes with the hot gas in which it bathes (simple calculations show that there are more electrons colliding with the dust particles than the number of atoms composing 1 μ dust grains and, since each electron has an energy of $\gtrsim 10$ eV, it is easy to see that dust grains will be completely destroyed in such an environment) and thus the short-lived nuclei will decay while in a gaseous state. Once the shocked wind reaches the globule region, some of it will get mixed with the globules. It is difficult to calculate the amount of gas that will effectively penetrate the globules. In order to estimate the amount of short-lived nuclei injected we will assume that the effective cross-section for wind capture is equal to the geometrical cross-section of the globules. Obviously this is an overestimate but it could give the right order of magnitude (detailed studies of the injection process should be done in the near future). With this last assumption, the amount of a short-lived nuclei injected in the globule depends only on the ejected mass of the short-lived nuclei, their half-life and the time they spend before being incorporated into the globule, the mass and radius of the globule (typical masses and radii would be $\sim 50\,M_\odot$ and a few tenths of parsec, respectively) and the distance between the globule and the massive star(s).

Let us apply these ideas to our own solar system. Wasserburg (1985) gives the abundance of short-lived species in the early solar system (the nucleus on the left is the short-lived one): ${}^{41}Ca/{}^{40}Ca \leq 8\cdot10^{-9}$, ${}^{26}Al/{}^{27}Al \sim 5\cdot10^{-5}$, ${}^{107}Pd/{}^{108}Pd \sim 2\cdot10^{-5}$, ${}^{146}Sm/{}^{144}Sm \sim 0.01$ (this ratio is still uncertain), ${}^{129}I/{}^{127}I \sim 1\cdot10^{-4}$, ${}^{244}Pu/{}^{238}U \sim 7\cdot10^{-3}$, ${}^{247}Cm/{}^{235}U \leq 5\cdot10^{-3}$. Now let us discuss the implications of these abundances: (i) ${}^{26}Al$: This nuclide is produced during central H-burning and ejected during the Of and WN stages of stellar evolution (it can also be produced during carbon burning and ejected during a supernova explosion). Prantzos et al. (1986) give estimates of the amount of ${}^{26}Al$ ejected in the winds of single massive stars. If we attribute the ${}^{26}Al$ in the early solar system to the winds of a massive star then this star was probably quite close (a few parsecs) because the transit time as to be quite short (a few milion years) and the dilution factor quite small if we want to explain the observations; (ii) ${}^{41}Ca$ and ${}^{107}Pd$: These 2 nuclides can be produced in a low fluence s-process (Prantzos et al. 1987; Langer et al. 1987; Arcoragi et al. 1987) and ejected during the WC and WO stages of stellar evolution (they can also be ejected in a supernova event). The fact that we do not detect ^{41}Ca implies that the ejecta that carried this nuclide decayed for more than 2 milion years before being incorporated in meteorites. On the other hand, ^{107}Pd is a much longer-lived nuclide (half-life $\sim$ 9.4 milion years). Its presence could be due to events not related to the formation of the solar system; (iii) ^{53}Mn (see Birck and Allègre 1985) and ^{146}Sm: These two nuclei are neutron deficient and can be produced by the p-process or by spallation (Arnould, this volume). The presence of ^{146}Sm is not confirmed yet (Wasserburg 1985). The half-life of ^{53}Mn is quite short ($\sim$ 5.5 milion years) so that its presence cannot be attributed to a very distant source (perhaps the same star that ejected the ^{26}Al); (iv) ^{129}I, ^{146}Sm and ^{247}Cm: These nuclides are r-process nuclei. They are fairly long-lived nuclides and could come from events not related to the formation of the solar system. They are probably synthesized during a supernova explosion.

The winds of nearby massive stars could explain the presence of ^{26}Al and ^{107}Pd in the early solar system. The other nuclides hint at some explosive event. If the presence of ^{53}Mn is related to such an event, then it seems that contamination by the ejecta of a *nearby* supernova is necessary in order to explain the presence of all the short-lived nuclei in the early solar system.

IV- DISCUSSION AND CONCLUSION

The Gum Nebula is a region where star formation occurs under simplified conditions. It is thus important to try to understand this region, and others like it, in some detail. The ideas presented in the last section have to be refined in order to get the real picture of what is going on in that nebula. Even thought the description is still partial, one can see that the injection of live short-lived nuclei is a natural consequence of induced star formation. There is a rapidly growing body of observations that give credibility to the scenario described in the last section. As an example let us mention the following observations:

(i) We have argued that dust should be destroyed in the shocked wind region due to the hot environment. Very low gas to dust ratios ($\lesssim 10^{-3} - 10^{-4}$) have been observed in many HII regions (see Dorland et al. 1986 and references given in their paper);

(ii) The study of ^{244}Pu fission tracks strongly hints at the in-situ decay of this nucleus (Drozd et al. 1977). But the very long half-life ($\sim$ 120 milion years) of ^{244}Pu does not exclude a non-live injection of shorter-lived nuclei. The need to have a source of heat in order to melt meteorites in the early solar nebula is a strong indication of in-situ decay of much shorter-lived nuclei than ^{244}Pu. The nuclide ^{26}Al is abundant enough to explain such an early melting process (it is interesting to note that Pudritz and Silk (1987) argue that live ^{26}Al could serve to transfer angular momentum and thus contribute to star formation by ionizing the gas in which it is embedded and coupling it to the magnetic field);

(iii) Molecular clouds show a clumpy structure on small scales ($\lesssim 10pc$, see Blitz and Thaddeus 1980) so that the density enhancements that give rise to cometary globules could be a normal component of molecular clouds.

We have argued that massive star induced low mass star formation can explain the presence of short-lived isotopic anomalies in the solar nebula. We have also argued that anomalies are likely to be injected *live* in the protosolar nebula due to the harsch environment that grains are likely to encounter before reaching the protosolar nebula. It should be emphasized that a lot of important questions remain to be answered before adopting such a model for the formation (and contamination) of the solar system, one of the most important being the frequency of such induced low mass star formation events. A very rough answer to the last point can be given by assuming that there are $\sim$ 10 low mass stars formed in the Gum Nebula, that there is always one such nebula within $\sim$ 500 pc of the solar system and that the lifetime of such a structure is $\sim$ 5 milion years. We can infer from these numbers that $\sim 10^8$ low mass stars have been formed in such a process during the lifetime of the Galaxy. This represents $\sim$ 1 star out of every 1000. This ratio should be considered as a lower limit and we cannot exclude for the time being ratios as high as 1 in 10. Careful study of the Gum Nebula and other nebulae, and more precise theoretical modeling of wind bubbles and the contamination process should help to clarify these questions in the near future.

ACKNOWLEDGEMENTS

Part of the work presented here was done while the author was in Brussels at the U.L.B. The autor wishes tc acknowledge helpful discussions with Marcel Arnould, Thierry Montmerle and Jim Truran. The author also wishes to thank R. Kuhfuß for a careful reading of the manuscript.

REFERENCES

Arcoragi, J.-P., Langer, N., Arnould, M., and Paulus, G. 1987, in preparation.

Beuermann, K.P. 1973, *Astrop. Space Sci.* **20**, 27.

Birck,J.L., and Allègre, C.J. 1985, *Meteoritics*, **20**, 609.

Blitz, L., and Thaddeus, P. 1980, *Astrop. J.* **241**, 676.

Brandt, J.C., Stecher, T.P., Crawford, D.L., and Maran, S.P. 1971, *Astrop. J.* **163**, L99.

Bruhweiler, F. C., Kafatos, M., and Brandt, J. C. 1983, *Comments Astrophys.* **10**, 1.

Cameron, A.G.W., and Truran, J.W. 1977, *Icarus* **30**, 447.

Cohen, M., and Schwartz, R. 1978, *Astrop. J.* **316**, 311.

Dorland, H., and Montmerle, T. 1987, *Astron. Astrophys.* **177** 243.

Dorland, H., Montmerle, T., and Doom, C. 1986, *Astron. Astrophys.* **160**, 1.

Drozd, R. J., Morgan, C. J., Podosek, F. A., Poupeau, G., Shirck., J. R., and Taylor, G. J. 1977, *Astrop. J.* **212**, 567.

Graham, J.A. 1986, *Astrop. J.* **302**, 352.

Graham, J. A., and Elias, J. H. 1983, *Astrop. J.*, **272**, 615. Gum, C.S. 1952, *Observatory* **72**, 151.

Klein, R. I., Whitaker, R. W., and Sandford II, M. T. 1985, in *Protostars and Planets II*, eds. D. C. Black and M. S. Mathews (Tucson: University of Arizona Press).

Langer, N., Arcoragi, J.-P., and Arnould, M. 1987, in preparation.

Morton, D. C. 1976, *Astrophys. J.*, **203**, 386.

Morton, D. C., and Wright, A.E. 1978, *Mont. Not. R. Soc.* **182**, 47P.

Olive, K.A., and Schramm, D.N. 1982, *Astrop. J.* **257**, 276.

Pettersson, B. 1987, *Astron. Astrophys.*, **171**, 101.

Prantzos, N., Arnould, M., and Arcoragi, J.-P. 1987, *Astrop. J.*, **315**, 209.

Prantzos, N., Doom, C., Arnould, M., and de Loore, C. 1986, *Astrop. J.*, **304**, 695.

Pudritz, R. E., and Silk, J. 1987, *Astron. Astrophys.*, **316**, 213.

Reeves, H. 1978, in *Protostars and Planets*, ed. T. Gehrels (Tucson: University of Arizona Press).

Reipurth, B. 1983, *Astron. Astrop.* **117**, 183.

Reipurth, B., and Bouchet, P. 1984, *Astron. Astrop.* **137**, L1.

Reynolds, R.J. 1976a, *Astrop. J.* **203**, 151.

—— 1976b, *Astrop. J.* **206**, 679.

Wallerstein, G., Silk, J., and Jenkins, E. B. 1980, *AStrop. J.*, **240**, 834.

Wasserburg, G. J. 1985, in *Protostars and Planets II*, eds. D. C. Black and M. S. Mathews (Tucson: University of Arizona Press).

Weaver, R., McCray, R., and Castor, J. 1977, *Astrop. J.* **218** 377.

Zealey, W. J., Ninkov, Z., Rice, E., Hartley, M., and Tritton, S. B. 1983, *Astrop. L.* **23**, 119.

III. Supernovae and SN 1987A

BINARY SYSTEMS AS SUPERNOVA PROGENITORS
(SOME FREQUENCY ESTIMATES)

A. Tornambè[1], F. Matteucci[2], I. Iben, Jr.[3],
and K. Nomoto[4]

(1) Istituto di Astrofisica Spaziale CNR, C.P. 67, I-00044 Frascati, Italy
(2) European Southern Observatory, Karl-Schwarzschild-Str. 2, D-8046 Garching, FRG
(3) University of Illinois, Urbana-Champaign, IL 61801, U.S.A.
(4) University of Tokyo, Meguro Ku, Japan

Abstract: We discuss a straightforward explanation of supernovae of type I in terms of binary systems evolving toward an explosive event according to the model of Iben and Tutukov (1984), and of Iben et al. (1987). The dependence of the estimated rates of such supernovae on free parameters is also discussed. The expected occurrence rates of type Ia, Ib and II SNe in elliptical galaxies and in our Galaxy are presented and compared with recent estimates of these rates by Van den Bergh, McClure, and Evans (1987). Agreement is within a factor of 2.

I. Introduction

The idea that a carbon-oxygen degenerate dwarf accreting matter from a hydrogen- and helium-rich envelope (Hoyle and Fowler 1964, Arnett 1968, 1969) or from a companion (Whelan and Iben 1973; Tutukov and Yungelson 1987, Webbink 1979a,b) could ignite carbon and then explode as a type I supernova has been encouraged by the success with which exploding white dwarf models (e.g. Nomoto, Thielemann, and Yokoi 1984, Southerland and Wheeler 1984, Woosley and Weaver 1986, and references therein) reproduce observed SN I supernova properties. Several models concerning the properties and the evolution of progenitor systems have been proposed in recent years. Starting in 1984, Iben and Tutukov in a series of papers formalized the idea that a close binary system composed of intermediate mass stars could evolve into a double degenerate C-O pair close enough to merge due to gravitational wave radiation thus giving rise to a type I supernova event.

One can summarize one of the several possible outcomes that their model can produce in the following steps:

i) The primary component with mass in the range 5-9 $M_{\odot}$ swells to fill its Roche lobe during its AGB phase when the external envelope is convective and a C-O core has already been developed in the centre. A dynamical mass transfer ensues giving rise to a common envelope phase during which the separation of the system is reduced due to drag forces. The mass transfer is not quenched until very little hydrogen and/or helium fuel remains above the burning shell, i.e. a nearly "naked" C-O core is left. Very little matter is accreted by the secondary.

ii) A similar history is followed by the secondary: once again a naked C-O core is left and the separation of the system has been shrunk further (down to 1-3 solar radii).

iii) Gravitational wave radiation abstracts angular momentum, causing further shrinkage in orbital separation until a merger of the two degenerate pairs becomes inevitable: a heavy disk is formed around the more massive white dwarf and accretion from this disk onto the WD triggers a carbon deflagration when the Chandrasekhar mass is attained (but see Iben 1987a,b).

Iben and Tutukov provided also the estimate of the expected rate of type I events which appeared to be in good agreement with the observations.

There are, however, some crucial aspects of the model which require a further exploration of this scenario.

First of all is the knowledge of the properties of the system, mainly separation following each common envelope phase. Iben and Tutukov used a crude semi-parametric formula based on the assumption that the energy required to eject common envelope material is nearly equal to the increase in the orbital binding energy of the system. This assumption could be shown by more careful analysis to be grossly in error. Another crucial point is the outcome of the merging phase itself of the double degenerate pairs.

Other problems are that the distribution of the binary systems with regard to component masses is poorly known, and the distribution of binaries with regard to separation function is also highly uncertain (see, however, Popova et al. 1982). It is thus obvious that the Iben and Tutukov model has to be further examined and probably considerably modified.

II. An easy attack point

From the previous short discussion it appears clear that the most crucial points to be understood in the future are the common envelope and the merging phases of degenerate structures. Both of these problems are very difficult and lead to the unavoidable conclusion that 3D hydrodynamical studies will be necessary to tackle these problems.

There is, however, an easy way to take, at present, in order to gain further insight into the problem: that is to understand how much the various parameters affect the final outcomes of a given population of binary systems.

This can be performed in a fast way by constructing an algorithm which transforms an assumed time-dependent distribution of binary parameters and birth rates into a time dependent distribution of various types of mergings. What one has to do is to analyze the fate of all the systems in a given mass and separation interval, by scanning the mass and the separation with narrow steps in component mass and orbital separation. Using the Roche geometry one can assume that when the currently evolving star swells to fill its Roche lobe during a convective envelope phase (first red giant branch phase or asymptotic giant branch phase) a common envelope will occur and the mass losing star will be completely deprived of the envelope up to the burning shell.

The properties of the bare stars (mass and composition) at the end of the common envelope phase can be derived by single star computations of the same mass and chemical composition. To compute the new separation of the system (A') one is still forced to make use of a semiparametric formula (for example, following Iben and Tutukov, $A' = A\alpha\ M_{1R}M_{2R}/M_1{}^2$, α being used as a free parameter).

With such a procedure the influence of the mass function of binary systems, of their q function and of separation function on the final outcome of merging systems can be tested.

Of course the first test to be performed is the one concerning the influence of the parameter α (or of other formulations of the effects of the common envelope) on the final behaviour of the merging events.

An extended analysis of all possible simulations is in preparation and will be published elsewhere (Tornambè 1987). Regarding the aim of this talk it is enough to mention that, as expected, the most crucial point of the whole picture is the efficiency of the common envelope phase in reducing the separation of the two components and it deserves a more detailed discussion.

Figure 1 shows the cumulative number of mergings of C-O plus C-O degenerate pairs versus time for a single burst of star formation at time t = 0.0 for several values of the parameter α. It appears clear from this figure that if the common envelope is not a disruptive event for the existence itself of the binary system, one has no substantial differences in the rate of the merging events even one Hubble time after the star forming event. However even in such a case, if the common envelope is too efficient in shrinking the components together (α = 0.5) no merging event will occur a few billion years after the forming event.

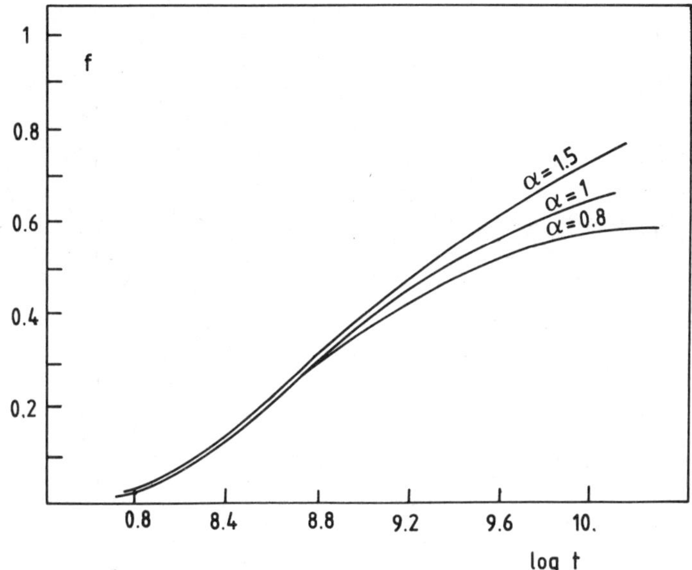

Fig. 1 - Cumulative numbers (in percent) of degenerate C-O C-O merging as a function of time according to various choices of the parameter α (see text).

Therefore, if we wish to justfy the existence of type I supernovae in elliptical galaxies by such a mechanism we must hope for a common envelope effect (as a mean) which is not too strong. If we restrict ourselves to larger values of α, we find that the outcome of binary evolution is not too sensitive to the precise value of α (say, in the range 0.8-2).

Other changes in the assumptions, such as a change of the mass function or the q distribution, inside reasonable limits, do not alter the total number of mergings after one Hubble time by more than a factor of two. The same occurs if some amount of mass is permitted to be accreted onto the secondary during the first common envelope event.

We can summarize our findings by saying that, if we accept a scenario in which the common envelope effect is such as to give rise to double degenerate mergings after one Hubble time, we should expect a merging rate substantially correct within a factor of 2.

III. Many Kinds of Mergers

One may easily incorporate into the analysis not only the outcomes of a generation of binary systems undergoing common envelope phases and forming degenerate dwarf binary pairs but also estimates of the frequeny of forming close binary pairs of many types other than the double degenerate C-O ones. As already discussed by Iben and Tutukov one finds among the resulting merging events also the following interesting ones:

a) C-O + He, where the He star is not degenerate
b) C-O + He, where both the components are degenerate
c) He + He, both degenerate components.

Since the merging of the two components is due to gravitational wave radiation, one can derive the expected rates of mass accretion during the mreging itself. Cases b) and c) should have quite large accretion rates (of the order of 10^{-5} solar masses per year). As a consequence of this, a straightforward supernova event is not expected.

Let us however analyze briefly case c), merging occurrence. Recently Saio and Nomoto (1987) studied the case of the accretion (at a rate of $\dot{M} = 10^{-7}\ M_{\odot}\ yr^{-1}$) over a 0.4 $M_{\odot}$ He WD. A preliminary report of these calculations has been given by Nomoto (1987). Saio and Nomoto find that when the accreting WD reaches 0.515 $M_{\odot}$, He ignites first off-center and then the burning front moves inward with a series of weak flashes. After having reached the center of the star, He-burning proceeds at a steady rate. The conclusion is that the He-He WD pair is at the end changed into a helium-main sequence star which will end its life as a CO WD.

An extremely interesting case of merging events is case a). In such a case the expected accretion rate of He matter onto the degenerate C-O dwarf is of the order of $10^{-8}\ M_{\odot}/yr$. This accretion rate is very near to the one required for an He detonation to take place after 0.1-0.4 $M_{\odot}$ of He have been accreted peacefully on the surface of the C-O dwarf (see, e.g., Nomoto 1982, Woosley, Taam and Weaver 1986). During the He detonation 0.1-0.3 $M_{\odot}$ of matter will be incinerated and, as a consequence, a type I supernova underluminous by a factor 3-6 with respect to a classical type I will be produced. Incinerated layers will be exposed earlier than in a classical type I supernova, and as a consequence this class of supernovae will resemble, in the spectrum, a type I SN born late. He and Fe are expected to be seen in the early time spectra. Oxygen is expected in the late time spectra if part of the unburned core is also ejected.

Figure 2 (from Tornambè and Matteucci 1987) gives the rate of such a type of merging for a single generation of binary systems born at time t = 0. The lifetime of these mergings does not extend over 1 billion years for the simple reason that the model requires a merging of an He-burning star, and hence the time delay after the formation of the He companion cannot exceed the maximum He lifetime in the obtained range of masses. The maximum rate is expected to be between $6 \cdot 10^7$ and 10^8 years after the star forming event of the progenitor binary system.

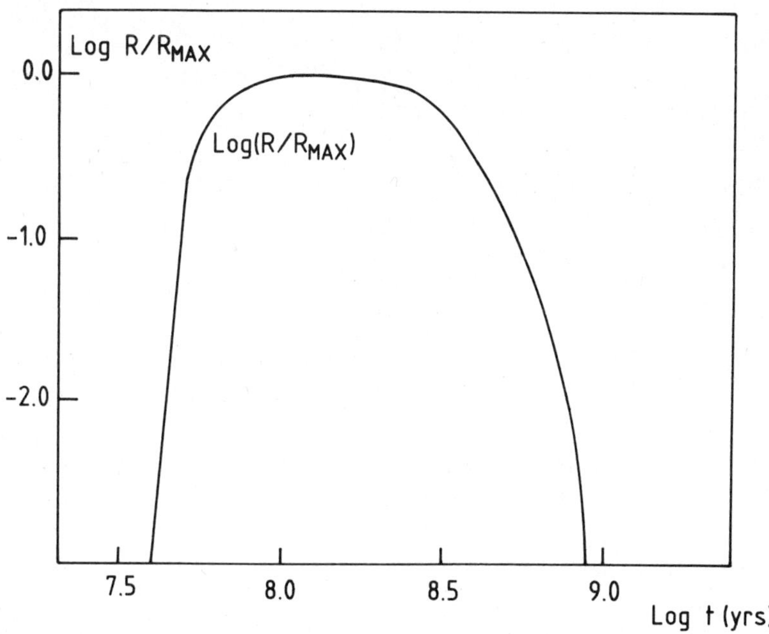

Fig. 2 - The rate of He star - CO dwarf merging as a function of the time for a single burst of star formation at t = 0.

Since the systems consisting of two degenerate white dwarfs can merge at any time after the star formation and those consisting of an He-star and a C-O white dwarf merge mostly in $6 \cdot 10^8$ years after the star formation, one expects that the SNe originating from the former systems are spread over the whole disk while the latter are more concentrated near the star formation regions.

Iben, Nomoto, Tornambè and Tutukov (1987) were tempted to identify the mergings of an He-star plus a C-O dwarf with the class of type Ib supernovae. In fact, most of the properties expected from such an event are the same as those observed in Type Ib supernovae (Wheeler and Levreault 1985, Uomoto and Kirshner 1985, Wheeler and Harkness 1986, Panagia, Sramek and Weiler 1986, Branch and Nomoto 1986).

IV. The Expected Rates in Spiral Galaxies and in Ellipticals

The previously discussed results for a single generation of stars can be easily extended to a continuous star formation rate. This can be done either for the case of a spiral galaxy where the star formation is an ongoing process, or for an elliptical galaxy where star formation presumably stopped several billion years ago. The galactic evolutionary models used for those two experiments are the one by Matteucci and Greggio (1986) and that described in Matteucci and Tornambè (1987). Figure 3 (from Tornambè and Matteucci 1987) shows the rates of double degenerate C-O mergings and those of He plus degenerate C-O mergings both for a typical elliptical galaxy and for our Galaxy. We are willing to identify the former with a classical type I supernovae (type Ia) and the latter with type Ib supernovae.

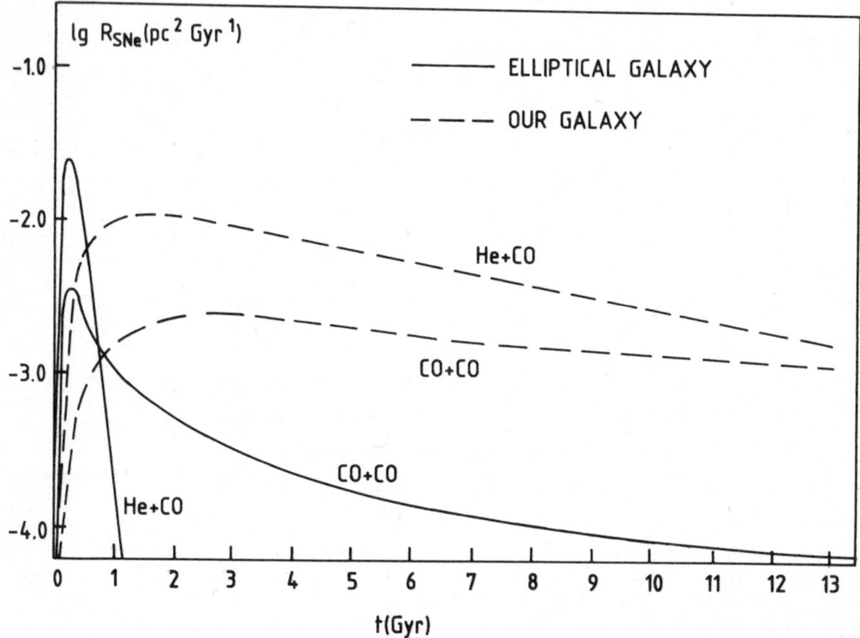

Fig. 3 - Total rates of mergings of various types as a function of time for a typical elliptical galaxy (full line) and for our Galaxy (dashed line). In the text it is suggested that He star + CO dwarf merging can give rise to type Ib SN events, while double CO dwarfs merging give rise to a type Ia SN event.

The rates at 13 billion years for our Galaxy, expressed in supernova units (1 SNu = 1 supernova every 100 years every 10^{10} $L_{B\odot}$), in order to be compared with the observations, are the following:

C-O + C-O mergings (type Ia SNe)	0.06 SNu
C-O + He mergings (type Ib SNe)	0.1 SNu
core collapsing massive stars (type II SNe)	0.75 SNu

A comparison with the observations is not easy and we will perform it in the light of the recent paper of van den Bergh, McClure and Evans, (VME, 1987). VME present the results of a 5 year monitoring of 1017 bright galaxies and find that the rates of Tammann (1982) are too high by a factor of at least 3. Their rates are (in supernova units):

Type Ia SNe	0.3 h^2	(i.e. 16%)
Type Ib SNe	0.4 h^2	(i.e. 22%)
Type II SNe	1.1 h^2	(i.e. 61%)

where h = H/100. Concerning our Galaxy, if one assumes a total blue luminosity of $2 \cdot 10^{10}$ $L_{B\odot}$ (see the discussion in VME for this point) the rate of 2.2 supernovae per century as given by van den Bergh (1983) translates into 1.1 SNu for our galaxy for all the supernova types.

Therefore, for our Galaxy the observed supernova rates for the 3 types of supernovae should be as follows:

Type Ia	0.17 SNu
Type Ib	0.24 SNu
Type II	0.67 SNu

These numbers should represent an upper limit since the galactic supernova rate of van den Bergh (1983) is somewhat larger with respect to that estimated by VME (see VME for a discussion). Theoretical rates (without any manipulation of the various parameters and assuming that the number of binary systems is equal to the number of single systems) are inside a factor of two from the observed rates. The best agreement is in the ratio between the type Ia and type Ib rates. The two rates are in fact affected by the same uncertainty sources in their absolute values; these uncertainty sources (mass function, separation function and so on) eliminate each other when the ratios are considered giving a surprisingly good agreement.

Even for elliptical galaxies a nice agreement is obtained if one considers that the Tammann rate may be overestimated by a factor of 3. In fact, observations by Tammann (1974) give 0.22 SN units for type Ia supernovae while theory is giving 0.04 (a M/L = 10 has been assumed). Finally, the current rate of type Ib supernovae in ellipticals is predicted to be zero, in agreement with the observations.

References

Arnett, W.D.: 1968, Nature 219, 1344.
Arnett, W.D.: 1969, Ap. Space Sci. 5, 180.
Branch, D., and Nomoto, K.: 1986, A&A 164, L13.

Hoyle, F., and Fowler, W.A.: 1960 Ap.J. 132, 565.
Iben, I., Jr., and Tutukov, A.V.: 1984, Ap.J. Suppl. 54, 335.
Iben, I., Jr., Nomoto, K., Tornambè, A., and Tutukov, A.V.: 1987, Ap.J. in press (June issue).
Iben, I., Jr.: 1987a, preprint.
Iben, I., Jr.: 1987b, preprint.
Matteucci, F., and Greggio, L.: 1986, A&A 154, 279.
Matteucci, F., and Tornambè, A.: 1987, A&A, in press.
Nomoto, K.: 1982, Ap.J. 257, 780.
Nomoto, K., Thielemann, F.K., and Yokoi, K.: 1984, Ap.J. 286, 644.
Nomoto, K.: 1987, in Progress in Particle and Nuclear Physics, Vol. 17, p. 249, ed. A. Faessler.
Panagia, N., Sramek, R.A., and Weiler, R.A.: 1986, Ap.J. Lett. 300, L55.
Popova, E.I., A.V. Tutkov, and L.R. Yungelson: 1982, Astroph. Space Sci. 88, 55.
Saio, H., Nomoto, K.: 1987, in preparation.
Southerland, P.G., and Wheeler, J.C.: 1984, Ap.J. 280, 282.
Tammann, G.A.: 1974, in "Supernovae and Supernova Remnants", eds. C.B. Cosmovici (Dordrecht: Reidel), p. 155.
Tammann, G.A.: 1982, in "Supernovae: A Survey of Current Research", eds. M.J. Rees and R.J. Stoneham (Dordrecht: Reidel), p. 371.
Tornambè, A., and Matteucci, F.: 1987, Ap.J. Lett., in press.
Tornambè, A.: 1987, in preparation.
Tutukov, A.V., and Yungelson, L.: 1979, Acta Astron. 29, 665.
Uomoto, A., and Kirshner, R.P.: 1985, A&A 149, L7.
van den Bergh, S.: 1983, P.A.S.P. 95, 388.
van den Bergh, S., McClure, R.D., and Evans, R.: 1987, January 1987 preprint of Dominion Astrophysical Observatory.
Webbink, R.F.: 1979a, in "Changing Trends in Variable Star Research, IAU Coll. 46, eds. F.M. Bateson, J. Smak and I.H. Urch (Hamilton, New Zealand: U. Waikato Press).
Webbink, R.F.: 1979b, in "White Dwarfs and Variable Degenerate Stars, eds. H.M. Van Horn and V. Weidemann (New York: University of Rochester), p. 426.
Wheeler, J.C., and Levrault, R.: 1985, Ap.J. Lett. 294, L17.
Wheeler, J.C., Harkness, R.: 1986, in "Distance to Galaxies and Deviation from the Hubble Flow", eds. B. Madore and R.B. Tully (Dordrecht: Reidel), in press.
Whelan, J.C., and Iben, I., Jr.: 1973, Ap.J. 186, 1007.
Woosley, S.E., Taam, R.E., and Weaver, T.A.: 1986, Ap.J. 301, 601.
Woosley, S.E., and Weaver, T.A.: 1986, Ann. Rev. Astr. Ap. 22, 205.

ON STELLAR MODELS FOR THE PROGENITOR OF SUPERNOVA 1987A

James W. Truran
Max-Planck-Institut für Astrophysik, 8046 Garching b. München, FRG
and
Dept. of Astronomy, University of Illinois, Urbana Ill.61801, USA

Achim Weiss
Max-Planck-Institut für Astrophysik
8046 Garching b. München, FRG

Abstract

We report the results of stellar evolutionary calculations for stars of initial masses, luminosities and metallicities compatible with the possible progenitor of Supernova 1987A in the Large Magellanic Cloud. We find, specifically, that the evolution of a star of initial main sequence mass in the range ~15-20 $M_{\odot}$ and initial metal composition $Z = 0.25\ Z_{\odot}$, compatible with the metallicity of the LMC, is quite consistent with the occurrence of a blue supergiant progenitor for Supernova 1987A: presumably, the B3 Ia supergiant Sanduleak -69 202. Some critical uncertainties associated with the stellar models are identified and discussed.

I. Introduction

Supernova 1987A was discovered in the Large Magellanic Cloud on February 24, 1987 (23). The optical features indicate that this event was a Type II Supernova. Astrometry (30) confirms the near-coincidence of the position of the supernova and that of star 1 of the Sanduleak -69 202 system. Additionally, UV observations (20) now suggest that Sanduleak -69 202 has indeed disappeared from the field. We will therefore proceed in this paper on the assumption that the stellar progenitor of Supernova 1987A was the B3 Ia blue supergiant Sanduleak -69 202. Our aim will be to explore the possible implications of this identification for models of the evolution of typical massive star ($M \gtrsim 10\ M_{\odot}$) progenitors of Type II supernovae.

In order to proceed with our study, we must first estimate the allowed range of luminosities and stellar masses. The identification of Sanduleak -69 202 as of spectral type B3 Ia implies an effective temperature of order 15,000 K. The apparent visual magnitude is m_V = 12.24 mag (16). Adopting a distance modulus m-M = 18.5 for the LMC and a reddening corection A_V = 0.6 mag, a photometric analysis of the several components of the Sanduleak -69 202 system by West et al. (30) indicates an absolute visual magnitude M_V = -6.8 mag for the B3 Ia star. If we further take a bolometric correction of 1.3, we obtain an absolute bolometric magnitude for the supernova progenitor M_{bol} = -8.1 mag and a corresponding luminosity of 1.3×10^5 $L_\odot$ or 5×10^{38} erg s^{-1}.

An estimate of the mass of the progenitor may be obtained from a survey of published models of massive star evolution (see, for example, the review by Chiosi and Maeder (17) and references therein). As we will learn from our subsequent discussion, it is necessary to give careful attention to results obtained for various assumptions concerning initial metallicity, the rate of mass loss, and the treatment of convection. The models of Brunish and Truran (4,5) indicate that a 15 $M_\odot$ star will reach the onset of carbon burning at a luminosity $\sim3.1\times10^{38}$ erg s^{-1} ($\sim8\times10^4$ $L_\odot$) quite independent of initial metallicity and of the presence of a moderate rate of stellar mass loss. Weaver et al. (28) found a 15 $M_\odot$ star of solar metallicity to have a luminosity of 3.7×10^{38} $L_\odot$, while the 20 $M_\odot$ model discussed by Wilson et al. (31) had a luminosity of 5.7×10^{38} erg s^{-1}. Similar results are reflected in the evolutionary tracks for massive stars undergoing mass loss, as described by Chiosi and Maeder (7). We therefore feel justified in proceeding on the assumption that the progenitor of Supernova 1987A had a mass in the range $\sim$15-20 $M_\odot$, although a somewhat higher mass is not excluded.

The issue we wish specifically to address in this paper is the identification of the progenitor with a blue supergiant. This behavior was not anticipated by stellar evolution and supernova theorists, since standard models for the expected massive star progenitors of Type II supernovae have generally predicted that they should evolve to red supergiants prior to the ignition of core carbon burning and thus should subsequently explode as red supergiants. The inclusion of mass loss effects encourages evolution to the red (7) and thus provides no easy solution to this problem. Metallicity effects, on the other hand, do seem to provide a possible explanation for the occurrence of a blue supergiant progenitor. Guided by the results of evolutionary calculations (5) which predict that stars of mass ~15-20 $M_\odot$ of low metallicity can reach the end of their evolution as blue supergiants, Truran et al. (26) suggested that the occurrence of a blue supergiant progenitor in a low metallicity population such as that which characterizes the Large Magellanic Cloud could be understood in this manner. Stellar evolution calculations reported by Hillebrandt et al. (10) and Arnett (1) indicate that a 15-20 $M_\odot$ star of metallicity Z = 1/4 $Z_\odot$ compatible with the LMC will reach the supernova stage as a blue supergiant. The compact structure of such a blue supergiant has also been shown (1,9,10,22,24,27,32,33) to be consistent with the subsequent development of the optical lightcurve and the observed spectral evolution of Supernova 1987A.

Our stellar evolution calculations for massive stars of low metallicity, relative to solar abundances, are presented in the next section, together with a review of results obtained by other researchers. We then briefly survey the dependences of massive star evolution on mass loss assumptions, opacity and the treatment of convection. A discussion of stellar evolution and its implications for the distribution of blue and red supergiants in metal-deficient populations follows, together with our conclusions.

II. Massive Star Evolution as a Function of Metallicity

The principal result of the calculations of Brunish and Truran (5, hereafter called BT) was the finding that stars of initial mass in the range 15 to 40 $M_\odot$ are to be found at effective temperatures of 10000 K or higher at the onset of central carbon burning, when their primordial metal contents are only a small fraction of the solar value. They investigated stars of mass 15, 20, 40 and 50 $M_\odot$ with metallicities of Z = 0.02, 0.01, 0.001 and 0.0002. From the observations of SK -69 202, we have already deduced a luminosity of $1.3 \cdot 10^5$ $L_\odot$. This corresponds to a stellar mass of 15 to 20 $M_\odot$, taking into account the uncertainties involved in the determination of M_{Bol} from m_V. Furthermore, a metallicity of 0.005 seems to be appropriate for the LMC (21), although some scatter exists and higher values for individual stars might be possible. Consequently, we have undertaken an investigation of the evolution of stars of masses M = 15,17,20,22, and 30 $M_{\odot}$ and compositions Z = 0.02, 0.01, 0.005, and 0.001. Our intention has been to confirm the principal result of BT for stellar masses and metal contents that seem to be appropriate for SK -69 202, and to show that, under plausible assumptions, the progenitor of a Type II supernova in a metal deficient stellar system like the LMC can be a blue supergiant.

The calculations were performed using the program of Kippenhahn et al. (17) in the version described by Weiss (29). As in BT, we used the Schwarzschild-criterion for convection and a very fine mass grid in the semiconvective regions of the models (for a discussion of the problems associated with the treatment of these layers, see the following section). Typically, evolved models consisted of 250 or more mass shells in the interior 99.9% of the stellar mass and another 50 grid points in the exterior regions. In the interior region, the stellar evolution equations are solved with the use of an implicit Henyey scheme, while in the exterior region they are explicitly integrated from the photosphere to the fitting point.

A major difference between our calculations and those of BT is the use of updated opacities. Our opacity tables were constructed by using the Astrophysical Opacity Library of Hübner et al. (13) for temperatures above 1 eV, together with additional values for the same mixtures of elements at lower temperatures, which kindly were provided by Hübner (11,12). For a more detailed description, see the discussion by Weiss (29). Seven tables for different compositions of hydrogen, helium, carbon, oxygen and all other elements ("metals") covering the temperature range from 3000 K to 10^9 K were available. Since only 5 of these tables are necessary for a linear interpolation in composition, it was possible to check the dependence of our results on the use of different combinations of tables. We found this dependence to be very weak for the case of M = 20 $M_{\odot}$, the one case we tested.

Our treatment of the nuclear reactions and energy generation is based upon the use of the analytic formulae provided by Kippenhahn et al. (17) and Thomas (25). The reaction rate for $^{12}C(\alpha,\gamma)^{16}O$ has been increased by a factor of three to take into account the recent experimental results. Our treatment of nucleosynthesis thus follows the abundances of only those five elements that are also taken into account in the opacities. Although this is a consistent treatment, it is clearly insufficient in the case of carbon burning, since all nuclear yields must be incorporated into the "metals". It was therefore not possible to continue our calculations past the carbon burning stage, and even this phase should be treated with some caution.

Mass loss, although most likely present in supergiants, has not been taken into account in our study, since BT showed that the evolution of metal-poor stars of 15 to 20 $M_{\odot}$ is not very sensitive to mass loss effects. The effects of convective overshooting were also neglected. A discussion of possible influences of mass loss and of alternative prescriptions for the treatment of convection will be given below.

The evolutionary tracks for stars of 15, 17, 20, 22, and 30 $M_{\odot}$ and a range of metallicities are presented in Figure 1.

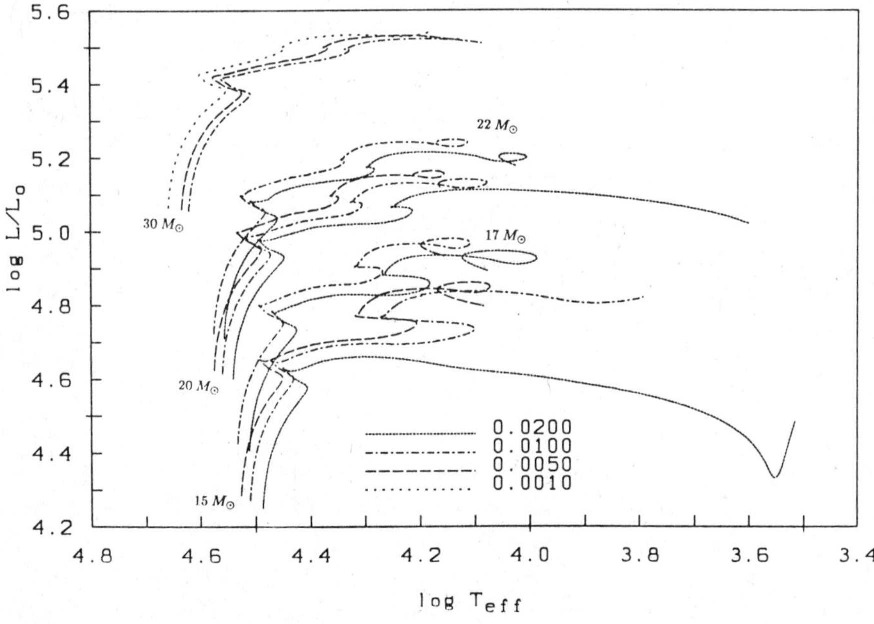

<u>Figure 1:</u> Evolutionary paths in the Hertzsprung-Russel Diagram for the stellar models described in the text. For each stellar mass, a set of tracks corresponding to different metal contents is illustrated.

In general, the calculations have been stopped when the model was in the phase of core carbon burning. The single exception, the model of $M = 15\ M_\odot$ and $Z = 0.02$, has been followed only to the ignition of helium, at which point it was already at an effective temperature cooler than 3000 K, the lower limit of our opacity tables. The tendency for models to move more slowly to the red at low metallicities is clear. In all cases with $Z = 0.005$, our adopted value for Sanduleak -69 202, the models were hotter than 10000 K during the phase of central carbon burning. For $M/M_\odot \geqslant 17$, even a metal content half of the solar value was sufficient to obtain a blue supergiant. Since the thermal time scale of the stellar envelope is of the order of 100 to 1000 years, and the nuclear time scale for core burning phases past carbon burning is always shorter than 100 years, we do not expect the star to change its position in the Hr diagram once carbon has been depleted in the core. We conclude from our calculations that a supernova progenitor of spectral type B3 for Supernova 1987A is plausible as a result of a reduced metal content. We thereby confirm the basic result of BT.

There are, however, several differences with respect to BT. Our $15\ M_\odot$ model with solar metallicity ignites helium as a red giant, whereas the corresponding model of BT remains a blue star at this phase. The model with $Z = 0.01$ has the same effective temperature and luminosity as that of BT at the beginning of central helium burning, but shows a redward loop between the end of core hydrogen burning and the beginning of core helium burning. This loop is present in most of our tracks, but tends to vanish for lower Z and higher masses. To our knowledge, this behaviour has not between encountered elsewhere, except in the recent work of Arnett (1) on the same subject. We attribute this behavior to changes in the envelope structure resulting from our use of updated opacities. Unfortunately, Arnett gives no details of his evolutionary calculations against which to check this argument. Nevertheless, we note a smooth transition from models that are very cool at the beginning of helium burning (which experience loops to the red in the phase of core contraction, but then return to higher effective temperatures) and those that always stay very hot. In general, our results resemble very much those of BT, although they have been obtained with somewhat different input physics and an entirely different stellar evolution code.

III. Dependences on Mass Loss, Opacity and Convection

An outstanding problem of stellar evolution calculations associated with the mass range of 15-40 $M_\odot$ is the treatment of semiconvective zones. An extensive review of different calculations and of the "contradictory and rather erratic" results obtained is given by Chiosi (6; see also the references given in this review for various computations). While all of these investigations have been performed for Population I stars, we expect the same problems also to occur for low metallicity stars, since we have identified semiconvective regions also in our models.

The important differences realized in the evolutionary sequences, when applying either the Schwarzschild or the Ledoux criterion, have been illustrated by Chiosi and Summa (8) for the case of a 20 $M_\odot$ star. The model calculated according to the Ledoux criterion ignited helium as a red supergiant, then performed a loop back to higher temperatures, and at the end of core helium burning finally became a red supergiant again. When the Schwarzschild criterion was used, the model performed no loop, but rather evolved gradually from blue to red. Chiosi and Summa noted that these differences they encountered result from different envelope structures and temperature gradients. It therefore is not surprising that changes in opacities also will modify the results of the calculations, since the temperature gradient in the radiative envelopes directly depends on the opacity. As far as we know, our calculations of massive star evolution are the first to use the Astrophysical Opacity Library, and we have mentioned the differences with respect to the results of BT, which we ascribe to these new opacities. Chiosi (6) also discussed the influence of different opacities in the case of the Carson opacities.

Bencivenni (2) presented calculations of the evolution of a 15 $M_\odot$ star with metallicity Z = 0.003, which also were performed using the Astrophysical Opacity Library. However, in contrast to our results, her star became a red supergiant already at the beginning of helium burning, and did not subsequently return to higher temperatures. Here, the different envelope structures can be understood on the basis of the fact that her models encountered so-called "convective breathing" of the convective core. Our models also showed semiconvective zones above the hydrogen-burning core that appeared, grew in mass, and then disappeared, but always remained detached from the convective core. Therefore, no mixing took place between these regions and the core. This example illustrates once more the uncertainties inherent in such calculations, and suggests that the sensitivities of evolutionary results to the underlying stellar parameters should be carefully examined.

One further process influencing the evolution of stars in the mass range of interesting for our models is mass loss. BT found that their adopted mass loss rates only slightly influenced the evolution of metal poor stars of 15 $M_\odot$. The models became somewhat cooler during helium and carbon burning. This effect was larger for 30 $M_\odot$ stars. As an example of the effects of mass loss on the evolution, consider the results of Maeder (19). A moderate mass loss rate (typically 10^{-6} $M_\odot$/yr during central hydrogen and helium burning) results in somewhat cooler models that nevertheless reach the end of their evolution as supernova progenitors on the Hayashi track. When a mass loss rate higher by a factor of ten was adopted (based on estimates of Wolf-Rayet mass loss rates), core helium burning in a 15 $M_\odot$ star took place entirely in the red supergiant region, but after the ignition of carbon the model (now having only 3.6 $M_\odot$ remaining) evolved to effective temperatures as high as 50000 K on a thermal time scale. It therefore seems to be possible to have blue supernova progenitors also from models with high mass loss rates (33). It seems somewhat unlikely, however, that such high wind-driven mass loss rates would be characteristic of lower mass stars ($\sim$15-20 $M_\odot$) of lower metallicities than the sun.

IV. Discussion

In the last two sections we have demonstrated that, for stars in the mass range of 15...30 $M_\odot$, a metal content of $< Z_\odot/2$ can result in blue supergiants as progenitors for Type II supernovae. This result, obtained with the use of updated opacities and a different stellar evolution code, generally confirms that of the earlier calculations of BT. Given the uncertainties discussed in the last section, this is not a trivial result and we find the agreement to be quite encouraging. We therefore offer our calculations as a straightforward explanation for the existence of a B3 I supergiant progenitor of SN 1987A. We cannot exclude, however, the possibility that a different treatment of convection, resulting e.g. in effects like "convective breathing" (2), or the inclusion of significant rates of mass loss might change our results considerably. The reviews of Chiosi and Maeder (7) and Chiosi (6) make it apparent that the question as to which result is to be preferred cannot yet be answered clearly. One possibility, for example, might be the existence of very high mass loss rates (19), which result in a return of the stellar models to higher effective temperatures. While such behavior might allow a reasonable interpretation of the observed concentrations of blue and red supergiants in diverse stellar populations, we nevertheless find it somewhat difficult to understand how a star can reach the stage of core collapse in such a configuration.

The question quite naturally arises here as to whether our stellar evolution calculations, and those of other researchers (1,5,10), for metal deficient stars imply a significant distortion of the Hertzsprung-Russell diagrams for low metallicity populations in general and for the Large Magellanic Cloud in particular. Humphreys and Davidson (15) have argued that massive star evolution has resulted in very similar supergiant populations in the Galaxy and the LMC, in spite of the small differences in the heavy element abundance levels. Of particular concern might be the fact that the observed M_{BOL} verses log T_e H-R diagrams for both the LMC (15) and the SMC (14) reveal the presence of a significant population of red supergiants at luminosities consistent with stars in the mass range $\sim$15-20 $M_{\odot}$. For the case of the LMC, we believe that this observed behavior might be simply a consequence of a spread in metallicity of a factor of a few, since our numerical results reveal that there exists a strong dependence of the extent of redward evolution on metallicity Z for values in the range $0.1\ Z_{\odot} \leq Z \leq Z_{\odot}$. A somewhat broader spread in metallicity might be necessary to allow an understanding of the H-R diagrams for the SMC, which is known to have a lower Z than the LMC. This could be a problem for our models. However, we believe the observations (14,15) of both the LMC and the SMC suggest a relatively larger population of supergiants in the temperature range $T_e \sim$ 10,000-20,000 K, compatible with the results of our calculations (3).

The alternative interpretation of these data is generally based on calculations of massive star evolution at high mass loss rates. Evolutionary tracks for massive stars of solar composition undergoing mass loss, reviewed by Chiosi and Maeder (7), reveal a more rapid evolution to the red subsequently followed by quite significant blueward excursions. For such models, the breadth of the blue supergiant region (and its relative population) is generally attributed to the presence of these blue loops. The occurrence of an immediate blue supergiant progenitor of Supernova 1987A is not easy to understand in this context, however. Given the fact that mass loss effects are not expected to be too significant for stars of mass ~15-20 $M_\odot$, we note that it is not easy to understand the quantitative and qualitative differences between our calculations (together with those of Brunish and Truran (4,5), Lamb et al. (18), and Arnett (1)) and those reviewed by Chiosi and Maeder (7). Further studies of the advanced stages of stars in this mass range seem essential.

V. Conclusions

We draw the following conclusions on the basis of the stellar evolution calculations reported and discussed in this paper:

(1) The evolution of a star of mass in the range ~15-20 $M_\odot$ and initial metal composition $Z = 1/4\ Z_\odot$, compatible with the metallicity of the LMC, is consistent with the occurrence of a blue supergiant progenitor for Supernova 1987A. Indeed, the predicted properties of such stars are quite in agreement with those observed for the presumed stellar progenitor Sanduleak -69 202.

(2) The envelope structure for such a stellar configuration is significantly more compact and exhibits a steeper mean density gradient and smaller photospheric radius. Such a structure has been found to be consistent with the spectral and light curve development of Supernova 1987A (1,9,10,22,24,27,32,33).

(3) A metallicity spread of a factor of a few for the young stellar component of the LMC is sufficient to allow an understanding of both (i) the presence of red supergiants at luminosities compatible with those of ~15-20 $M_\odot$ stars (14,15) and (ii) a blue supergiant progenitor of mass ~15-20 $M_\odot$ for Supernova 1987A. It seems to us, generally, to be more difficult to understand how blue loops can yield a blue supergiant presupernova configuration. The presence of

a significant population of red supergiants in the critical luminosity range in the SMC (14), in a population of somewhat lower average metallicity, suggests however that the problem may be more complicated.

Acknowledgements

This research was supported in part by the United States National Science Foundation under grant ASI 86-11500 at the University of Illinois. J.W.T. wishes to express his thanks to the Alexander von Humboldt Foundation for support by a U.S. Senior Scientist Award and to Professor R. Kippenhahn for the hospitality of the Max-Planck-Institut für Astrophysik, Garching bei München.

References

1. Arnett, W.D., 1987, preprint
2. Bencivenni, D., 1987, this volume and private communication
3. Brunish, W.M., Gallaher, J.S., and Truran, J.W., 1986, Astron. J. 91, 598
4. Brunish, W.M., and Truran, J.W., 1982a, Astrophys. J. 256, 247
5. Brunish, W.M., and Truran, J.W., 1982b, Astrophys. J. Suppl. 49, 447
6. Chiosi, C., 1978, in Philip and Hayes (eds.), The HR Diagram, IAU Symp. No. 80, Reidel
7. Chiosi, C., and Maeder, A., 1986, Ann. Rev. Astron. Astrophys. 24, 329
8. Chiosi, C., and Summa, C., 1970, Astrophys. Space Sci. 8, 478
9. Grassberg, E.K., Imshennik, V.S., Nadyozhin, D.K., and Utrobin, V.P., 1987, preprint
10. Hillebrandt, W., Höflich, P., Truran, J.W., and Weiss, A., 1987, Nature, in press
11. Hübner, W.F., 1979, private communication
12. Hübner, W.F. and Magee, N.H., Jr., 1983, private communication
13. Hübner, W.F., Merts, H.L., Magee, N.H. Jr., and Argo, M.F., 1977, Los Alamos Sci. Lab. Rept. LA 6760 M
14. Humphreys, R.M., 1983, Astrophys. J. 265, 176
15. Humphreys, R.M. and Davidson, K., 1979, Astrophys. J. 232, 409
16. Isserstedt, J., 1975, Astron. Astrophys. Suppl. 19, 259
17. Kippenhahn, R., Weigert, A., and Hofmeister, E., 1967, in Alder, Fernbach and Rotenberg (eds.), Methods in Computational Physics, Academic Press, New York, vol. 7, p. 129
18. Lamb, S.A., Iben. I., Jr., and Howard, W.M., 1976, Astrophys. J. 207, 209
19. Maeder, A., 1981, Astron. Astrophys. 102, 401

20. Panagia, N., Gilmozzi, R., Clavel, J., Barylak, M., Gonzalez Riesta, R., Lloyd, C., Sanz Fernandez, de Cordoba, L., and Wamsteker, W., 1987, Astron. Astrophys. 177, L25
21. Peimbert, M., and Torres-Peimbert, S., 1974, Astrophys. J. 193, 327
22. Shaeffer, R., Casse, M., Mochkovitch, R., and Cahen, S., 1987, preprint
23. Shelton, I., 1987, IAU Circular Number 4316
24. Shigeyama, T., Nomoto, K., Hashimoto, M., and Sugimoto, D., 1987, preprint
25. Thomas, H.-C., 1967, Z. f. Astrophys. 67, 420
26. Truran, J.W., Höflich, P., Weiss, A., and Meyer, F., 1987, Messenger (ESO) , 47, 26
27. Wampler, E.J., Truran, J.W., Lucy, L.B., Höflich, P., and Hillebrandt, W., 1987, preprint
28. Weaver, T.A., Zimmerman, G.B., and Woosley, S.E., 1978, Astrophys. J. 225, 1021
29. Weiss, A., 1987, Astronomy and Astrophysics, in press
30. West, R.M., Lauberts, A., Jørgensen, H.E., and Schuster, H.-E., 1987, Astron. Astrophys. 177, L1
31. Wilson, J.R., Mayle, R., Woosley, S.E., and Weaver, T.A., 1986, Ann. N.Y. Acad. Sci., in press
32. Woosley, S.E., Pinto, P.A., and Ensman, L., 1987, preprint
33. Woosley, S.E., Pinto. P.A., Martin, P.G., and Weaver, T.A., 1987, preprint.

A FEW COMMENTS ON THE EVOLUTIONARY HISTORY OF SN 1987a BEFORE EXPLOSION

Alvio Renzini
European Southern Observatory, Garching b. München

Although not yet certainly proved, I will adopt here the view that Sanduleak –69 was a single star, and has exploded to become SN 1987a. This connection has generated a little surprise in some people, as it was thought that a B3 supergiant could not explode. It has then been argued that the lower metallicity prevailing in LMC should force $\sim 15M_{\odot}$ stars to spend their whole lifetime in the blue side of the HR diagram, and then, indeed, explode as blue supergiants (Hillebrandt et al. 1987; Arnett 1987).

I have two comments concerning the former reaction. The first, very general and *philosophical*, is that SN explosions are set up in the deep core of single stars, and the physical state of the deep core has very little connection with the surface layers, i.e. with either the stellar radius or the spectral type. Moreover, particularly in massive stars the model stellar radius turns out to be extremely sensitive to virtually any change in either the input physics or composition. There is a deep physical reason for this inclination of stellar envelopes towards exaggerated reactions to minor stimuli. Indeed, just out of the main sequence band, radiative stellar envelopes are subject to thermal instabilities caused primarily by the metal contribution to opacity around 10^6K (Renzini 1984). Major manifestations of these instabilities are the runaway expansion from the main sequence band towards red giant dimensions, and the runaway retreats from such dimensions which can occur during the core helium burning phase. But otherwise these radiative envelope are in nearly neutral equilibrium, which explains why the slightest change (e.g. in the luminosity released by the burning core) can trigger a runaway expansion or contraction. This is to say that, when allowance is made for the current uncertainties in e.g. mass loss, opacity, mixing, etc., the theoretician cannot predict from first principles the stellar radius at the moment of the SN explosion, and therefore he cannot be surprised if on occasion a B3 supergiant blows off.

The second point that I would like to make starts by drawing your attention to Figure 1 in Maeder (1984). Having parametrised mass loss in some reasonable way, Maeder finds that (for solar metallicity) stars less massive than $\sim 15M_{\odot}$ explode as red supergiants (RSG), those in the mass range between roughly 30 to $60M_{\odot}$ also become RSGs, but severe mass loss drives them back to the blue where they explode as Wolf-Rayet stars, and finally stars above $\sim 60M_{\odot}$ never become RSGs, but evolve directly to the Wolf-Rayet stage, and then explode. This scenario is in fair agreement with what we know about the evolution of galactic massive stars. One will notice that the transition between RSG and WR exploders is somewhere between 15 and $30M_{\odot}$. Again, there is no wonder if core collapse hits a $\sim 20M_{\odot}$ star midway in its journey from the RSG to the (missed!) WR configurations. The objection that this should be a rare event can just be paired with the widespread comment that "this supernova is different from all others observed so far".

More serious is perhaps the objection that Maeder's figure applies to the Milky Way, while LMC is known for being more poor in metals. For $Z = Z_{\odot}/4$, $\sim 15M_{\odot}$ stars may indeed spend their whole lifetime in the blue side of the HR diagram (cf. Hillebrandt et al. 1987; Arnett 1987). This follows from the reduced contribution of metal opacity around 10^6K, which tends to suppress the mentioned thermal instability of the envelope. A closer look to LMC massive stars may however help in better understanding the past history of the supernova. This can be done by ideally placing Sanduleak -69 in the HR diagram of LMC bright stars given by Humphreys (1984, her Figure 2). Adopting $M_{bol} \simeq -7.8$ and $Log\, T_{eff} \simeq 4.1$ for Sanduleak -69, we can then see this star surrounded by quite a few of similar temperature and luminosity, we also see that a mass of around $20M_{\odot}$ looks appropriate, and last (but not least) we see that RSGs populate the diagram up to $M_{bol} \simeq -9.2$. From this we can conclude that in LMC the bulk of stars less massive than $\sim 40M_{\odot}$ do indeed become RSGs, and then the circumstantial evidence favours the idea that Sanduleak -69 has also experienced her RSG phase, before migrating back to the blue, most likely thanks to severe mass loss in the RSG phase itself. This argument cannot completely exclude the *always-blue* interpretation for the past history of Sanduleak -69, as one could still argue that this star was more metal poor than the average.

Is there any way of assessing whether Sanduleak -69 did actually experience a RSG phase before evolving to her pre-explosion configuration? I can see two possibilities, namely: i) light echos of the SN light on the remnant RSG wind, and ii) a prompter appearance of helium rich layers than mass conservative models would predict. Concerning i), the remnant RSG wind, perhaps now in form of Rayleigh-Taylor knots thanks to the action of the B3 wind, is expected to lie at a distance

$$D = \sim 1pc \left(\frac{v_{RSGW}}{10\, km/s}\right)\left(\frac{\Delta t}{10^5 yr}\right)$$

form the SN, where v_{RSGW} is the RSG wind velocity, and Δt is the time elapsed since the star left the RSG region. Clearly, the chance of observing some faint light echos in a few years from now is crucially dependent on Δt. Concerning ii), apart from modelling difficulties, an early disappearance of Balmer lines from the SN spectrum could just indicate that the precursor lost some fraction of its mass, without preference for the evolutionary phase at which this could have taken place. However, a very massive hydrogen envelope might suggest that substantial mass loss during a RSG phase did not take place.

REFERENCES

Arnett, W.D. 1987, preprint

Hillebrandt, W., Höflich, P., Truran, J.W., Weiss, A. 1987, *Nature*, in press

Humphreys, R.M. 1984, *Observational Tests of Stellar Evolution Theory*, ed. A. Maeder and A. Renzini (Dordrecht: Reidel), p. 279

Maeder, A. 1984, *ibid*, p. 299

Renzini, A. 1984, *ibid*, p. 21

MODEL CALCULATIONS FOR SCATTERING DOMINATED ATMOSPHERES AND THE USE OF SUPERNOVAE AS DISTANCE INDICATORS

P.Höflich
Max-Planck-Institut für Physik und Astrophysik, Institut für Astrophysik, Karl Schwarzschild Str. 1, 8046 Garching, FRG

Summary

We present results of calculations of scattering dominated atmospheres in order to interpret the spectra of type II supernovae. We assume spherical geometry and a density profile which is either a power law or given by the expansion of a stellar structure. For hydrogen up to 8 energy levels are allowed to deviate from LTE. Effects of sphericity are most important for the continuum forming regions and are responsible for much higher colour temperatures in the optical wavelength range than the corresponding effective temperature would imply. Whereas non-LTE effects are small for the continuum flux in the optical range, they become most important for the hydrogen lines. The strong influence of line blanketing on the observed spectra is demonstrated. This effect can lead to significant errors in the interpretation of the spectra and in the Baade Wesselink method for using supernovae as distance indicators. The application of our model on the SN1987a imply a distance of 46 $\pm$ 4 kpc for the LMC.

1. Introduction

Supernova explosions are spectacular events which have called attention of of astronomers since a long time. This objects are divided in mainly two subclasses on the basis of spectral criteria. Type II supernova are those in which the lines of the Balmer series are clearly seen and dominate the optical spectra and which individually show wide spectral variations, whereas hydrogen cannot be observed in type I supernova (Panagia, 1985). In the following only type II supernovae are considered. These objects are expected a consequence of the evolution of massive stars with masses $\geq 8M_{\odot}$ and the initial event is believed to start with a core collaps.

Supernovae are the brightest single celestial objects and may reach the same luminosity at maximum as a whole galaxy. Therefore they can be used as candles (Kirshner and Kwan 1974) to determine the distances of galaxies by the Baade Wesselink method (Baade 1926, Wesselink 1946) and they can be applied to calibrate other methods for the distance determination of galaxies, especially to fix the constant H_o of Hubble's law which gives the relation between the observable expansion velocity and the distances of galaxies.

The Baade-Wesselink method uses ratios of observed fluxes at different wavelengths or the slope of the spectrum and the velocities at different times to determine stellar quantities such as the photospheric radius and the effective temperature by comparison with model predictions. Using the calculated intrinsic fluxes the distance of an object can be derived from the observed brightness. In principle the same atmospheric models may be applied as for normal stars or as for HII regions. However there are a number of difficulties due to the properties which distinguish the supernova envelopes substantially as well from stellar atmospheres and as from "classical" HII regions: i) The density structure cannot be assumed to be constant or given by the hydrostatic equation and ii) the dominance of scattering over absorption through a large fraction of the atmosphere, even in the continuum forming region. In addition the typical particle densities (10^9 *to* 10^{12} cm^{-3}) are much higher than in "classical" HII regions (10^1... 10^4 cm^{-3}) and lower than in stellar photospheres ($10^{13...17}$ cm^{-3}). Therefore one can neither make the common assumptions for atmospheres (i.e. local thermodynamical equilibrium (LTE), plane geometry) nor those for "classical" HII regions (occupation numbers given by the cases of Menzel and Baker (Menzel 1937), constant temperature, ionisation due only to photoionisation from the ground state, etc.). On the contrary, collisional excitation and photoionisation due to Balmer continuum photons are the most important processes for hydrogen in supernova envelopes as has already been shown for envelopes with about the same density (Höflich and Wehrse 1987).

To address these problems we have carried out calculations for models of low-density scattering dominated atmospheres. The assumptions of the models are described in the following section. The results of the computations are discussed and the relevants of such models for the use of supernovae as distance indicators is demonstrated in section 3. In the last section we give a short conclusion of the main results.

2. Model characteristics

In order to get more detailed information we have applied a modified computer code (Höflich et al. 1986, Höflich and Wehrse 1987) for the construction of spherical extended non-LTE models. It was tried to use as few approximations in view of the physical treatment as possible and to reduce the number of free parameters.

We assume stationarity, which is a a reasonable approximation because the radiative timescales (on the order of a few minutes) are much shorter than the hydrodynamic timescales soon after the initial increase of the luminosity. Spherical symmetry is assumed and the density profile is taken to be either a power law (Weaver and Woosley, 1980)

$$\rho(r) \propto r^{-n} \qquad n \approx 5...20,$$

or a self-similar expansion of the stellar structure of the assumed progenitor

$$\rho(r) = \lambda^{-3}\, \rho(R) \text{ and } r = \lambda\, R,$$

where R is the radial distance in the progenitor. Because the kinetic energy of matter in the envelope remains nearly constant a homologous expansion is assumed. Consequently the velocity is a linear function of the distance r. Note that the second density law corresponds to a power law in which n is a function of the distance in the envelope corresponding to a time dependent n of the seen photosphere. The envelope is assumed to consist of hydrogen. Eight levels are allowed to deviate from LTE. Bound-bound and bound-free transitions have been included in the rate equations. Bound-bound opacities of the same transitions are also included in the radiation transport equation. In addition the opacities due to the higher members of the Balmer series are treated as part of the Balmer continuum. This is a reasonable approximation because higher levels than 6 are near to LTE. Continuum opacities from hydrogen (bound-free and free-free) and Thomson scattering are taken into account. Radiative equilibrium is assumed for the whole photosphere to determine the temperature profile.

3. Discussion of the model calculations

The calculations discussed here have been carried out in collaboration with G. Shaviv and R. Wehrse (Höflich et al. 1986) and more recently in collaboration with A.Weiss, J.Truran and W. Hillebrandt in order to develope a model for SN1987a in the Large Magellanic Cloud (LMC) (Hillebrandt et al. 1987, Höflich et al. 1987).

Several different definitions of the optical depths are relevant for this problem. For scattering dominated atmospheres they are the following: (i) τ_{abs} is the optical depth for true absorption. $\tau_{abs} \approx 1$ corresponds to the innermost layers from which photons can be observed. (ii) τ_{sc} is the optical depth in scattering and it is about equal the extinction optical depth τ_{ext}. $\tau_{sc} \approx 1$ occurs much higher in the atmosphere. The corresponding diameter would be measured as the diameter of the supernova if the envelope is resolved in angle. (iii) $\tau_{gen} \approx \sqrt{\tau_{sc} * \tau_{abs}}$. In the layers where $\tau_{gen} \approx 1$ most of the photons which are observed in the emitted continuum are generated.

In Table 1 free model parameters and some derived physical quantities are listed for models, which have power law density gradient $\propto r^{-10}$ and an effective temperature T_{eff} of 8000 K. In the second column we give the photospheric radii R_{5000} as input parameter for the models, the corresponding bolometric luminosities L_{bol} and the mass densities $\rho(R_{5000})$. R_{5000} is defined as the distance at which the optical depth from outside equals one at 5000 Å for true absorption. Note that this photospheric radius and therefore T_{eff} is only a definition for extended atmospheres but that the use of τ_{abs} results an upper limit for T_{eff} in respect to the other definitions of the optical depths (see above).

Table 1: Free model parameters and some derived physical quantities of supernova envelopes for $T_{eff} = 8000$ K, a density profile $\rho(r) \propto r^{-10}$ and $v(R_{5000}) = 8000$ km/sec. (R_{5000}: photospheric radius; L_{bol}: bolometric luminosity; $\rho(R_{5000})$: mass density; τ_{sc}: optical depth for Thomson scattering; $v(R_{5000})$: velocity; optical depth for true absorption; T_C(5000 Å) : colour temperature at 5000 Å).

Model number	R_{5000} (cm)	L_{bol} (erg/sec)	$\rho(R_{5000})$ (g/cm^3)	τ_{sc} (R_{5000})	T_C(5000 Å) (K)
1	$1\ 10^{14}$	$2.92\ 10^{40}$	$1.83\ 10^{-12}$	7.9	9000
2	$5\ 10^{14}$	$7.30\ 10^{41}$	$9.13\ 10^{-13}$	20.3	9200
3	$1\ 10^{15}$	$2.92\ 10^{42}$	$7.64\ 10^{-13}$	32.9	9400
4	$3.5\ 10^{15}$	$3.58\ 10^{43}$	$5.81\ 10^{-13}$	94.5	9700

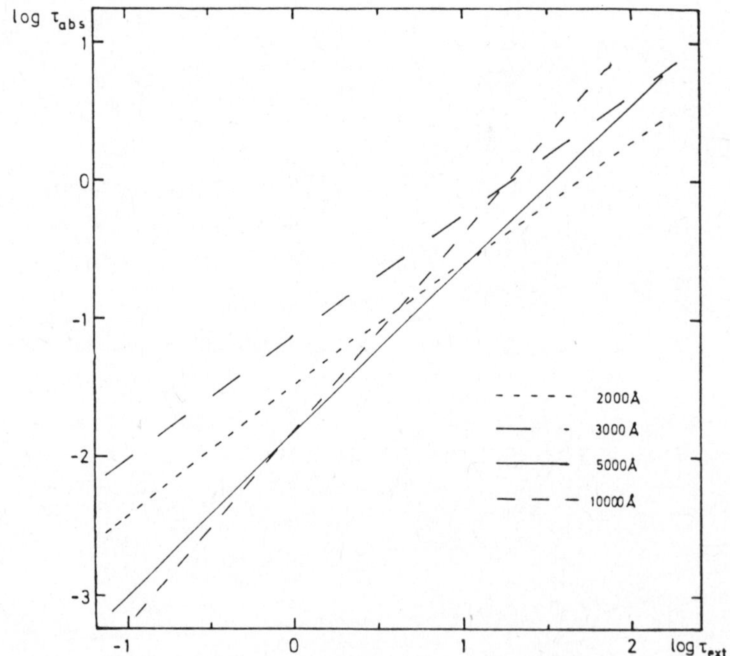

Figure 1: The run of τ_{abs} as a function of $\tau_{ext} \approx \tau_{sc}$ of the model with $R_{5000} = 1\ 10^{15}$ cm (see Table 1)

The scattering optical depths dominates τ_{abs} as can be seen from column 5 of Table 1 and Figure 1. In this figure the depths for true absorption at several wavelengths as a function of the optical extinction depth τ_{ext} of model 3 is shown (see Table 1). Most of the emitted photons are generated at depths between 0.3 and 0.8 in true absorption corresponding to scattering depths of about more than 10 depending on the wavelength.

The temperature profiles of the models 1 and 3 and the grey temperature slope are given in Figure 2. The temperature gradients differ significantly from the grey solution for all models but show an increasing departure with the stellar radius mainly due to the extension effect. The change in the temperature of the continuum forming region between 3000 Å and 5000 Å is about 1.15 in model 3.

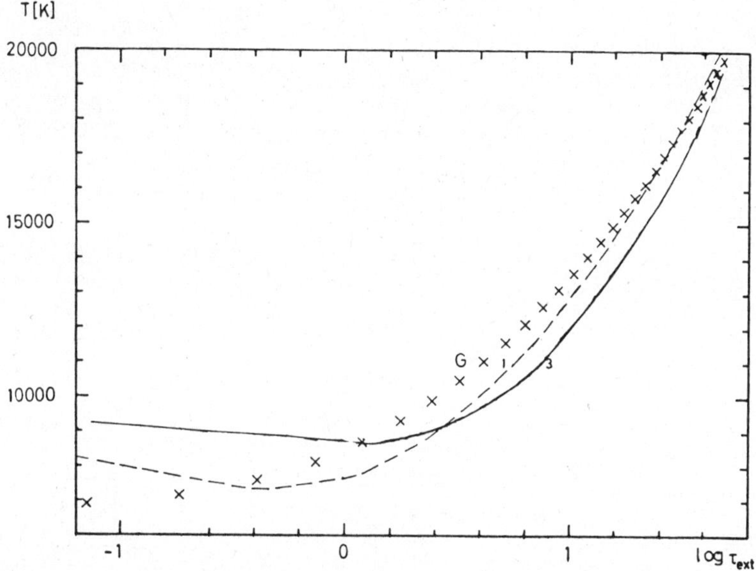

Figure 2: The temperature profile as a function of $\tau_{ext} \approx \tau_{sc}$ of the models 1 and 3 (see Table 1) in comparison with the grey solution (x) marked by G.

To demonstrate the influence of non-LTE effects on the continuum forming region we have given the maximum departure coefficient of hydrogen as a function of τ_{ext} for the models of Table 1 (see Figure 3). In addition the values of the maximum departure coefficient at various optical depths (0.1 and 0.5) for true absorption at several wavelengths are marked. Obviously the non-LTE effects fanish outside the continuum forming region. This behaviour can be understood as due mainly to the high optical depth in scattering which cause a nearly isotropic radiation field and to a lesser extent by the flatter temperature profile due to sphericity. These effects restore LTE for scattering dominated atmospheres. Note that non-LTE becomes most important for the lines because they are formed above the continuum forming region.

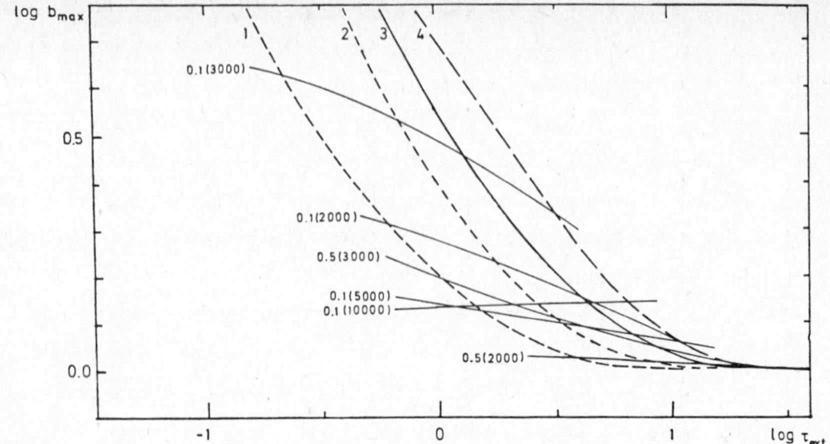

Figure 3: The run of the maximum departure coefficient as a function of $\tau_{ext} \approx \tau_{sc}$ for the models given in table 1. The additional curves give various optical depths for true absorption as a function of wavelength and marked as (τ_{abs}/wavelength).

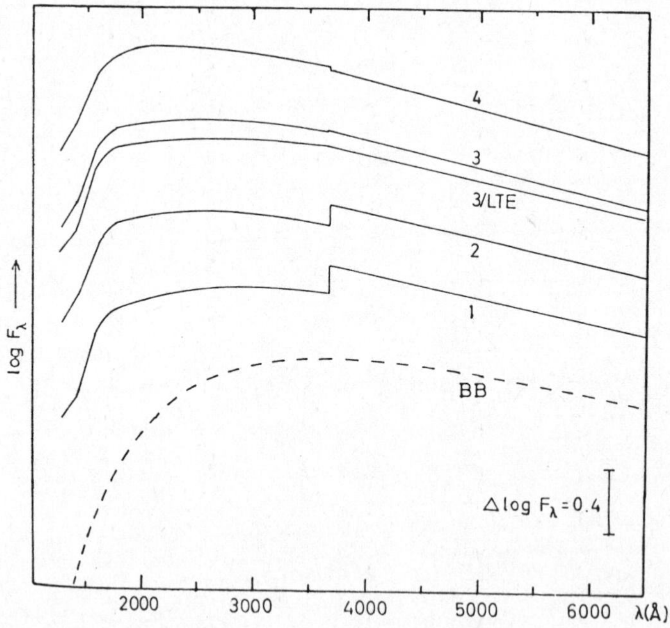

Figure 4: The flux distribution as a function of wavelength of the models given in table 1. The spectrum 3/LTE refers to model 3 calculated under the assumption of LTE. The curve marked BB is the slope of the black body function for the same T_{eff}.

In Figure 4 the scans of the calculated continuum fluxes are shown. The basic effect found by Shaviv et al. (1984), that the spectra appear to have higher colour temperatures in the UV, is clearly seen. But even in the optical wavelength range the colour temperature is much higher than would be expected if the slope of the spectrum is fitted by a black body (see Table 1; Kirshner and Kwan 1974). This implies that the monochromatic luminosity as derived from the colour temperature would be significantly overestimated of about a factor of 1.4 to 2 by using a black body function. This corresponds to an error in using the Baade-Wesselink method to determine distances of about 20 to 40 % . Thus the Hubble constant H_o would be underestimated by using the observed brightness of supernovae in galaxies as a distance indicator.

Therefore sophisticated models have to be used to yield correct intrinsic luminosities for an observed supernovae. Because type II supernova show strong differences in their spectra every supernova observation has to be interpreted individually.

But we want to point out that the spectra in the UV, the optical and IR wavelength range are strongly effected by line blanketing. In the optical range the blanketing is due to hydrogen lines mainly, even in very early stages of the evolution. For demonstration a calculated spectrum of a very early stage of about 0.9 days after the initial event of SN1987a is shown in Figure 5.

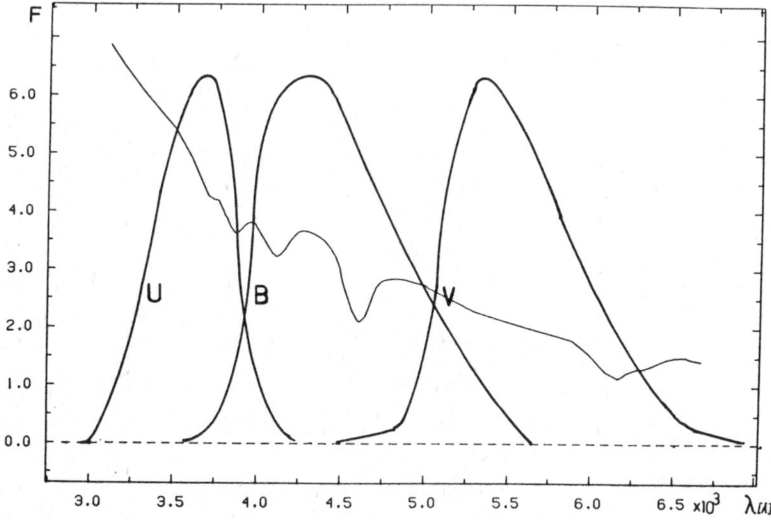

Figure 5: The relative flux as a function of wavelength as calculated by a hydrogen line blanketed model ($T_{eff} = 14000\ K$; $R_{5000} = 1\ 10^{14}\ cm$; $v(R_{5000}) = 21500\ km/sec$). The density profile is determined by the homologous expansion of a B3I star with an expansion factor of 50. In addition we give the transmission functions in the UBV colour system (Johnson, 1966)

To interpret observations by continuum slope of the spectra, only the very small wavelength range between 5200 and 5700 Å can be used. This clearly demonstrates the importance of line blanketing, even if the measured colours in broad band filters such as the UBV system of Johnson (1966) are used. Because the hydrogen lines are formed in layers further out than the continuum non-LTE becomes most important as mentioned above. In addition this is true for lines of heavier elements which contamine the flux in the UV as Lucy has shown (Lucy 1987).

To fit this early stage of SN1987a we have used the observed UBV colours (IAU Circular 4316, Hillebrandt et al. 1987). The distance of SN1987a has been derived by the calculated luminosity in the V filter. Assuming an interstellar reddening of $A_V = 0.45^m$ (Wampler et al. 1987), and $m_V = 4.81...5.1^m$ (IAU Circular 4316) we get an distance of 46 ± 4 kpc for this supernova which is about the distance of the LMC. This value is in agreement with distances of the LMC (42...56 kpc) as derived by other methods (Andersen et al. 1985, Walker,A.R. 1985) but clearly supports to the lower distances.

4. Conclusion

The results of our calculations are the following. The continuum forming region in a supernova atmosphere is mainly influenced by extension effects and the run of temperature. Non-LTE effects are less important for the formation of the continuum, but they become very important for the hydrogen lines because they are formed in outer photospheric layers than the continua.

The effective temperature is much lower than the observed colour temperature in the optical wavelength range would imply. The determination of the effective temperature by using a black body function would yield a strong overestimate of the luminosity. This results an underestimate of the Hubble constant by about 20 ... 40 % .

Non-LTE effects and line blanketing have to be taken into account for the interpretation of the observed spectra of supernovae if they are used as distance indicators. Because the type II supernovae are a very heterogeneous group with respect to the observed luminosity and to the spectral behaviour, supernova spectra have to be interpreted individually to be usefull as distance indicators. With this model we determined the distance of SN1987a and the LMC as 46 ±4 kpc.

References

Andersen,J., Blecha,A., Walker,M.F., *Astron. Astrophys.* **150** L12 (1985)

Baade,W. *Astron. Nachrichten* **228** 359 (1926)

Hillebrandt,W., Höflich,P., Truran,J.W., Weiss,A. submitted to *Nature* (1987)

Höflich,P., Wehrse,R., Shaviv,G. *Astron. Astrophys.* **163** 105 (1986)

Höflich,P., Wehrse,R. *Astron. Astrophys.* in press (1987)

Höflich,P., Weiss,A., Hillebrandt,W., and Truran,J.W. in preparation (1987)

Johnson,H.L. *Ann.Rev.Astron.Astrophys.* **4** 197 (1966)

Kirshner,R., Kwan,J. *Astrophys. J.* **193** 27 (1974)

Lucy,L.B private communication (1987)

Menzel,D.H. *Astrophys.J.* **85** 330 (1937)

Panagia,N. in "Supernovae as Distance indicators", *Lecture notes in physics* **224**, Springer Verlag, Berlin Heidelberg New York Tokyo (1985)

Shaviv,G., Wehrse,R., Wagoner,R.V. *Astrophys. J.* **289** 198 (1984)

Walker,A.R. *Mon.Not.Roy.astr.Soc.* 212 343 (1985)

Wampler, E.J., Truran, J.W., Lucy, L.B., Höflich, P., and Hillebrandt, W. *Nature* in press (1987)

Weaver,T.A., Woosley in "Supernova Spectra" *A.I.P.Conf.Proc.No.* **63** (1980)

Wesselink,A.J. *Bull.Astron.Inst.Neth.* **368** 91 (1946)

SYNTHETIC SPECTRA FOR SUPERNOVAE II

W. Spies, P. Hauschildt, R. Wehrse, B. Baschek
Institut für Theoretische Astrophysik, Universität Heidelberg
Im Neuenheimer Feld 561, D-6900 Heidelberg

G. Shaviv
Dept. of Physics, Technion, Israel Institute of Technology
IL-32000 Haifa

ABSTRACT

Model atmospheres for supernovae of type II have been calculated taking into account the effects of sphericity and velocity fields. We obtain a good fit for the energy distribution of the recent supernovae SN 1980 K and SN 1987 A.

1. INTRODUCTION

The spectrum emitted from the photosphere of a supernova contains a large amount of information on the nature of the explosion and on the metal enrichment and the heating of the interstellar matter. For a quantitative interpretation of supernova spectra a detailed determination of the photospheric properties (effective temperature, velocity field, density profile, chemical composition) is essential. However, due to the large geometric extension of the photosphere and the high velocity fields involved, the spectra are very complex and cannot be analyzed in a simple way (as e.g. by means of a standard analysis).

In this paper we report new results of an attempt to study the spectra by means of detailed photospheric models following Shaviv et al. (1984).

2. PHYSICAL MODEL AND ASSUMPTIONS

The supernova photosphere is calculated with the following assumptions:

(i) the configuration is spherical,

(ii) the density follows a power law with an exponent $n = 10$,

(iii) LTE (including scattering) holds,

(iv) no energy is generated in the atmosphere, i.e. radiative equilibrium holds,

(v) the expansion velocity increases proportional to the radius (free coasting atmosphere).

All relevant continuous opacity sources and (in some models) several thousands of the most important spectral lines are taken into account.

The radiative transfer equation for the moving atmosphere is solved for about 4000 wavelength points, most of which are in the UV and the visible.

3. EXAMPLES OF CALCULATED ENERGY DISTRIBUTIONS AND COMPARISON WITH OBSERVATIONS

In Fig. 1 the UV spectrum of SN 1980 K (Benvenuti et al., 1982) is shown together with two continuum models with effective temperatures T_{eff} = 10500 K and T_{eff} = 12000 K. For both energy distributions a radius of 10^{15} cm and a velocity $v_{exp} = 10^4$ km s^{-1} at an optical depth of absorption $\tau_{abs}(\lambda 5000$ Å$) = 1$ are assumed. The composition is solar. The calculated energy distribution is reddened corresponding to E(B-V) = 0.34.

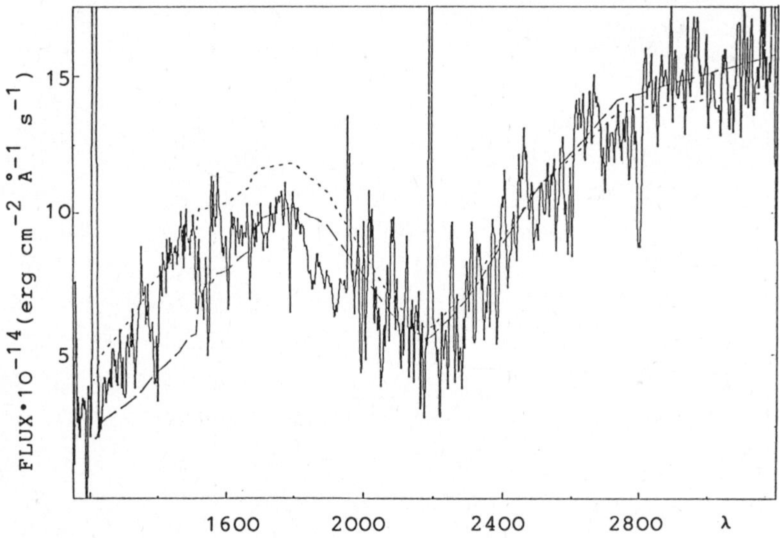

Fig. 1. Comparison of the observed IUE spectrum of the SN 1980 K on October 30, 1980, with calculated continuous energy distributions for T_{eff} = 10500 K (broken curve) and T_{eff} = 12000 K (dotted curve).

It is seen that the calculated spectra follow the slopes of the observed distribution remarkably well and that the effective temperature of the supernova is certainly bracketed by 10500 and 12000 K. Fig. 2 shows the UV fluxes of the recent LMC supernova 1987 A (Kirshner and Sonneborn, 1987) and the energy distribution of a blanketed model with T_{eff} = 8000 K and $v_{exp} = 1.5.10^4$ km s^{-1}. The reddening corresponds to E(B-V) = 0.20. Note the agreement with the broad dip around 1700 Å and the peaks in the range 1300 to 1600 Å. It is not possible to attri-

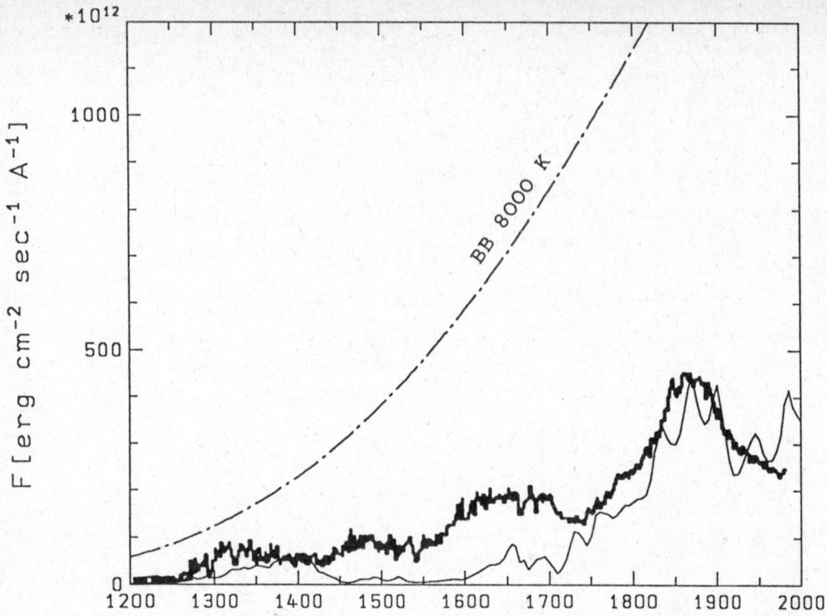

Fig. 2. Comparison of the observed IUE spectrum of the SN 1987 A on February 26, 1987, with the calculated energy distribution of a line-blanketed model for T_{eff} = 8000 K (thin line).

bute an observed feature to a single species or even line because of the large velocity gradient in the atmosphere and the corresponding smearing out of lines. It is clear, however, from the line statistics that the observed depressions are due to lines from iron group elements.

Additional details are given in Hauschildt et al. (1987) and Spies et al. (1987).

ACKNOWLEDGMENT

We thank Dr. R.P. Kirshner for allowing us to use his observational data of the SN 1987 A in advance of publication. This work was supported in part by the Deutsche Forschungsgemeinschaft (Sonderforschungsbereich 328).

REFERENCES

Benvenuti, P., Sanz Fernandez de Cordoba, L., Wamsteker, W., Macchetto, F., Palumbo, G.C., Panagia, N.: 1982, An Atlas of UV Spectra of Supernovae, ESA SP-1046

Hauschildt, P., Wehrse, R., Shaviv, G.: 1987, submitted to Astrophys.J.

Kirshner, R.P., Sonneborn, G.: 1987, Astrophys. J. (Letters), in press

Shaviv, G., Wehrse, R., Wagoner, R.V.: 1984, Astrophys. J. 289, 198

Spies, W., Hauschildt, P., Wehrse, R., Shaviv, G.: 1987, submitted to Astrophys. J.

MONTE CARLO METHODS FOR NEUTRINO TRANSPORT IN TYPE-II SUPERNOVAE

Hans-Thomas Janka
Max-Planck-Institut für Physik und Astrophysik , Institut für Astrophysik
Karl-Schwarzschild-Str. 1 , D-8046 Garching bei München
Federal Republic of Germany

Abstract : *Neutrinos play an important role in the type-II supernova scenario. Numerous approaches have been made in order to treat the generation and transport of neutrinos and the interactions between neutrinos and matter during stellar collapse and the shock propagation phase. However, all computationally fast methods have in common the fact that they cannot avoid simplifications in describing the interactions and, furthermore, have to use parameterizations in handling the Boltzmann transport equation. In order to provide an instrument for calibrating these treatments and for calculating neutrino spectra emitted from given stellar configurations, a Monte Carlo transport code was designed. Special attention was payed to an accurate computation of scattering kernels and source functions. Neutrino spectra for a hydrostatic stage of a 20* $M_\odot$ *supernova simulation were generated and conclusions drawn concerning a late time revival of the stalled shock by neutrino heating.*

1. Neutrinos and Supernovae

During the collapse of the central Fe-Ni- core of a massive star, neutrinos are emitted mainly by electron captures on protons. They leave the core — thus generating only a small entropy increase — as long as densities above about $10^{12}\ gcm^{-3}$ are not yet reached. However, the final value of the electron fraction $Y_e = n_e/n_B$, which emerges from this deleptonization process at the moment when neutrino trapping occurs, determines crucially the subsequent evolution of the collapsing star (see e.g. *Bruenn* 1985, 1986a, 1986b).

Over 97% of the gravitational binding energy of the stellar core (more than $10^{53} ergs$) is stored within the neutrino gas that leaks out of the dense center over time scales of a few seconds and transports energy, momentum, and lepton number through the outer layers of the collapsed object.

Although the principal physical picture seems to be understood quite well, most numerical simulations are unsuccessful in producing supernova explosions by the prompt bounce shock mechanism (*Hillebrandt* 1985). The initial shock is damped due to nuclear dissociations and additional neutrino losses. It thus changes to a standing accretion shock at a radial position between 100 and 200 kilometers instead of heating the outer layers and giving escape velocities to them. Indeed, an alternative explosion scenario is represented by the so called 'delayed explosion' mechanism (*Bethe* and *Wilson* 1985), which finds the weakened shock being revived by the hot neutrino flux that transfers energy from the neutrino sphere up to the region behind the shock. The efficiency of this heating process, however, depends sensitively on the neutrino temperature (*Lattimer* and *Burrows* 1984). It follows that only a thorough calculation of the neutrino interactions and transport through those stellar regions where neutrinos and matter gradually decouple as a function of the neutrino energy and where the surface neutrino spectrum is formed can yield evidence for the possibility of the shock being able to resume propagation and to move out successfully.

2. Characteristics and Problems of Neutrino Transport

Typical values for temperatures and densities in the core collapse scenario range from $10^{10}K$ to about $2...3 \cdot 10^{11}K$ and from some $10^9 gcm^{-3}$ up to more than $10^{14} gcm^{-3}$. Whereas in regions of matter density below around $3...4 \cdot 10^{10} gcm^{-3}$ neutrinos essentially stream freely, and at densities over about $5...6 \cdot 10^{11} gcm^{-3}$ (equilibrium) diffusion models neutrino transport quite well, the neutrino flux in the intermediate range cannot be described by so simple a picture. As neutrino scattering and absorption cross sections are of the order of $10^{-44} cm^2$, neutrino- nucleon interactions can neither be neglected completely nor are they capable of establishing local equilibrium. Moreover, the total cross sections are roughly proportional to the square of the neutrino energy, so the opacity of the matter is not the same for the whole population of neutrinos. As a consequence of this there is actually no unique 'neutrino sphere' in the sense of a well defined photosphere. Significant deviations from local equilibrium neutrino distributions must be expected under such circumstances.

The correct treatment of the general problem of neutrino transport through dense stellar matter requires the application of a complex set of equations (see e.g. *Schinder* and *Shapiro* 1986). Besides equations describing the hydrodynamical behaviour of the star, one needs *Poisson's* equation, an equation of state, conservation laws for baryon number, lepton number, energy, and momentum in the stellar gas, and finally the *Boltzmann* transport equation, which expresses the total change of the local neutrino distribution function as a result of all the interactions neutrinos undergo with the particles in the stellar gas. The latter relation reads in the case of spherical symmetry in an *Euler*ian coordinate frame:

$$\left(\frac{1}{c}\frac{\partial}{\partial t} + \frac{(1-\mu^2)}{r}\frac{\partial}{\partial \mu} + \mu\frac{\partial}{\partial r}\right) f_i(\epsilon,\mu,r,t) \;=\; \Gamma_i - \Lambda_i f_i \;, \qquad (2.1)$$

where i denotes the different species of neutrinos, $\mu = \hat{r}\cdot\hat{\nu}$ gives the cosine of the angle between the radial direction and the direction of neutrino motion, and $\epsilon = p \cdot c$ is the neutrino energy. Γ_i stands for the neutrino emissivity due to all sources and Λ_i for the total neutrino absorptivity. Both the much more complicated form in a comoving frame and the explicit dependence of the interaction kernels on the neutrino distribution function make it a numerically cumbersome, slowly converging integro- differential equation, which cannot be used in connection with full hydrodynamical computations of the supernova event. Instead, the usual procedure (see e.g. *Castor* 1972, *Arnett* 1977, *Mihalas* 1978, *Bludman* and *Van Riper* 1978, *Bowers* and *Wilson* 1982, *Hillebrandt* 1985) is to derive moment equations by integration over the angles and then to use a closure condition for constructing a diffusion equation. The diffusion coefficient has to be adjusted adequately so as to reproduce correctly the well known limits of diffusion and free streaming. For the local neutrino occupation functions, either equilibrium distributions or simple nonequilibrium assumptions (e.g. via definition of a special 'temperature' for the neutrino gas different from the local matter temperature) are common, or — in a computationally more expensive treatment — a multigroup representation is used. Moreover, the interaction kernels and source functions also need simplifications. These can involve a conservative handling of neutrino- nucleon interactions and a *Fokker- Planck*- approximation of the neutrino- electron scattering, in addition to the treatment of the electrons as extremely relativistic particles. Most crucial, however, seem to be the use of the flux- limiter in the expression for the diffusion constant and the representation of the neutrino distribution function, although there is no general agreement in the literature with regard to the validity and accuracy of the other approximations (e.g. *Lichtenstadt* et al. 1978, *Myra* et al. 1986, *Mayle* 1985, *Tubbs* 1978, 1979, *Tubbs* et al. 1980). As a manifestation of these uncertainties, neutrino spectra for the various supernova stages show strong dependence on the applied method of transport (see e.g. *Sato* and *Suzuki* 1987).

In order to be able to check individual influences, to calibrate the flux limiting parameter, and to perform thorough and accurate calculations of the neutrino spectra for given model situations, a Monte Carlo transport scheme was developed within the course of this work. It is a generalization of a work by *Tubbs* (1978), who used the method for equilibration simulations on an infinite background medium.

3. The Monte Carlo Method for Fermion Transport

In contrast to an integration of the *Boltzmann* transport equation in time, which means the persecution of the temporal changes of the statistical ensemble average represented by the particle distribution function, the Monte Carlo method seeks to *generate* the ensemble average by following the individual trajectories of a great sample of particles. The latter is achieved by deriving 'probability laws' for all the physical events the particles face on their ways through the stellar gas. In general the physical processes involved are statistical by nature (e.g. neutrino generation, scattering, absorption) and allow modelling of individual particle destinies via sampling the appropriate probability density functions with sequences of random numbers. For instance, if one has to select among several different competing interactions with the rates $R_1, R_2, R_3, ..., R_n$ (interactions per particle per second) this can be done by using the discrete distribution

$$\{p_i \ , \ i = 1, 2, 3, ..., n\} \quad = \quad \left\{ \frac{R_1}{R}, \frac{R_2}{R}, ..., \frac{R_n}{R} \right\} \quad , \tag{3.1}$$

where $R = \sum_{i=1}^{n} R_i$ gives the total interaction rate and p_i describes the probability of the particle being involved in process i. In an analogous fashion, the probability of a particle with initial energy ϵ scattering into the energy interval $[\epsilon', \epsilon' + d\epsilon']$ is represented by the continuous density function

$$p(\epsilon, \epsilon')d\epsilon' \ = \ \left(\frac{1}{R_S} \cdot \frac{dR_S}{d\epsilon'} \right) d\epsilon' \quad , \tag{3.2a}$$

where

$$R_S \ = \ \int \frac{dR_S}{d\epsilon'} \, d\epsilon' \tag{3.2b}$$

is the total scattering rate of particles of energy ϵ from a specified kind of target.

The application of interaction rates represents a characteristic procedure in handling the two interchanging gases. The transported neutrinos are followed explicitly on their ways through the stellar medium, while, instead of treating the stellar gas particles in a similar manner, the stellar gas can be considered as retaining the equilibrium configuration all the time, because the much faster electromagnetic and strong interactions are capable of damping perturbations very quickly. Conservation laws are either trivially fulfilled in the framework of a Monte Carlo scheme (e.g. particle numbers) or their validity is guaranteed in the limit of large test particle numbers as a consequence of statistical averaging (*Tubbs* 1978).

The Monte Carlo method shows some special characteristics when applied to the transport of fermions. If degeneracy is important, phase space blocking effects cannot be neglected. In order to generate the local phase space occupation function, it is therefore necessary to follow the whole ensemble of physical particles simultaneously. This can be accomplished by introducing 'sample' or 'test' particles, which have to be identified with values for energy, radial position, direction angle of motion, and a weighting factor counting the real number of physical particles represented by the test particle. The implicit assumption here is that the average behaviour of a bundle of many particles can be simulated by the destiny of the test particle. This submicroscopic averaging is valid as long as the distribution of real particles is sufficiently localized around the sample particle's properties (energy etc.), i.e. the bundle is tiny (*Tubbs* 1978).

In order to ensure that *Pauli*'s exclusion principle is obeyed, the neutrino phase space occupation function, which is used in determining the final state inhibition factor of the interaction rates (and influences the Monte Carlo transport *only* on that way), must be constructed properly. *Tubbs* (1978) proposed a formulation that can be generalized for the purposes of this work. The indices k, j shall label an energy- angle cell $[\epsilon_k, \epsilon_{k+1}] \times [\mu_j, \mu_{j+1}]$.

$$B_{kj} = n_{kj}/n_{kj}^{EQ} \tag{3.3}$$

with

$$n_{kj} = V^{-1} \cdot \sum_{i=1}^{N} W_i \cdot \Delta_{i,kj} \tag{3.4}$$

$$\Delta_{i,kj} = \begin{cases} 1\,, & \text{if particle } i \text{ is in cell } kj\,; \\ 0\,, & \text{otherwise} \end{cases}$$

and

$$n_{kj}^{EQ} = \frac{2\pi}{(hc)^3} \cdot (\mu_{j+1} - \mu_j) \cdot \int_{\epsilon_k}^{\epsilon_{k+1}} d\epsilon\; \epsilon^2 \cdot f_\nu^{EQ}(\epsilon) \tag{3.5}$$

then expresses the fraction of the equilibrium concentration to which the group is filled. Here

$$f_\nu^{EQ} = \frac{1}{1 + \exp[(\epsilon - \mu_\nu^{EQ})/k_B T]}\;, \tag{3.6}$$

with $\mu_\nu^{EQ} = \mu_e + \mu_p - \mu_n$, describes the chemical equilibrium occupation function, n_{kj} is the actual neutrino number density in group kj (W_i weight factor of individual test particle, V spatial volume), and n_{kj}^{EQ} stands for the according equilibrium value. Then the relation

$$[f_\nu(\epsilon)]_{kj} \equiv f_{kj}(\epsilon) = B_{kj} \cdot f_\nu^{EQ}(\epsilon) \tag{3.7}$$

defines an appropriate distribution function, which exhibits the correct equilibrium behaviour in the limit of $B_{kj} \longrightarrow 1$. Note that f_{kj} should not be hindered from slightly exceeding unity under some conditions. The competing effects of absorption and emission of neutrinos will guarantee the accurate value *on the average*. Nevertheless, in order to avoid negative values of the blocking factors $(1 - f_{kj})$, B_{kj} must be replaced by

$$B_{kj} \longrightarrow \min \left\{ B_{kj}\,,\; B_{kj}^{max} \equiv 1 + \exp\left[(\epsilon_k - \mu_\nu^{EQ})/k_B T\right] \right\}\;, \tag{3.8}$$

which is consistent with the exclusion principle.

4. Supernova Model and Transport Simulations

For the purpose of evaluating the possibility that the stalled shock might be revived by energy deposition due to neutrinos in the stellar gas behind it, the neutrino flux from the collapsed center of a $20M_\odot$ star (*Hillebrandt* 1985) was investigated. This particular stage of the supernova event — around $12ms$ after core bounce — is characterized by a standing accretion shock at a radius of approximately $140\,km$, which corresponds to a mass shell of $1.3M_\odot$. The matter inside has achieved almost hydrostatic conditions and remains electron rich, so that positrons can be neglected and electron neutrinos dominate all other kinds by a significant factor. Because of the high temperatures generated in the shock, nuclei are fully dissociated into free nucleons. For the transport simulation, a window between an inner radius of $30km$ and the shock position as outer boundary was chosen. The density falls smoothly in this region from about $10^{12}gcm^{-3}$ to about $3...4 \cdot 10^{11}gcm^{-3}$ at $r \approx 75km$, and then exhibits a steeper slope down to approximately $5 \cdot 10^{9}gcm^{-3}$ at the position of the shock. Table 1 summarizes the most important quantities as a function of the radial position. Note that the electron fraction shows a rapid increase just outside the mean neutrino sphere. These special features of thermodynamic and composition parameters of the stellar medium explain the characteristics of the behaviour of the weak reaction rates and mean free flight times listed in table 2, especially the rapid decreases of the neutrino emission rate and of the absorption and neutrino- electron scattering mean free flight times.

At the inner edge neutrinos diffuse into the region of interest here at a given constant rate during the course of the calculation. Because of the low densities, no incoming flux had to be considered at the outer boundary. The Monte Carlo runs were performed with a radial partition into 10 zones; a calculation with 15 zones revealed no differences. The sufficiency of this zoning is clear from the fact that the length scales of steep changes in the composition parameters are significantly smaller than the local mean free paths.

The neutrino distribution function was represented with the use of an energy- angle mesh of $(45-60) \times 10$ cells; a cut-off energy of $90MeV$ meant a neglect of not more than 1.2% of the neutrino spectrum, even in the regions of highest densities.

The calculations were performed on a background of spherical geometry (and symmetry) and the composition and thermodynamic quantities of the stellar gas were altered due to the exchange of energy and lepton number with the streaming neutrinos. The net neutrino number and energy flux at the inner boundary had values of about $1.6 \cdot 10^{58}s^{-1}$ and $8.3 \cdot 10^{53}ergs\,s^{-1}$, respectively. The stellar gas contained only electrons, protons, and neutrons; the weak interactions that therefore had

RZ	$\overline{R}$	ρ	Y_e	T	n_n	μ_e
	$[10^6 cm]$	$[10^{10} \frac{g}{cm^3}]$		$[MeV]$	$[10^{33} \frac{1}{cm^3}]$	$[MeV]$
1	3.79	82.80	0.233	10.185	379.31	18.472
2	4.93	55.55	0.188	8.296	269.40	15.066
3	6.07	52.05	0.128	5.193	271.06	16.631
4	7.21	39.79	0.123	3.075	208.40	17.164
5	8.34	19.65	0.191	2.277	94.94	16.232
6	9.48	5.58	0.221	2.427	25.94	10.262
7	10.62	2.74	0.296	2.366	11.53	8.577
8	11.76	1.33	0.386	1.939	4.89	7.509
9	12.89	1.21	0.447	1.804	4.01	7.869
10	14.03	0.771	0.467	1.557	2.45	6.889

Table 1 : Thermodynamic and composition parameters of the initial model ($t = 0\ s$). $\overline{R}$ denotes the mean radii of the zones, and the electron chemical potentials μ_e include the electron rest mass. The gas is neutral according to the relation $n_e = n_p$.

RZ	R_{Em}	$\overline{\epsilon_\nu}$	t_{Ab}	$t_{\nu n}$	$t_{\nu p}$	$t_{\nu e}$
	$[s^{-1} cm^{-3}]$	$[MeV]$	$[s]$	$[s]$	$[s]$	$[s]$
1	$2.45 \cdot 10^{40}$	33	$1.02 \cdot 10^{-6}$	$4.83 \cdot 10^{-6}$	$2.19 \cdot 10^{-5}$	$1.31 \cdot 10^{-5}$
2	$4.94 \cdot 10^{39}$	27	$2.10 \cdot 10^{-6}$	$1.02 \cdot 10^{-5}$	$6.01 \cdot 10^{-5}$	$3.60 \cdot 10^{-5}$
3	$9.93 \cdot 10^{38}$	18	$5.94 \cdot 10^{-6}$	$2.29 \cdot 10^{-5}$	$2.11 \cdot 10^{-4}$	$1.45 \cdot 10^{-4}$
4	$2.95 \cdot 10^{38}$	15	$1.59 \cdot 10^{-5}$	$4.34 \cdot 10^{-5}$	$4.17 \cdot 10^{-4}$	$4.24 \cdot 10^{-4}$
5	$1.27 \cdot 10^{38}$	12	$1.02 \cdot 10^{-4}$	$1.49 \cdot 10^{-4}$	$8.49 \cdot 10^{-4}$	$1.06 \cdot 10^{-3}$
6	$8.32 \cdot 10^{36}$	12	$1.08 \cdot 10^{-4}$	$5.45 \cdot 10^{-4}$	$2.58 \cdot 10^{-3}$	$2.29 \cdot 10^{-3}$
7	$2.99 \cdot 10^{36}$	12	$2.14 \cdot 10^{-4}$	$1.23 \cdot 10^{-3}$	$3.92 \cdot 10^{-3}$	$3.40 \cdot 10^{-3}$
8	$8.12 \cdot 10^{35}$	9	$9.12 \cdot 10^{-4}$	$5.12 \cdot 10^{-3}$	$1.10 \cdot 10^{-2}$	$9.23 \cdot 10^{-3}$
9	$8.55 \cdot 10^{35}$	9	$1.14 \cdot 10^{-3}$	$6.26 \cdot 10^{-3}$	$1.05 \cdot 10^{-2}$	$9.38 \cdot 10^{-3}$
10	$2.72 \cdot 10^{35}$	9	$1.64 \cdot 10^{-3}$	$1.02 \cdot 10^{-2}$	$1.59 \cdot 10^{-2}$	$1.51 \cdot 10^{-2}$

Table 2 : Characteristic time scales of the initial model (calculated for empty neutrino phase space). The values $t_{\nu a}$ are the mean free flight times between scatterings of neutrinos of energy $\overline{\epsilon_\nu}$ at particles of kind a, the t_{Ab} give the analogous numbers for neutrino capture on free neutrons. Hereby the average neutrino energy within each radial zone is approximately equal to $\overline{\epsilon_\nu}$. R_{Em} denotes the total emission rates of neutrinos with energies less than 90 MeV.

to be considered were neutrino generation by electron capture reactions, neutrino absorption on free neutrons, and neutrino scattering from electrons, neutrons, and protons. These were calculated without any limitations concerning energy and momentum exchange in the single event, i.e. the (nondegenerate) thermal distributions of the nucleons as well as the energy and momentum transfer to them were taken into account and the electrons were treated as an arbitrarily degenerate *Fermi* gas without neglecting their rest mass. The general expression for the weak interaction rate of a process $\nu + a \longrightarrow b + c$ reads

$$R(\epsilon,\mu,r,t) \;=\; g_a \cdot (2\pi)^{-9} \cdot \int d^3p_a\, f_a^{EQ} \int d^3p_b\,(1-f_b^{EQ}) \int d^3p_c\,(1-f_c^{EQ})\; |M_{a\nu\rightarrow bc}|^2 \quad , \quad (4.1)$$

where g_a is the spin degeneracy factor of particle a and $|M|^2$ is the square of the matrix element of the process, summed over final spins and averaged over the initial ones. If scattering is to be described ($b = a', c = \nu'$), f_c^{EQ} has to be replaced by f_ν. Note that f_ν will not necessarily be an equilibrium occupation function, while the other particles are assumed to be distributed according to the local equilibrium configurations. The evaluation of equation 4.1 proceeded essentially as described in the publications by *Tubbs* (1978), *Tubbs* and *Schramm* (1975), and *Yueh* and *Buchler* (1976a,b).

5. Results and Conclusions : Neutrino Spectra and Delayed Explosions

The neutrino flux and the phase space distributions had finally achieved a quasistationary state after a calculation period of about $6.5 \cdot 10^{-4}s$. This is suggested by the characteristic weak interaction time scales of the system (table 2) as well as by the typical transport time scale $\tau_T = (R_{out} - R_{in})/c \approx 3.8 \cdot 10^{-4}s$, which is the critical quantity for those zones for which local absorption and emission time scales are longer. The development of the neutrino number densities and the evolution of the surface properties — neutrino luminosity and mean neutrino energy (figs. 2a,2b) — show a rapid increase, followed by a relaxation towards certain values. A comparison of the number fluxes (see table 3) at both boundaries also confirms the accurate conservation of the flux, indicating a quasistationary behaviour. Furthermore, a simple three zone model of the stellar layers identifies the sufficient length of the simulation runs. Imagine that, in the interesting window of the star, there is an inner region of high densities ($\rho \gtrsim 4...5{\cdot}10^{11}gcm^{-3}$) where the extremely fast local emission and absorption processes are able to establish a stationary phase space occupation very rapidly, so that the number flux is conserved locally and is equal to the (net) inner boundary value J_+. Further out, the zone of moderate matter densities ($4{\cdot}10^{11}gcm^{-3} \gtrsim \rho \gtrsim 3...4{\cdot}10^{10}gcm^{-3}$) is characterized by the competition of the incoming diffusion flux and the offstream to lower density regions. The latter show no significant rates of local weak interactions. So flux conservation is

RZ	n_ν	$\langle \epsilon_\nu \rangle$	$\langle \epsilon_\nu^2 \rangle$	μ_ν^{EQ}	$\widetilde{T}_\nu$	$\widetilde{\mu}_\nu$	$J_{\nu,D}$
	$[10^{32} cm^{-3}]$	$[MeV]$	$[MeV^2]$	$[MeV]$	$[MeV]$	$[MeV]$	$[s^{-1}]$
0	340.50	34.63					$15.62 \cdot 10^{57}$
1	181.02	32.10	1304.22	4.291	9.966	4.793	$15.50 \cdot 10^{57}$
2	94.54	26.57	910.55	2.753	8.294	3.063	
3	50.91	18.17	415.07	6.485	5.409	6.192	
4	32.49	13.40	211.16	9.280	3.710	7.493	
5	21.36	11.90	161.80	9.352	3.336	6.286	
6	11.85	11.13	142.82	6.639	3.306	3.862	
7	7.25	10.48	127.03	5.272	3.216	2.197	
8	5.08	9.936	113.72	5.315	3.095	1.261	
9	3.80	9.714	108.12	5.946	3.069	0.343	
10	2.90	9.522	104.32	5.307	3.040	−0.466	
S		9.50					$15.96 \cdot 10^{57}$

Table 3 : Neutrino properties of the developed stellar model ($t = 6.53 \cdot 10^{-4}\ s$). The μ_ν^{EQ} are the chemical potentials supposing local chemical equilibrium were achieved. $\widetilde{T}_\nu$ and $\widetilde{\mu}_\nu$ mean artificial thermodynamic parameters of the neutrino gas calculated by claiming that the actual local neutrino number densities n_ν and the mean neutrino energies $\langle \epsilon_\nu \rangle$ should be reproduced by *Fermi-* distributions $f_\nu(\epsilon_\nu; \widetilde{\mu}_\nu, \widetilde{T}_\nu)$. The given values of $J_{\nu,D}$ figure the effective *diffusion* flux over the zone boundaries,which is caused by the different neutrino number densities within neighbouring radial cells ($J_{\nu,D} = \frac{1}{4} \cdot c \cdot 4\pi R_z^2 \cdot (n_{\nu,z} - n_{\nu,z+1})$ with R_z as the radius of the boundary between zone z and zone $z+1$). Under RZ $= 0$ the properties of the incoming diffusion flux are listed and under RZ $=$ S those of the neutrino stream leaving the star.

RZ	T	n_n	Y_e	μ_e	$\Delta\varepsilon_{acc}$
	$[MeV]$	$[10^{33} cm^{-3}]$		$[MeV]$	$[10^{32} \frac{MeV}{cm^3}]$
1	10.185	381.81	0.228	18.086	−1246
2	8.375	264.61	0.202	15.615	1915
3	5.328	266.29	0.143	17.330	1941
4	3.163	209.26	0.119	16.884	210.9
5	2.421	99.51	0.152	14.790	−490.7
6	2.502	25.00	0.249	10.701	175.7
7	2.397	11.46	0.301	8.592	19.83
8	1.978	4.86	0.391	7.492	11.07
9	1.819	4.14	0.429	7.737	−7.56
10	1.576	2.47	0.463	6.838	0.571

Table 4 : Thermodynamic and composition parameters for the developed model at a time of calculation $t = 6.53 \cdot 10^{-4}\ s$. $\Delta\varepsilon_{acc}$ is the time integrated net change of the local energy density of the stellar gas.

rapidly achieved here, too. It is then mainly the mean neutrino number density $n_\nu = Y_\nu \cdot n_B$ in the intermediate shell that determines the outward stream of particles according to

$$J_- = a \cdot Y_\nu \quad , \tag{5.1}$$

where a is roughly a constant in time and can be fixed from the Monte Carlo transport results. The neutrino fraction Y_ν shows a comparatively slow relaxation towards a quasistationary value due to the competing fluxes J_+ and J_- according to the net balance

$$\dot{Y}_\nu + \dot{Y}_e + \frac{J_-}{n_B \cdot V_\nu} = \frac{J_+}{n_B \cdot V_\nu} \tag{5.2}$$

(V_ν is the volume of region 2, n_B the local baryon number density). If one assumes the net rate of change of the electron fraction $\dot{Y}_e$ (governed by neutrino absorption and emission processes) to be neglegible compared to the effects of transport on Y_ν (that means local absorption and emission must be almost in balance), then a differential equation for the time evolution of Y_ν follows

$$\dot{Y}_\nu = \frac{1}{n_B \cdot V_\nu} \cdot (J_+ - a\,Y_\nu) \quad , \tag{5.3}$$

which has the solution

$$Y_\nu(t) = \frac{J_+}{a} \cdot \left\{ 1 - (1 - \frac{a}{J_+} Y_{\nu,i}) \cdot \exp\left[-(t - t_i)/\tau_\nu\right] \right\} \quad , \tag{5.4}$$

with

$$\tau_\nu = \frac{V_\nu \cdot n_B}{a} \tag{5.5}$$

being a typical 'flux relaxation' time scale, which can be determined from the Monte Carlo results to be

$$\tau_\nu \approx 4.1 \cdot 10^{-4}\ s \quad . \tag{5.6}$$

This supports our assumption that the adopted calculation period of at least $6.5 \cdot 10^{-4} s$ should ensure quasistationarity in this respect, too.

Which information do the Monte Carlo simulations yield about the neutrino properties in the corresponding supernova state ? The local neutrino energy space distributions (fig. 1) reveal the expected behaviour. While the inner zones (1-3) with densities greater than about $5 \cdot 10^{11} gcm^{-3}$ essentially achieve chemical equilibrium between neutrinos and stellar matter, the overlying zones (4-7) with densities in the range from $4 \cdot 10^{11} gcm^{-3}$ to $3 \cdot 10^{10} gcm^{-3}$ exhibit a growing shift of the actual neutrino distribution to higher energies, in comparison to the local equilibrium distribution. A 'hole' in the lower phase space develops due to the large mean free paths of the low energy neutrinos, whereas the higher energy states are overpopulated because of the hot flux from deeper regions. Further out (zones 8-10), the neutrinos are by far hotter than the stellar matter and the

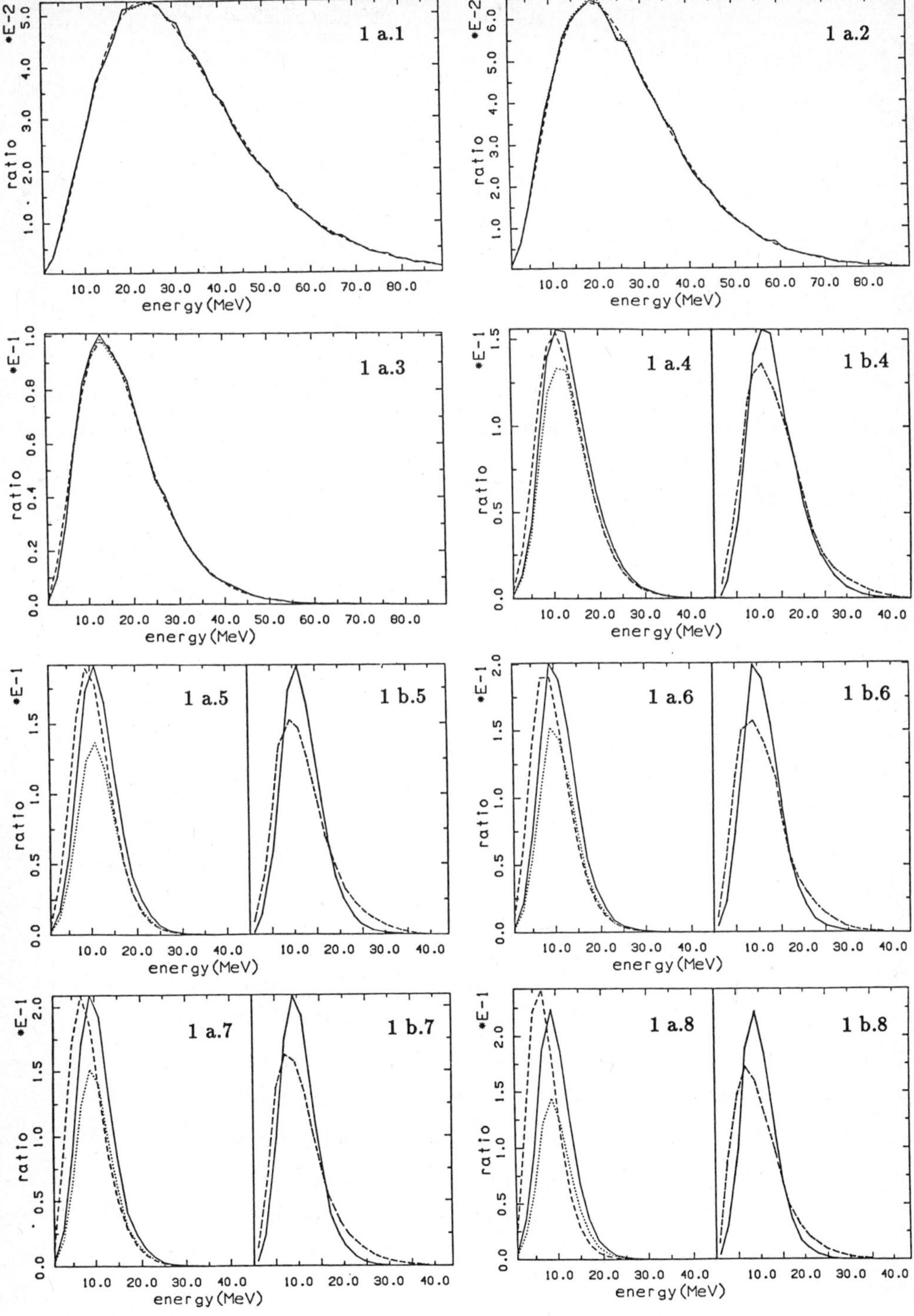
1 a.1
1 a.2
1 a.3
1 a.4
1 b.4
1 a.5
1 b.5
1 a.6
1 b.6
1 a.7
1 b.7
1 a.8
1 b.8
ratio
*E-2
*E-1
energy (MeV)

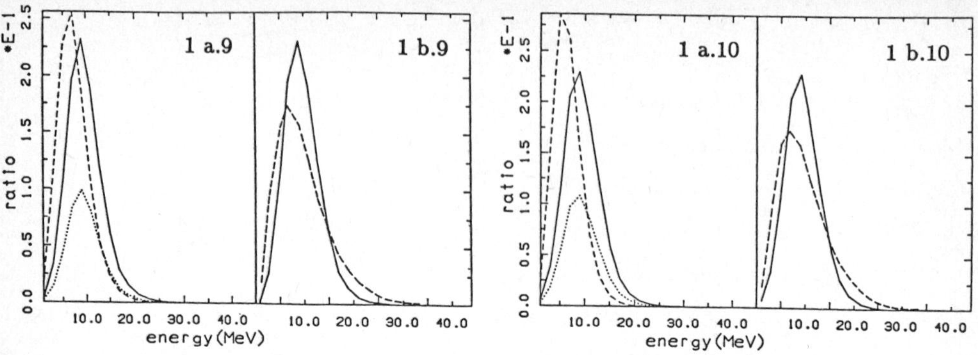

Figs. 1 a.1-a.10 : Neutrino energy space distributions for the 10 radial zones at a time of calculation $t = 6.53 \cdot 10^{-4}\,s$.

– – – – – : Distribution of momentary chemical equilibrium n_k^{EQ}/n^{EQ} ,

———— : actual neutrino distribution at the given moment n_k/n ,

·········· : n_k/n^{EQ} as a measure for the deviation from equilibrium.

The n_k denote the number densities in the energy cells k (attributed to the cell centres), n means the total number density ; the equilibrium values are marked by the superscript 'EQ'.

Figs. 1 b.4-b.10 : Comparison between the actual nonequilibrium neutrino energy distributions and the (isotropic) thermal distributions for parameters $\widetilde{T}_\nu$ and $\widetilde{\mu}_\nu$ such that local neutrino number densities and neutrino energy densities are reproduced correctly ($t = 6.53 \cdot 10^{-4}\,s$).

energy distribution remains nearly unchanged with increasing radius, which is a hint that the free stream limit is reached, i.e.: local weak processes are of minor influence on the spectrum and the neutrino number density decreases solely by reason of geometry. The flux in this region is caused by the anisotropy of the angular distributions of the particles; this differs from the behaviour in the innermost shells, which realize neutrino number concentrations such that the diffusing flux is maintained by the gradients over the zone boundaries. The corresponding numbers can be found in table 3. The change of the mean neutrino energy from about $34.6 MeV$ at the inner edge down to $\langle \epsilon_{\nu,S} \rangle = 9.5 MeV$, when the neutrinos leave the star, is also shown there. Together with a number flux of $1.59 \cdot 10^{58} s^{-1}$, this yields a surface neutrino luminosity of $23.7 \cdot 10^{52} ergs\, s^{-1}$. Note that the values of $\langle \epsilon_{\nu,S} \rangle$ and $L_{\nu,S}$ as well as their temporal behaviour do not change, if a thermal distribution for the incoming flux at the inner boundary is replaced by a totally degenerate distribution with a mean neutrino energy of $18.7 MeV$ (compare figs. 2a,2b). This means that the spectrum at the surface (fig. 3 gives the time integrated form) is completely determined by the events within the investigated window.

Figs. 2 a,b : Surface neutrino luminosity $L_{\nu,S}$ and corresponding mean neutrino energy $\langle\epsilon_{\nu,S}\rangle$ as functions of calculation time. The transport computations started with empty neutrino phase space. Crosses ($\times$) mark the simulation run with totally degenerate energy distribution and points ($\bullet$) that with thermally distributed neutrinos in the diffusion flux entering the region of interest at the inner boundary. (Some values of the luminosity curve are attributed with the 1σ-error limits.)

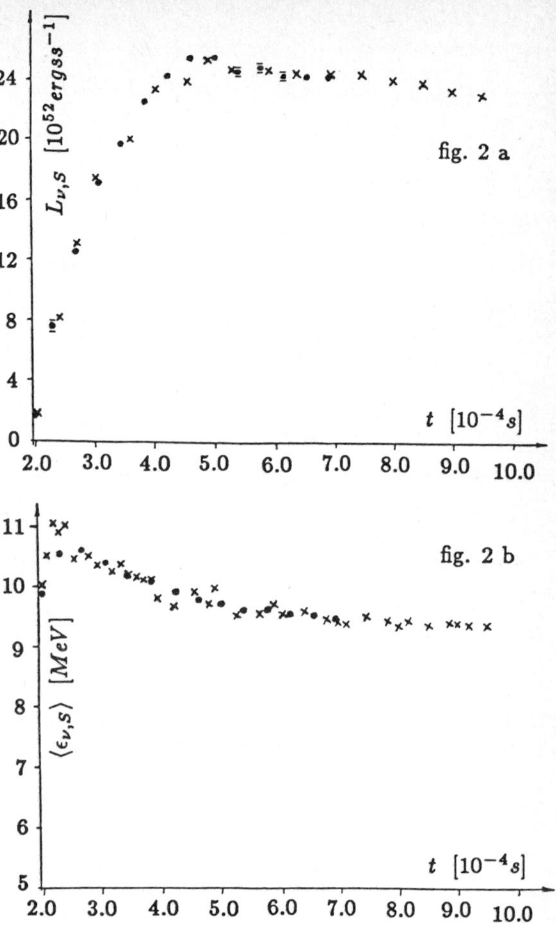

Fig. 3 : Time integrated spectra of the surface neutrino fluxes (time: $t = 6.53 \cdot 10^{-4}\, s$).

———— : Differential energy flux density (in MeV per cm^2 and per MeV),

.... : differential number flux density ($\times$ 5) (in particles per cm^2 and per MeV).

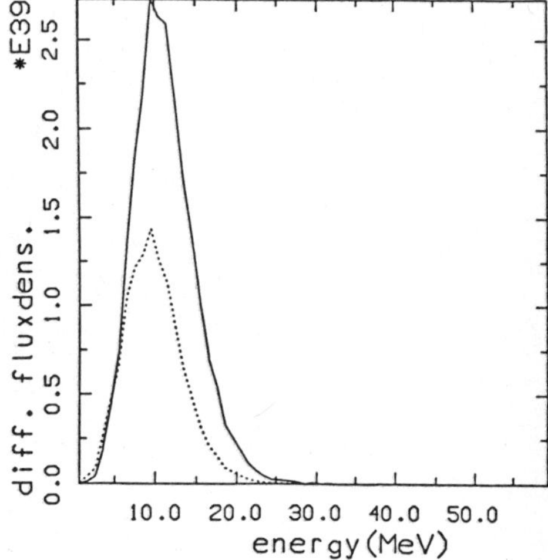

An interesting question is how accurately the local neutrino distributions can be reproduced with an (isotropic) *Fermi*- kind occupation function. If one demands that the latter should give correct values for the number and energy densities, one gets two equations which can be solved for an artificial neutrino 'chemical potential' $\widetilde{\mu}_\nu$ and an artificial neutrino 'temperature' $\widetilde{T}_\nu$. The numbers are listed in table 3. They show good agreement with μ_ν^{EQ} and the matter temperature T where chemical equilibrium prevails, whereas in the free stream region $\widetilde{T}_\nu$ remains radially constant and $\widetilde{\mu}_\nu$ changes consistent with the geometrical emptying of the phase space. Nevertheless, the energy dependence of the distributions shows (fig. 1b) that they do not describe the real spectra very well. The shape of the actual distributions is much narrower and its peak therefore higher. Qualitatively, the same result is obtained if the condition of isotropy is dropped and all the particles are assumed to move radially outward in the case of free streaming.

We now ask whether the features we have described provide any information about a possible delayed explosion in the supernova model we are considering ? A critical parameter yielding evidence about the efficiency of the neutrino heating is the heating time scale defined by

$$\tau_H = \frac{E_G}{\dot{E}} \tag{5.7}$$

(*Lattimer* and *Burrows* 1984) as the time it will take the neutrinos to deposit an energy equivalent to the local gravitational binding energy per unit mass in the stellar material behind the shock. Therefore $\dot{E}$ represents the energy transfer rate from the neutrinos to the matter and includes the contributions from energy gain due to neutrino absorptions on neutrons ($\dot{E}_{A,\nu_e}$), from energy loss caused by neutrino reemission ($\dot{E}_{E,\nu_e}$), and from energy exchanges via neutrino- electron scattering events ($\dot{E}_{S,\nu_e}$). Approximate expressions can be derived for all three of these, on the base of the interaction rates given by *Tubbs* and *Schramm* (1975). Assuming extremely relativistic electrons and neglecting both energy transfer to the nucleons and final state blocking for the electrons and neutrinos, they may be written in the free stream limit (units: $ergs\,g^{-1}s^{-1}$)

$$\dot{E}_{A,\nu_e} \cong 5.2\cdot 10^{-20} Y_n \cdot \frac{L_\nu}{4\pi R_m^2}\cdot \left[\langle\epsilon_\nu^2\rangle + 2\Delta m\langle\epsilon_\nu\rangle + (\Delta m)^2\right] \tag{5.8a}$$

$$\begin{aligned}\dot{E}_{E,\nu_e} \cong 1.7\cdot 10^{16}\cdot Y_p\cdot T_m^6\cdot\Big[\, & F_5(\eta_e) - 3\left(\frac{\Delta m}{T_m}\right)F_4(\eta_e) + \\ & + 3\left(\frac{\Delta m}{T_m}\right)^2 F_3(\eta_e) - \left(\frac{\Delta m}{T_m}\right)^3 F_2(\eta_e)\Big]\end{aligned} \tag{5.8b}$$

$$\begin{aligned}\dot{E}_{S,\nu_e} \approx \rho^{-1}\cdot\Big[\, & 3.1\cdot 10^{28}\cdot \widetilde{T}_\nu^5 T_m^4\cdot F_3(\eta_e)\cdot F_4(\widetilde{\eta}_\nu)\ + \\ & -\ 2.0\cdot 10^{23}\cdot \widetilde{T}_\nu^4 T_m^5\cdot F_3(\eta_e)\cdot F_3(\widetilde{\eta}_\nu)\Big]\end{aligned} \tag{5.8c}$$

where T_m and R_m are temperature and radial position of the matter, ρ is the local mass density, L_ν denotes the neutrino energy flux in $ergs\,s^{-1}$, $\eta_e = \mu_e/T_m$ and $\widetilde{\eta}_\nu = \widetilde{\mu}_\nu/\widetilde{T}_\nu$ are degeneracy parameters for electrons and neutrinos, and $\Delta m = 1.3\,MeV$. The F_n's mean the usual (relativistic) *Fermi* integrals and all energies and temperatures are measured in MeV. Taking the neutrino flux, neutrino energy averages ($\langle\epsilon_\nu\rangle$ and $\langle\epsilon_\nu^2\rangle$) and neutrino parameters $\widetilde{T}_\nu$ and $\widetilde{\eta}_\nu$ from the Monte Carlo calculations, a positive net heating of

$$\dot{E}_{\nu_e} = \dot{E}_{A,\nu_e} + \dot{E}_{S,\nu_e} - \dot{E}_{E,\nu_e} = 3.3 \cdot 10^{20} \quad erg s\, g^{-1} s^{-1} \tag{5.9}$$

results. This implies a heating time of

$$\tau_H \approx 40 \;\; ms \tag{5.10}$$

immediately behind the shock, where $E_G = |-(G \cdot M_R)/R_m| \approx 1.2 \cdot 10^{19} ergs\, g^{-1}$, $T_m \approx 0.9 MeV$, $\eta_e \approx 3.8 MeV$, $\rho \approx 2 \cdot 10^9 g\, cm^{-3}$, and $Y_n \approx 1 - Y_p \approx 0.53$.

This very small value can be understood in terms of the high neutrino luminosity and the comparatively low matter temperature, which reduces reemission losses considerably. Clearly, from this kind of brief snap shot of at most $1\,ms$ duration, no conclusions can be drawn concerning the long time evolution of the luminosity, which must certainly decrease very soon due to the deleptonisation of the stellar matter (what then would demand the choice of a different inner boundary flux). But — as long time hydrodynamical calculations confirm (*Mayle* 1985) — the spectral shape of the neutrino stream will not be subjected to such drastic changes and the mean neutrino energy of about $9.5 - 10\,MeV$ should subsequently guarantee heating times significantly lower than the typical hydrodynamical time scales of $200 - 300\,ms$ (*Lattimer* and *Burrows* 1984), because antineutrinos will contribute in the heating as soon as the depletion of the electron phase space allows thermal neutrino production to occur. Convective processes in the deeper regions might also provide very high neutrino luminosities over periods of some $100\,ms$, while neutrino annihilation reactions can possibly yield additional energy deposition in the zones behind the shock, as recently emphasized by *Goodman* et al. (1987). Ultimately, since all hydrodynamical processes tend to support the revival of the shock (*Bethe* and *Wilson* 1985; *Lattimer* and *Burrows* 1984), 'time will be on the supernova's side'.

6. Summary

A Monte Carlo transport code for neutrinos was developed, which allows one to calculate neutrino spectra for given stellar configurations very accurately. Although it is not suitable for direct application in connection with hydrodynamical calculations, it can be used for calibration of faster transport methods. An investigation of a $20 M_\odot$ supernova model in a stage about $12\,ms$ after core bounce shows that the local neutrino spectra cannot be described satisfactorily by thermal distributions. The (electron) neutrinos have mean energies of about 10 MeV when they leave the star. This favors the revival of the stalled shock and the occurrence of a delayed explosion, producing a supernova event with the decisive aid of the enormous energy reservoir present in form of neutrinos.

Acknoledgements

I would like to thank W. Hillebrandt for his valuable advice in many detail questions, and J. W. Truran for improving the manuscript.

References

Arnett, W.D., 1977, *Ap. J.* **218**, 815.

Bethe, H.A., and Wilson, J.R., 1985, *Ap. J.* **295**, 14.

Bludman, S.A., and Van Riper, K.A., 1978, *Ap. J.* **224**, 631.

Bowers, R.L., and Wilson, J.R., 1982a, *Ap. J. Suppl.* **50**, 115.

Bruenn, S.W., 1985, *Ap. J. Suppl.* **58**, 771.

Bruenn, S.W., 1986a, *'A Comparison of Recent Numerical Calculations of Stellar Core Collapse'*, preprint.

Bruenn, S.W., 1986b, *Ap. J.* **311**, L69.

Castor, J.I., 1972, *Ap. J.* **178**, 779.

Goodman, J., Dar, A., and Nussinov, S., 1987, *Ap. J.*, **314**, L7.

Hillebrandt, W., 1985, in *High Energy Phenomena Around Collapsed Stars*, NATO-ASI **C195**, ed. F. Pacini, (Dodrecht: D. Reidel), p.73.

Lattimer, J.M., and Burrows, A., 1984, in *Problems of Collapse and Numerical Relativity*, NATO-ASI **C134**, ed. D. Bancel and M. Signore, (Dodrecht: D. Reidel), p.147.

Lichtenstadt, I., Ron, A., Sack, N., and Wagschal, J.J., and Bludman, S.A., 1978, *Ap. J.* **226**, 222.

Mihalas, D., 1978, *Stellar Atmospheres*, (San Francisco: W.H. Freeman and Company).

Mayle, R.W., 1985, *Ph. D. Thesis*, to be published in *Astroph. Journal.*

Myra, E.S., Bludman, S.A., Hoffman, Y., Lichtenstadt, I., Sack, N., and Van Riper, K.A., 1986, *'The Effects of Neutrino Transport on the Collapse of Iron Stellar Cores'*, preprint.

Sato, K., and Suzuki, H., 1987, submitted to *Phys. Rev. Lett.*

Schinder, P.J., and Shapiro, S.L., 1986, in *Radiation Hydrodynamics in Stars and Compact Objects*, Proc. IAU Colloq. **89**, ed. D. Mihalas and K.H. Winkler, p. 121.

Tubbs, D.L., 1978, *Ap. J. Suppl.* **37**, 287.

Tubbs, D.L., 1979, *Ap. J.* **237**, 846.

Tubbs, D.L., and Schramm, D.N., 1975, *Ap. J.* **201**, 467.

Tubbs, D.L., Weaver, T.A., Bowers, R.L., Wilson, J.R., and Schramm, D.N., 1980, *Ap. J.* **239**, 271.

Yueh, W.R., and Buchler, J.R., 1976a, *Astrophys. and Space Science* **39**, 429.

Yueh, W.R., and Buchler, J.R., 1976b, *Astrophys. and Space Science* **41**, 221.

NEUTRINOS FROM SN 1987A: REMARKS ON POSSIBLE INTERPRETATIONS

Wolfgang Hillebrandt
Max-Planck-Institut für Physik und Astrophysik
Institut für Astrophysik
Karl-Schwarzschild-Straße 1
D-8046 Garching b. München, FRG

Abstract

An attempt is made to interpret the neutrino events observed a few hours before the optical outburst of Supernova 1987A in the Large Magellanic Cloud (LMC) on February 23, 1987, in the framework of existing theoretical models. It will be shown that: 1. The neutrinos observed by the KAMIOKA and IMB detectors can be explained by standard core collapse scenarios, provided the neutrino pulse seen by the Mont Blanc experiment represented background events only. 2. If the neutrinos detected by the Mont Blanc experiment are real a more complicated scenario has to be invented. 3. From the widths of the observed neutrino pulses a model independent upper limit on neutrino rest masses can be derived. 4. Model dependent interpretations indicate a finite neutrino rest mass.

1. Neutrino Detections

Two neutrino pulses have been detected prior to the optical outburst of SN 1987A at Feb. 23.12 (UT) (Aglietta et al., 1987a) and at Feb. 23.32 (UT) (Hirata et al., 1987; Bionta et al., 1987), respectively. Two photographs taken at about Feb. 23.44 (UT) (McNaught, I.A.U. Circular No. 4389) show the supernova at a visual magnitude of about $6^m_.4$ only 2.8×10^4 sec and 1.1×10^4 sec, respectively, after the neutrino events were discovered. This close correlation indicates that the neutrinos were indeed emitted from the exploding star.

The pulse seen in the Mont Blanc experiment consisted of 5 neutrinos spread over $\Delta t \simeq 7$sec with measured positron energies between 7 and 11

Table 1: Properties of neutrino events detected from SN 1987A

Experiment	Event No.	Time (UT) (Feb. 23)	Electron energy (MeV)	Electron angle (degrees)
Mont Blanc	1	$2^h52^m36^s.79$	7(±30%)	----
	2	$40^s.65$	8	----
	3	$41^s.01$	11	----
	4	$42^s.70$	7	----
	5	$43^s.80$	9	----
KAMIOKA	1	$7^h35^m41^s.00$(?)	20.0±2.9	18±18
	2	$41^s.11$	13.5±3.2	15±27
	3	$41^s.30$	7.5±2.0	108±32
	4	$41^s.32$	9.2±2.7	70±30
	5	$41^s.51$	12.8±2.9	135±23
	6	$41^s.69$	6.3±1.7	68±77
	7	$42^s.54$	35.4±8.0	32±16
	8	$42^s.73$	21.0±4.2	30±18
	9	$42^s.92$	19.8±3.2	38±22
	10	$50^s.22$	8.6±2.7	122±30
	11	$51^s.43$	13.0±2.6	49±26
	12	$53^s.44$	8.9±1.9	91±39
IMB	1	$7^h35^m41^s.37$	38±10	74±11
	2	$41^s.79$	37± 9	52± 8
	3	$42^s.02$	40±10	56± 9
	4	$42^s.52$	35± 9	63±10
	5	$42^s.94$	29± 7	40± 6
	6	$44^s.06$	37± 9	52± 8
	7	$46^s.38$	20± 5	39± 6
	8	$46^s.96$	24± 6	102±15

MeV (see table 1). From the second pulse, KAMIOKA detected 12 neutrino events spread over $\Delta t \simeq 13$sec, whereas IMB saw 8 neutrinos with $\Delta t \simeq 6$sec. The electron (positron) energies in the second pulse ranged from 6 to 35 MeV (KAMIOKA) to 20 to 40 MeV (IMB) (see table 1). It is interesting to note that the angular distribution of the electrons detected by the KAMIOKA experiment is not quite isotopic, as one would expect from the reaction $\bar{\nu}_e + p > e^+ + n$. In particular, the high energy events ($E_e \gtrsim 20$MeV) seem to be strongly forward peaked away from the LMC. Information on the angular distribution of events is also available for the IMB data but they are less conclusive because 25% of the photomultipliers were inoperative when the neutrinos were detected.

2. Significance and Consistency of the Neutrino Detections

Unfortunately, only a computer clock was used by the KAMIOKA-group. The arrival time of the neutrinos in their detector, therefore, is uncertain by about ±1 minute. Since the IMB detection was within this time interval and an accidental coincidence is extremely unlikely, we will assume that the second neutrino pulse began at 7:35:41 (UT) given by the IMB experiment. In a 1 minute window around this time the Mont Blanc detector had no signal above background (Aglietta et al., 1987a).

According to Hirata et al. (1987), the rate at which events like the one observed in the KAMIOKA detector can arise from statistical fluctuations is of the order of one event every 7×10^7 yrs. Although the statistical significance of the IMB detection is much less and gives a rate of one event of multiplicity 9 every 3 yrs from background noise, from both experiments we can safely conclude that the second neutrino burst was indeed real.

With respect to the first neutrino events, the conclusions are less certain. Aglietta et al. (1987a) state that their background rate of finding 5 positrons above threshold in a 10 sec interval is one every 1.5 yrs. This was concluded from an average background trigger rate of 0.12 per 10 sec in the run including the Feb. 23.12 event. This translates into a probability of 3×10^{-4} to find 5 events in a 10 seconds bin in a 4 hrs window around the assumed time of the supernova explosion. We want to note that the statistical significance of the Mont Blanc detection is about the same as that of the IMB detection. However, since at the time of the Mont Blanc events KAMIOKA recorded 4 positrons only (M. Koshiba, private communication), just slightly above their typical background rate of 2 events in 10 sec, we cannot exclude the possibility that the first pulse was not real but was due to random fluctuations.

In order to check the consistency of the three detections we can proceed as follows. It is likely that the Mont Blanc neutrino detector sees all events above 7 MeV. From table 1 it then follows that the number of events "seen" is $3 \pm \sqrt{3} = 3 \pm 1.7$. By taking into account the ratio of free protons in both experiments and the detection efficiency of the KAMIOKA detector, we can predict that KAMIOKA should have seen 17±10 events at that time. A more detailed analysis (Aglietta et al.,

1987b) gives a somewhat lower number (12±8). It is obvious, therefore, that the expected number of events is marginally consistent with the observed 4±2 events. IMB and KAMIOKA can be compared in a similar way. Taking into account the detection efficiency of IMB for events above 20 MeV and the fact that 25% of the photomultiplier tubes did not work at the time of the second neutrino pulse, the total number of events above 20 MeV in the IMB detector would be 22±5. Since the volume of the KAMIOKA detector is about a factor of 2.3 smaller one would expect an event number of 10±3 in KAMIOKA. The actual number of events, however, is 4±2, again marginally consistent with the expectation. A similar conclusion is reached if events above 30 MeV are considered for which the trigger efficiency of the IMB experiment is 56%.

So in conclusion, from statistics alone one cannot rule out the possibility that the first neutrino pulse seen by the Mont Blanc neutrino observatory may indeed have been real. Although its statistical significance is certainly much lower than that of the second pulse, the absence of a clear signal in the KAMIOKA detector would still be consistent if the average neutrino energy in the first pulse would have been sufficiently low, lower than 6 to 7 MeV, say, in which case the higher detection efficiency of the Mont Blanc experiment at low energies could explain the observed number of events.

3. Neutrino Energies

We will now proceed on the assumption that both neutrino pulses were real detections and will investigate the energetics of the events in order to check whether or not they are in conflict with astrophysical expectations.

We will first discuss the implications from the Mont Blanc detection. As can be seen from table 1, two of the detected events had energies near threshold and the first of them preceded the others by four seconds. We, therefore, will assume that the number of events was 4±2. Because of the much higher cross-section of the $(\bar{\nu}_e,p)$-reaction as compared to (ν,e)-scattering we will assume that the incident reaction was $\bar{\nu}_e+p\to e^{+}+n$, for which the cross-section is 9.5×10^{-44} $(E_\nu/\mathrm{MeV})^2$ cm^2, where E_ν is the incident neutrino energy in MeV. Note that for this reaction the incident neutrino energy on the average will only be about 10 to 20% higher than the observed positron energy. By taking the number of protons in the liquid scintillator of the Mont Blanc experiment ($\simeq 8\times10^{30}$) we can estimate the integrated flux of neutrinos at the detector and find $(5\pm3)\times10^{12}$ $(E_\nu/\mathrm{MeV})^{-2}$ cm^{-2}. For the distance to the LMC, this corresponds to a total number of anti-neutrinos emitted from the source of about $(1.2\pm0.7)\times10^{60}$ $(E_\nu/\mathrm{MeV})^{-2}$ anti-neutrinos. The total energy in the burst then is $E_{tot}\simeq(1.2\pm0.7)\times10^{60}$ $(E_\nu/\mathrm{MeV})^{-1}$ (MeV) or $E_{tot}\simeq(1.9\pm1.1)$ $(E_\nu/\mathrm{MeV})^{-1}\times10^{54}$ ergs. By assuming an average neutrino energy of 6 MeV in order to be consistent with the (non-)detection of KAMIOKA we end up with a total anti-neutrino energy of $(3\pm2)\times10^{53}$erg. Since about the same energy will be emitted in neutrinos and anti-neutrinos the total energy in electron-neutrinos has to be of the order of $(6\pm4)\times10^{53}$erg. This number has to be compared with the binding energy of a neutron star which is at most about 5×10^{53}erg for reasonable equations of state (Sato and Suzuki, 1987). Therefore, within the statistical errors the energy required in order to explain the Mont Blanc events is consistent with the assumption that a significant fraction of the binding energy of a neutron star is radiated away by neutrinos in a few seconds, provided only electron neutrinos were emitted.

One can argue that about the same energy, as in ν_es will be radiated in ν_μs and ν_τs, in which case the total energy in neutrinos would exceed the binding energy of a neutron star significantly. This, however, is not true if the temperature at the neutrino sphere is as low as we have assumed. The time-scale to fill the neutrino phase space is about $13T^{-5}$ sec for ν_es and a factor of 5 longer for ν_μs and ν_τs for the reaction $e^-+e^+\rightarrow\nu+\bar{\nu}$. Here T is the temperature at the ν-sphere in MeV. For a temperature of 2 MeV (consistent with the assumed average neutrino energy of 6 MeV) this gives time-scales of about 0.4 sec for ν_es and 2 sec for ν_μs and ν_τs. Therefore, it is conceivable that many fewer thermal μ- and τ-neutrinos were emitted from the source.

An analysis similar to the one here has been performed by Sato and Suzuki (1987) and many others for the KAMIOKA and IMB events, assuming again that all neutrinos detected were $\bar{\nu}_e$s. Sato and Suzuki (1987) find that the KAMIOKA data can be fitted by a thermal neutrino spectrum corresponding to a temperature of (2.8±0.3) MeV, whereas the IMB data require a significantly higher temperature of (4.6±0.7) MeV. The total energy in three neutrino flavours is then $(2.9\pm0.6)\times10^{53}$ erg for KAMIOKA and $(1.5^{+1.2}_{-0.6})\times10^{53}$ erg for the IMB pulse. If some of the events detected by both experiments are considered to be noise (in particular events no. 10, 11, and 12 of KAMIOKA, see table 1), the total energy is reduced by about a factor of 2. If we compare the energy required to explain both the first and the second neutrino pulse it becomes obvious that the energy in the first pulse has to be higher by about a factor of 2, and we also find that the energy in the second pulse is certainly consistent with the assumption that a neutron star of about 1.5 $M_{\odot}$ has formed.

As we have mentioned in section 2, several events detected by KAMIOKA were strongly forward directed indicating (ν_e,e)-scattering rather than $(\bar{\nu}_e,p)$-reactions. If this would be true the energy estimates given above would have to be revised. Sato and Suzuki (1987) find that due to the small scattering cross-sections the integrated luminosity corresponding to the first 2 events would already be 1.5×10^{53}ergs even if a scattering angle of 0^0 is assumed.

So in conclusion again, we cannot rule out any one of the two neutrino pulses from energy arguments alone, since both are still consistent with the assumption that a neutron star has formed in SN 1987A and has radiated away a large fraction of its binding energy in form of thermal neutrinos during the first few seconds of its life. It is also apparent, however, that this conclusion is rather uncertain since in the case of the Mont Blanc detections we have to rely on the poor statistics of a few events and in the case of KAMIOKA and IMB we do not know for sure which (if any) of the detections were due to (ν,e)-scattering. Note also that due to statistical errors the total neutrino energy cannot be determined to better than about a factor of two (see also de Rujula (1987) for an extended discussion).

4. Astrophysical Scenarios

Most attempts to interpret the neutrinos seen from SN 1987 start from the assumption that the first pulse was not real but rather was due to noise in the Mont Blanc detector. The main arguments given in favour of this interpretation are:

1. KAMIOKA should have seen a clear signal.
2. The energy required to explain this pulse is too high.

But, as we have tried to show in the previous two sections, both arguments are not conclusive, the main uncertainty being the statistics of only a few events. There is also some dispute going on whether the early evolution of the visual light curve of SN 1987 is best fitted by an event starting at the time of the KAMIOKA detection (Arnett, 1987; Woosley et al., 1987) or a few hours earlier (Wampler et al., 1987).

If one dismisses the first pulse the interpretation of the second one is straightforward and in good agreement with theoretical predictions, in particular if one assumes that the last three events seen by KAMIOKA and the last two events observed by IMB were just noise. The time spread of the KAMIOKA signal is then 2 sec only and IMB gives a pulse width $\Delta t \simeq 3$sec. Burrows and Lattimer (1986) have shown that about 80% of the thermal neutrinos above 7 MeV are emitted from a newly born neutron star within the first 2 sec and 95% are emitted during the first 4 sec. This is in very good agreement with the pulse width given above. Also the neutrino temperatures obtained from numerical models for the first 1.4 sec after core-collapse (Mayle, 1985; Mayle, Wilson, and Schramm, 1987) seem to agree fairly well with the observations, although the neutrino luminosity seems to have been somewhat higher than was predicted by the models. If all 12 events seen by KAMIOKA and all 8 events found by IMB came from SN 1987A this causes a problem since theoretical models did not predict that after about 10 sec the mean neutrino energy could still be as high as 11 MeV or more. We will come back to this question later in section 5.

The main predictions made by these models are that in a year or two a neutron star should become visible either as a thermal x-ray source, a pulsar, or a synchrotron nebulae after the supernova envelope has become transparent. Moreover, γ-ray lines from the decay of radioactive ^{56}Co should be detectable (Woosley et al., 1987).

If, on the contrary, we assume that both neutrino pulses were real, a more complicated and speculative scenario has to be invented. Such a scenario has to explain the following facts:

1. The neutrino temperature of the first pulse was low ($\lesssim$2MeV). Nevertheless, at least 2×10^{53}ergs have been emitted in form of thermal electron neutrinos and anti-neutrinos.

2. The second pulse was delayed by about 4.5 hours relative to the first pulse.
3. The total energy emitted in both pulses was equal within factors of about 2.

Hillebrandt et al. (1987) have suggested that the first pulse signaled the formation of a neutron star, whereas the second pulse came from the further collapse to a black hole. We will have to see whether or not this scenario can explain the observations mentioned above.

An upper limit of the electron neutrino luminosity can be obtained from the exclusion principle (Bludman and Ruderman, 1975), and we find in the black body limit

$$L_{\nu} \lesssim 10^{49} \left[\frac{T}{\mathrm{MeV}}\right]^4 \left[\frac{R}{10\ \mathrm{km}}\right]^2 \ (\mathrm{erg\ sec^{-1}}) \qquad (1)$$

From the observations we get $L_{\nu} \gtrsim 4\times10^{52}$ erg s^{-1} and $T \lesssim 2$ MeV. Consequently the radius of the neutrino sphere has to be at least about 150 km. For a non-thermal neutrino spectrum the luminosity may exceed the limit given by eq.1 and thus the radius of the neutrino sphere may be somewhat smaller. In any case, it seems that the required radius of the neutrino sphere is significantly larger than that obtained from non-rotating core-collapse models (~30-80 km; Wilson et al., 1986; Hillebrandt, 1986). One way out of this problem would be to assume that the stellar core was rapidly rotating in which case the neutrino "sphere" would become anisotropic, its "radius" might be larger and neutrinos could be transported much more efficiently by large scale circular motions (Müller and Hillebrandt, 1981). An anisotropy of the emitted neutrino pulse might also help to reduce the amount of energy needed in order to explain the Mont Blanc events, provided the pulse was beamed towards us. These difficulties, therefore, may not be insurmountable.

The progenitor of SN 1987A, Sanduleak-69 202, was a blue supergiant with a mass around 15 $M_{\odot}$ to 20 $M_{\odot}$. The collapse of the core of such a star may indeed lead to the formation of a rather massive neutron star (Wilson et al., 1986), and for some models it was found that the mass of the compact remnant exceeded that of a cold non-rotating neutron star. The question then is whether or not the further collapse to a black hole can be delayed by several hours. One can think of two effects which might stabilize the neutron star temporarily, its thermal energy content and its angular momentum. Cooling time scales can be obtained from eq.1. For a "surface" temperature of about 10^{10} K they will be around one hour. Loss of rotational energy due to dipole radiation can be estimated to be on typical time scales of about

$$\tau \simeq 6\times10^{4} \left[\frac{M}{M_{\odot}}\right] \left[\frac{P}{10^{-3}\mathrm{sec}}\right]^{2} \left[\frac{B}{10^{13}\ \mathrm{Gauss}}\right]^{-2} \left[\frac{R}{30\ \mathrm{km}}\right]^{-4} \ \mathrm{sec}, \qquad (2)$$

where P is the period of rotation, B the magnetic field strength, and M and R mass and radius, respectively, of the newly born neutron star. Again we obtain the right time scale within an order of magnitude.

A more difficult problem is to estimate the amount of neutrinos and their energy which may be emitted during the collapse to a black hole. Again we can use eq.1 to estimate the duration of the second neutrino pulse. Since we have an energy $E_{\nu}\gtrsim 4\times10^{52}$ergs in electron neutrinos and anti-neutrinos and the neutrino temperature was $\lesssim$4MeV we find $\tau\gtrsim$2sec, which is much longer than the expected time-scale for the collapse to a black hole. On the other hand, a neutron star collapsing to a black hole may not radiate neutrinos like a black body and we should replace eq.1 by the true Pauli-limit (Straumann, private communication) which gives

$$L_{\nu} \lesssim 3.2\times10^{53} \left[\frac{<\nu>}{20\ \mathrm{MeV}}\right]^{4} \left[\frac{R}{10\ \mathrm{km}}\right]^{2} \ (\mathrm{erg\ sec^{-1}}) \ , \qquad (3)$$

where now $<\nu>$ is the average neutrino energy. From eq.3 it is obvious that the observed amount of energy can, in principle, be emitted in about a tenth of a second. Time-scales of this order may result if a rapidly spinning object is collapsing to a black hole (Eardly, 1983). The observed dispersion of the neutrino signal should then be caused by other effects, e.g. a finite neutrino rest mass (see section 5).

We have to note, however, that it is very unlikely that neutrino diffusion out of a collapsing neutron star can give such a high neutrino luminosity, because diffusion times are of the order of seconds. Again, large scale convective motions may provide a way out of this difficulty.

So in conclusion, it seems possible, though not easy, to invent an astrophysical scenario that can explain that two neutrino pulses may have been observed from SN 1987A. It is also obvious, however, that the most easy explanation of the neutrino events is to dismiss the Mont

Blanc events as noise and to assume that the second pulse signaled the formation of the neutron star. Future observations will reveal which interpretation is the correct one.

5. Fundamental Properties of Neutrinos

The fact that neutrinos have been seen from SN 1987A tells us that life-time of the neutrino has to be at least 10^5 years, or, to be more precise $t_{1/2} \geq 40h\ ((m_0c^2)/(eV))$ in the co-moving frame of the neutrinos, and this upper limit can be improved for decay modes involving γ-rays by at least six orders of magnitude from the absence of a γ-ray burst associated with SN 1987A (Chupp, IAU-circular no. 4365, 1987; v. Feilitzsch, private communication).

Moreover, the observed widths of the neutrino pulses can be used to obtain a model-independent upper limit for neutrino rest masses, and if certain model assumptions are made, the data may even indicate a finite neutrino rest mass.

In order to obtain such estimates we can proceed as follows. If neutrinos are emitted from a source at distance d with a certain energy distribution on time-scales short compared with the spread of the signal seen by a distant detector a rest mass can be inferred from the relation

$$m_0 = 4.28 \left[\frac{E_\nu^{min}}{7\ \mathrm{MeV}}\right] \left[\frac{d}{52\ \mathrm{kpc}}\right]^{-1/2} \sqrt{\Delta t\ \frac{\alpha^2}{\alpha^2-1}} \quad (\mathrm{eV}/c^2)\ , \qquad (4)$$

where $\alpha = E_\nu^{max}/E_\nu^{min}$ is the observed energy band width of neutrinos (Hillebrandt et al., 1987). If we do not use a model for the intrinsic time spread of the signal, eq.4 gives an upper limit on the rest mass. From the Mont Blanc data (table 1) we obtain (including possible errors in the energy determination) $m_0 \leq 23\ \mathrm{eV}/c^2$. The IMB and KAMIOKA data give similar estimates, namely 33 eV/c^2 and 15 eV/c^2, respectively (see e.g. Bahcall and Glashow, 1987; Arnett and Rosner, 1987; Burrows and Lattimer, 1987). It is obvious that these upper limits are reduced considerably if it is assumed that the first Mont Blanc event as well as the last three KAMIOKA events and the last two IMB events are background noise. In particular, the data then are compatible with a zero rest mass since the intrinsic spread of the neutrino signal from a forming neutron star is expected to be of the order of a few seconds (Burrows and Lattimer, 1986).

Therefore, only if it should turn out that SN 1987A left behind a black hole rather than a neutron star can we hope to get model-independent information on finite neutrino rest masses. Based upon the assumption that the second neutrino pulse signaled the formation of a black hole Hillebrandt et al. (1987) have analysed the KAMIOKA data. They found that the data can be fitted by two groups of neutrinos with different masses. The first group included events 1 through 6 and the second one events 7 through 8. The rest mass obtained for the first group was $(7.6^{+4}_{-3})(eV/c^2)$ whereas the second gave $(69^{+24}_{-45})(eV/c^2)$. If this interpretation is correct one would have to attribute the second group of events to μ- and τ-neutrinos. Because of the much lower cross-section ($\sigma \simeq 1.3 \times 10^{-45}$ $(E_\nu/MeV)^{-1}$ cm^2) the total energy in μ- and τ-neutrinos then has to be of the order of $(1.5\pm1.0)\times10^{54}$ergs, significantly higher than the binding energy of a neutron star. Note that the rest mass of a $2M_\odot$ object is about 4×10^{54}ergs only. This interpretation, therefore, seems to face an energy problem unless either the neutrino emission was anisotropic and beamed towards us, or the number of scattering events was accidentally high in the KAMIOKA detector.

6. Summary

We have demonstrated that both from statistical arguments and from the energetics of the neutrino events observed by various neutrino detectors the possibility cannot be ruled out that SN 1987A emitted two neutrino bursts prior to the optical outburst. Moreover, it has been shown that an astrophysical scenario can be invented which can explain the presence of two neutrino pulses without leading to principle difficulties. If, however, future observations should reveal the presence of a neutron star remnant in SN 1987A the first neutrino pulse would have to be dismissed. In this latter case the properties of the observed second neutrino pulse would be consistent with theoretical predictions of core-collapse models and the upper limit for the electron neutrino rest mass would agree with those obtained from terrestrial experiments.

References

1) Aglietta, M., et al.; 1987a, Europhysics Letters, in press.
2) Aglietta, M., et al.; 1987b, Europhysics Letters, in press.
3) Arnett, W.D.; 1987, Ap. J., in press.
4) Arnett, W.D., and Rosner, J.L.; 1987, Phys. Rev. Lett. 58, 1906.
5) Bahcall, J.N., and Glashow, S.L.; 1987, Nature 326, 476.
6) Bionta, R.M., et al.; 1987, Phys. Rev. Lett. 58, 1494.
7) Bludman, S.A., and Ruderman, M.A.; 1975, Ap.J. 195, L19.
8) Burrows, A., and Lattimer, J.M.; 1986, Ap. J. 307, 178.
9) Burrows, A., and Lattimer, J.M.; 1987, Ap. J., submitted.
10) De Rujula, A.; 1987, CERN preprint.
11) Eardly, D.M.; 1983, in Gravitational Radiation, N. Deruelle and T. Piran, eds., North Holland (Amsterdam), p.257.
12) Hillebrandt, W.; 1986, NATO-ASI.
13) Hillebrandt, W., Höflich, P., Kafka, P., Müller, E., Schmidt, H.U., and Truran, J.W.; 1987, Astron. Astrophys., in press.
14) Hillebrandt, W., Höflich, P., Kafka, P., Müller, E., Schmidt, H.U., Truran, J.W., and Wampler, E.J.; 1987, Astron. Astrophys. 177, L41.
15) Hillebrandt, W., Höflich, P., Truran, J.W., and Weiss, A.; 1987, Nature, in press.
16) Hirata, K., et al.; 1987, Phys. Rev. Lett. 58, 1490.
17) Mayle, R.W.; 1985, Ph.D. thesis, Lawrence Livermore National Laboratory.
18) Mayle, R.W., Wilson, J.R., and Schramm, D.N.; 1987, Ap.J., in press.
19) Müller, E., and Hillebrandt, W.; 1981, Astron. Astrophys. 103, 358.
20) Sato, K., and Suzuki, H.; 1987, Phys. Rev. Lett., submitted.
21) Wampler, E.J., Truran, J.W., Lucy, L.B., Höflich, P., and Hillebrandt, W.; 1987, Astron. Astrophys., submitted.
22) Wilson, J.R., Mayle, R., Woosley, S.E., and Weaver, T.A.; 1986, Ann. N.Y. Acad. Sci., in press.
23) Woosley, S.E., Pinto, P.A., and Ensman, L.; 1987, Ap. J., submitted.

List of Participants

J.-P. Arcoragi	(MPI f. Astrophysik, Garching, D)
M. Arnould	(Universite Libre de Bruxelles, B)
J. Audouze	(Institut d'Astrophysique, Paris, F)
B. Baschek	(Universität Heidelberg, D)
W. Becker	(Universität Münster, D)
D. Bencivenni	(Inst. Astrofisica Spaziale, Frascati, I)
J.B. Blake	(Aerospace Corporation, Los Angeles, USA)
H. Boffin	(Universite Libre de Bruxelles, B)
A. Burkert	(Universität München, D)
A. Chieffi	(Inst. Astrofisica Spaziale, Frascati, I)
I.J. Danziger	(ESO, Garching, D)
P. Descouvemont	(Universite Libre de Bruxelles, B)
M.F. El Eid	(Universität Göttingen, D)
B. A. Fryxell	(University of Chicago, USA, guest of MPA)
H.J. Haubold	(Sternwarte Babelsberg, Potsdam, DDR)
G. Hensler	(Universität München, D)
W. Hillebrandt	(MPI f. Astrophysik, Garching, D)
P. Höflich	(MPI f. Astrophysik, Garching, D)
T. Janka	(MPI f. Astrophysik, Garching, D)
F. Käppeler	(Kernforschungszentrum Karlsruhe, D)
K.L. Kratz	(Universität Mainz, D)
R. Kuhfuß	(MPI f. Astrophysik, Garching, D)
N. Langer	(Universität Göttingen, D)
F. Matteucci	(ESO, Garching, D)
G. Meynet	(Observatoire de Geneve, Sauverny, CH)
E. Müller	(MPI f. Astrophysik, Garching, D)
J.M. Pearson	(Universite de Montreal, Canada)
N. Prantzos	(Institut d'Astrophysique, Paris, F)
M. Rayet	(Universite Libre de Bruxelles, B)
H. Rebel	(Kernforschungszentrum Karlsruhe, D)
A. Renzini	(Osservatorio Astronomico, Bologna, I)
C. Rolfs	(Universität Münster, D)
M.M. Sharma	(Ludwig-Maximilians-Universität, München, D)
O. Straniero	(Inst. Astrofisica Spaziale, Frascati, I)
A. Tornambè	(Inst. Astrofisica Spaziale, Frascati, I)
J.W. Truran	(MPI f. Astrophysik, Garching, D)
F. Voss	(Kernforschungszentrum Karlsruhe, D)
F.B. Waanders	(Universität Münster, D)
J. Wampler	(ESO, Garching, D)
A. Weiß	(MPI f. Astrophysik, Garching, D)
M. Wiescher	(University of Notre Dame, USA)

Lecture Notes in Physics

Vol. 257: Statistical Mechanics and Field Theory: Mathematical Aspects. Proceedings, 1985. Edited by T.C. Dorlas, N.M. Hugenholtz and M. Winnink. VII, 328 pages. 1986.

Vol. 258: Wm. G. Hoover, Molecular Dynamics. VI, 138 pages. 1986.

Vol. 259: R.F. Alvarez-Estrada, F. Fernández, J.L. Sánchez-Gómez, V. Vento, Models of Hadron Structure Based on Quantum Chromodynamics. VI, 294 pages. 1986.

Vol. 260: The Three-Body Force in the Three-Nucleon System. Proceedings, 1986. Edited by B.L. Berman and B.F. Gibson. XI, 530 pages. 1986.

Vol. 261: Conformal Groups and Related Symmetries – Physical Results and Mathematical Background. Proceedings, 1985. Edited by A.O. Barut and H.-D. Doebner. VI, 443 pages. 1986.

Vol. 262: Stochastic Processes in Classical and Quantum Systems. Proceedings, 1985. Edited by S. Albeverio, G. Casati and D. Merlini. XI, 551 pages. 1986.

Vol. 263: Quantum Chaos and Statistical Nuclear Physics. Proceedings, 1986. Edited by T.H. Seligman and H. Nishioka. IX, 382 pages. 1986.

Vol. 264: Tenth International Conference on Numerical Methods in Fluid Dynamics. Proceedings, 1986. Edited by F.G. Zhuang and Y.L. Zhu. XII, 724 pages. 1986.

Vol. 265: N. Straumann, Thermodymamik. VI, 140 Seiten. 1986.

Vol. 266: The Physics of Accretion onto Compact Objects. Proceedings, 1986. Edited by K.O. Mason, M.G. Watson and N.E. White. VIII, 421 pages. 1986.

Vol. 267: The Use of Supercomputers in Stellar Dynamics. Proceedings, 1986. Edited by P. Hut and S. McMillan. VI, 240 pages. 1986.

Vol. 268: Fluctuations and Stochastic Phenomena in Condensed Matter. Proceedings, 1986. Edited by L. Garrido. VIII, 413 pages. 1987.

Vol. 269: PDMS and Clusters. Proceedings, 1986. Edited by E.R. Hilf, F. Kammer and K. Wien. VIII, 261 pages. 1987.

Vol. 270: B.G. Konopelchenko, Nonlinear Integrable Equations. VIII, 361 pages. 1987.

Vol. 271: Nonlinear Hydrodynamic Modeling: A Mathematical Introduction. Edited by Hampton N. Shirer. XVI, 546 pages. 1987.

Vol. 272: Homogenization Techniques for Composite Media. Proceedings, 1985. Edited by E. Sanchez-Palencia and A. Zaoui. IX, 397 pages. 1987.

Vol. 273: Models and Methods in Few-Body Physics. Proceedings, 1986. Edited by L.S. Ferreira, A.C. Fonseca and L. Streit. XIX, 674 pages. 1987.

Vol. 274: Stellar Pulsation. Proceedings, 1986. Edited by A.N. Cox, W.M. Sparks and S.G. Starrfield. XIV, 422 pages. 1987.

Vol. 275: Heidelberg Colloquium on Glassy Dynamics. Proceedings, 1986. Edited by J.L. van Hemmen and I. Morgenstern. VIII, 577 pages. 1987.

Vol. 276: R.Kh. Zeytounian, Les Modèles Asymptotiques de la Mécanique des Fluides II. XII, 315 pages. 1987

Vol. 277: Molecular Dynamics and Relaxation Phenomena in Glasses. Proceedings, 1985. Edited by Th. Dorfmüller and G. Williams. VII, 218 pages. 1987.

Vol. 278: The Physics of Phase Space. Proceedings, 1986. Edited by Y.S. Kim and W.W. Zachary. IX, 449 pages. 1987.

Vol. 279: Symmetries and Semiclassical Features of Nuclear Dynamics. Proceedings, 1986. Edited by A.A. Raduta. VI, 465 pages. 1987.

Vol. 280: Field Theory, Quantum Gravity and Strings II. Proceedings, 1985/86. Edited by H.J. de Vega and N. Sánchez. V, 245 pages. 1987.

Vol. 281: Ph. Blanchard, Ph. Combe, W. Zheng, Mathematical and Physical Aspects of Stochastic Mechanics. VIII, 171 pages. 1987.

Vol. 282: F. Ehlotzky (Ed.), Fundamentals of Quantum Optics II. Proceedings, 1987. X, 289 pages. 1987.

Vol. 283: M. Yussouff (Ed.), Electronic Band Structure and Its Applications. Proceedings, 1986. VIII, 441 pages. 1987.

Vol. 284: D. Baeriswyl, M. Droz, A. Malaspinas, P. Martinoli (Eds.), Physics in Living Matter. Proceedings, 1986. V, 180 pages. 1987.

Vol. 285: T. Paszkiewicz (Ed.), Physics of Phonons. Proceedings, 1987. X, 486 pages. 1987.

Vol. 286: R. Alicki, K. Lendi, Quantum Dynamical Semigroups and Applications. VIII, 196 pages. 1987.

Vol. 287: W. Hillebrandt, R. Kuhfuß, E. Müller, J.W. Truran (Eds.), Nuclear Astrophysics. Proceedings. IX, 347 pages. 1987.